Glencoe Earth Science
Features and Benefits

Feature	Description	Pages
Dynamic Instructional Strategies	**... present clear and comprehensive coverage of middle school science.**	
	■ Each engaging chapter opener includes a *Launch Lab* and *Foldables*™.	7, 39
	■ *National Geographic Visualizing* features illustrate important concepts in middle school science.	19, 52
A Strong Reading Strand	**... encourages active reading and learning for students of all reading levels. In the *Student Edition:***	
	■ as you read gives students a preview of learning objectives and vocabulary for each section;	8, 40
	■ *Reading Checks* help students check their reading comprehension; and	11, 41
	■ *Caption questions* ask students to interpret visuals.	9, 43
	■ *Reading Essentials, An Interactive Student Textbook* is designed to help struggling readers comprehend core content. It is written at a reading level of about two to three grades below the *Student Edition.*	
Meeting the Needs of All Students	**... facilitates understanding of science concepts for students of all learning levels. In the *Teacher Wraparound Edition:***	
	■ *Differentiated Instruction* strategies help meet the needs of students with learning disabilities and physical challenges, or create opportunities to enrich and extend students' learning;	9, 41
	■ *Daily Intervention* provides intervention strategies for struggling students; and	13, 45
	■ *Identifying Misconceptions* helps uncover and address common science misconceptions.	38F
	■ The *English/Spanish Glossary*, also in the *Student Edition,* helps English-language learners comprehend science terms.	176–181
Extensive Standardized Test Practice	**... gives students the opportunity to practice for state and national exams.**	
	■ Each chapter ends with a variety of standardized test practice questions, including *Multiple Choice, Short Response/Grid In*, and *Open Ended.*	66–67
A Variety of Labs	**... gets students excited about being involved in science. The *Student Edition* provides:**	
	■ *MiniLABs,* traditional labs, and *Design Your Own, Model and Invent,* and *Use the Internet* labs; and	12, 42
	■ *Extra Try at Home Labs* provide opportunities for students to practice their science skills at home with adult supervision using materials from the kitchen, junk drawer, or backyard.	148–149
	■ *Virtual Labs* CD-ROM contains an interactive virtual lab for each chapter.	17, 53
	■ *Video Labs* (VHS) reinforce lab techniques and safety skills, offer troubleshooting tips, and give expected outcomes.	12, 55
	■ The *Science Lab Manual, Probeware Lab Manual,* and *Science Inquiry Lab Manual* provide additional opportunities to practice laboratory techniques.	
Multi-Level Review	**... presents multiple opportunities for all students to review and master content.**	
	■ Each section ends with a review that contains a *Summary* of the section's major concepts and a *Self Check* that has questions to assess student learning and practice math or science skills.	22, 54
	■ The *Study Guide* at the end of each chapter can preview, review, summarize, and visualize the chapter's main ideas.	33, 63
	■ *Study Guide* and *Reinforcement* help students grasp core content.	
Teacher Resources	**... provide innovative strategies to help new and experienced teachers.**	
	■ *Chapter Resources Fast File* ™ contains important reproducible masters.	38B
	■ Section Focus, Assessment, and Teaching transparencies accompany each chapter.	68C
	■ *Performance Assessment in the Science Classroom* has assessment guidelines, strategies, sample rubrics, and more.	
Online Resources	**... enrich the learning experience with the click of a mouse.**	
	■ For prescreened Web links, standardized test practice, self-check quizzes, chapter tests, *Vocabulary PuzzleMaker,* extra math practice, science career information, current science news, and *WebQuest* interactive projects, visit **bookj.msscience.com**.	
	■ The complete interactive *Student Edition* is available at The McGraw-Hill Learning Network Web site, **mhln.com**.	
Technology	**... provides timesaving products to help teachers creatively engage their students.**	
	■ *MindJogger Videoquizzes* (VHS & DVD) provide a game-show style interactive quiz for each chapter.	
	■ Easy to edit *Interactive Chalkboard* Microsoft® PowerPoint® presentations include step-by-step lessons, an image bank, chapter and section review questions, standardized test practice, and transparencies.	
	■ *ExamView® Pro Testmaker* CD-ROM in English or Spanish allows you to customize assessments.	
	■ *TeacherWorks* CD-ROM is your all-in-one resource center that helps you plan and organize lessons.	
	■ *StudentWorks* CD-ROM solves the heavy backpack problem.	

SAFETY SYMBOLS	HAZARD	EXAMPLES	PRECAUTION	REMEDY
DISPOSAL	Special disposal procedures need to be followed.	certain chemicals, living organisms	Do not dispose of these materials in the sink or trash can.	Dispose of wastes as directed by your teacher.
BIOLOGICAL	Organisms or other biological materials that might be harmful to humans	bacteria, fungi, blood, unpreserved tissues, plant materials	Avoid skin contact with these materials. Wear mask or gloves.	Notify your teacher if you suspect contact with material. Wash hands thoroughly.
EXTREME TEMPERATURE	Objects that can burn skin by being too cold or too hot	boiling liquids, hot plates, dry ice, liquid nitrogen	Use proper protection when handling.	Go to your teacher for first aid.
SHARP OBJECT	Use of tools or glassware that can easily puncture or slice skin	razor blades, pins, scalpels, pointed tools, dissecting probes, broken glass	Practice common-sense behavior and follow guidelines for use of the tool.	Go to your teacher for first aid.
FUME	Possible danger to respiratory tract from fumes	ammonia, acetone, nail polish remover, heated sulfur, moth balls	Make sure there is good ventilation. Never smell fumes directly. Wear a mask.	Leave foul area and notify your teacher immediately.
ELECTRICAL	Possible danger from electrical shock or burn	improper grounding, liquid spills, short circuits, exposed wires	Double-check setup with teacher. Check condition of wires and apparatus.	Do not attempt to fix electrical problems. Notify your teacher immediately.
IRRITANT	Substances that can irritate the skin or mucous membranes of the respiratory tract	pollen, moth balls, steel wool, fiberglass, potassium permanganate	Wear dust mask and gloves. Practice extra care when handling these materials.	Go to your teacher for first aid.
CHEMICAL	Chemicals can react with and destroy tissue and other materials	bleaches such as hydrogen peroxide; acids such as sulfuric acid, hydrochloric acid; bases such as ammonia, sodium hydroxide	Wear goggles, gloves, and an apron.	Immediately flush the affected area with water and notify your teacher.
TOXIC	Substance may be poisonous if touched, inhaled, or swallowed.	mercury, many metal compounds, iodine, poinsettia plant parts	Follow your teacher's instructions.	Always wash hands thoroughly after use. Go to your teacher for first aid.
FLAMMABLE	Flammable chemicals may be ignited by open flame, spark, or exposed heat.	alcohol, kerosene, potassium permanganate	Avoid open flames and heat when using flammable chemicals.	Notify your teacher immediately. Use fire safety equipment if applicable.
OPEN FLAME	Open flame in use, may cause fire.	hair, clothing, paper, synthetic materials	Tie back hair and loose clothing. Follow teacher's instruction on lighting and extinguishing flames.	Notify your teacher immediately. Use fire safety equipment if applicable.

Eye Safety
Proper eye protection should be worn at all times by anyone performing or observing science activities.

Clothing Protection
This symbol appears when substances could stain or burn clothing.

Animal Safety
This symbol appears when safety of animals and students must be ensured.

Handwashing
After the lab, wash hands with soap and water before removing goggles.

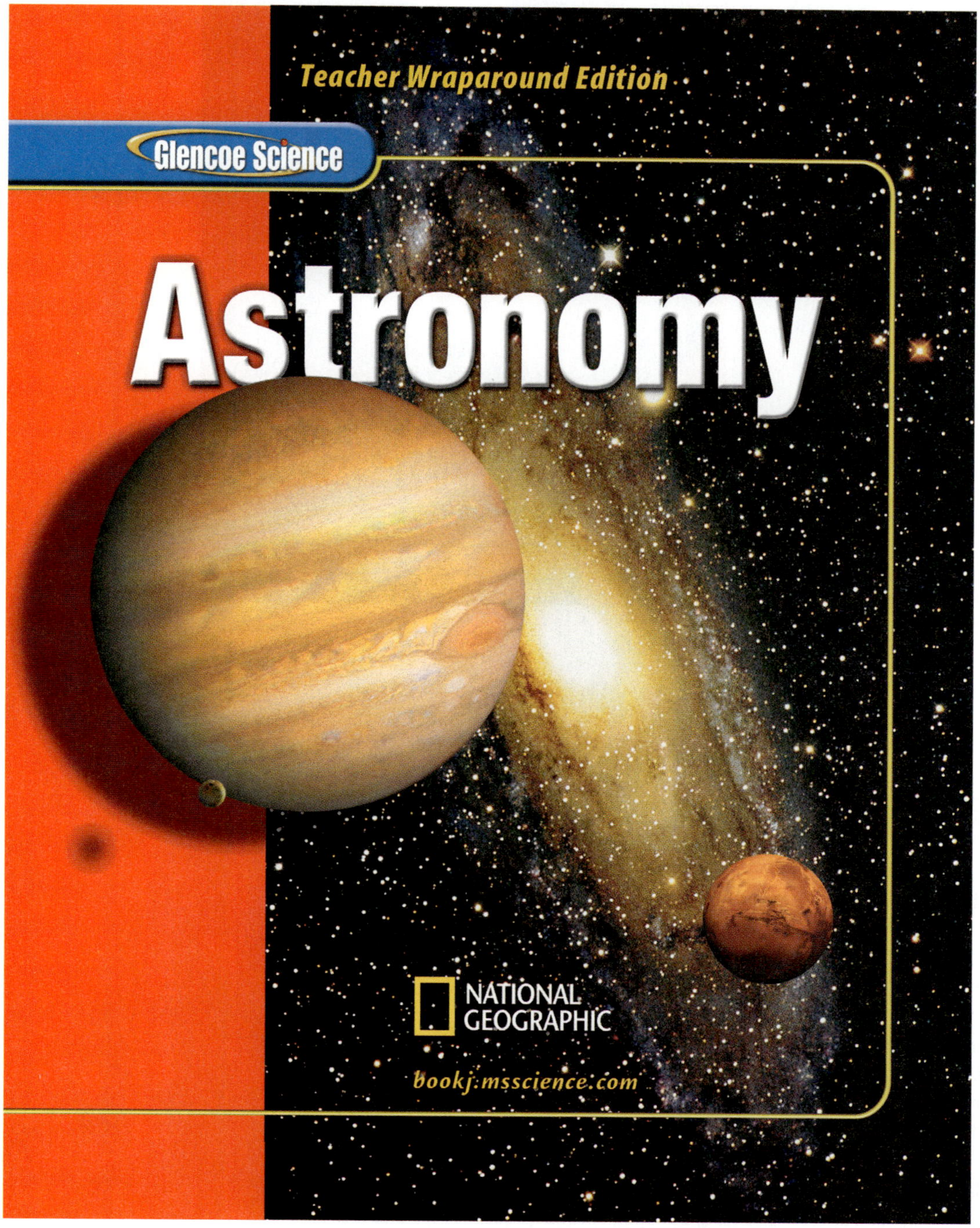

McGraw Hill Glencoe

New York, New York   Columbus, Ohio   Chicago, Illinois   Peoria, Illinois   Woodland Hills, California

Glencoe Science

Astronomy

This collection of images is of Jupiter, Io (one of its moons), Mars, and the Andromeda Galaxy. The Andromeda Galaxy is the most distant object visible to the human eye. At a distance of 2.2 million light years, it appears as a fuzzy patch of light in the night sky.

McGraw Hill Glencoe

The McGraw·Hill Companies

Send all inquiries to:
Glencoe/McGraw-Hill
8787 Orion Place
Columbus, OH 43240-4027

ISBN 0-07-861761-8 (Student Edition)
ISBN 0-07-861762-6 (Teacher Wraparound Edition)

Printed in the United States of America.

1 2 3 4 5 6 7 8 9 10 027/043 09 08 07 06 05 04

Authors

NATIONAL GEOGRAPHIC
Education Division
Washington, D.C.

Ralph M. Feather Jr., PhD
Assistant Professor
Geoscience Department
Indiana University of Pennsylvania
Indiana, PA

Dinah Zike
Educational Consultant
Dinah-Might Activities, Inc.
San Antonio, TX

Series Consultants

CONTENT

William C. Keel, PhD
Department of Physics and Astronomy
University of Alabama
Tuscaloosa, AL

MATH

Teri Willard, EdD
Mathematics Curriculum Writer
Belgrade, MT

READING

Carol A. Senf, PhD
School of Literature, Communication, and Culture
Georgia Institute of Technology
Atlanta, GA

SAFETY

Aileen Duc, PhD
Science 8 Teacher
Hendrick Middle School, Plano ISD
Plano, TX

Sandra West, PhD
Department of Biology
Texas State University-San Marcos
San Marcos, TX

ACTIVITY TESTERS

Mary Helen Mariscal-Cholka
William D. Slider Middle School
El Paso, TX

Science Kit and Boreal Laboratories
Tonawanda, NY

Series Reviewers

Mary Ferneau
Westview Middle School
Goose Creek, SC

Annette D'Urso Garcia
Kearney Middle School
Commerce City, CO

Nerma Coats Henderson
Pickerington Lakeview Jr. High School
Pickerington, OH

Michael Mansour
Board Member
National Middle Level Science Teacher's Association
John Page Middle School
Madison Heights, MI

Mary Helen Mariscal-Cholka
William D. Slider Middle School
El Paso, TX

Sharon Mitchell
William D. Slider Middle School
El Paso, TX

Glencoe Science 15-Book Series

Teach science your way!

With the 15 Life, Earth, and Physical Science titles in our modular series, you can select the science topics you want to cover and customize your science curriculum in any way you choose.

F Earth Materials and Processes

1 Minerals
2 Rocks
3 Earth's Energy and Mineral Resources
4 Plate Tectonics
5 Earthquakes
6 Volcanoes

G The Changing Surface of Earth

1 Views of Earth
2 Weathering and Soil
3 Erosional Forces
4 Water Erosion and Deposition
5 Clues to Earth's Past
6 Geologic Time

H The Water Planet

1 Water
2 Freshwater at Earth's Surface
3 Groundwater Resources
4 Ocean Motion
5 Oceanography

I The Air Around You

1 Atmosphere
2 Weather
3 Climate
4 Air Pollution

J Astronomy

1 Exploring Space
2 The Sun-Earth-Moon System
3 The Solar System
4 Stars and Galaxies

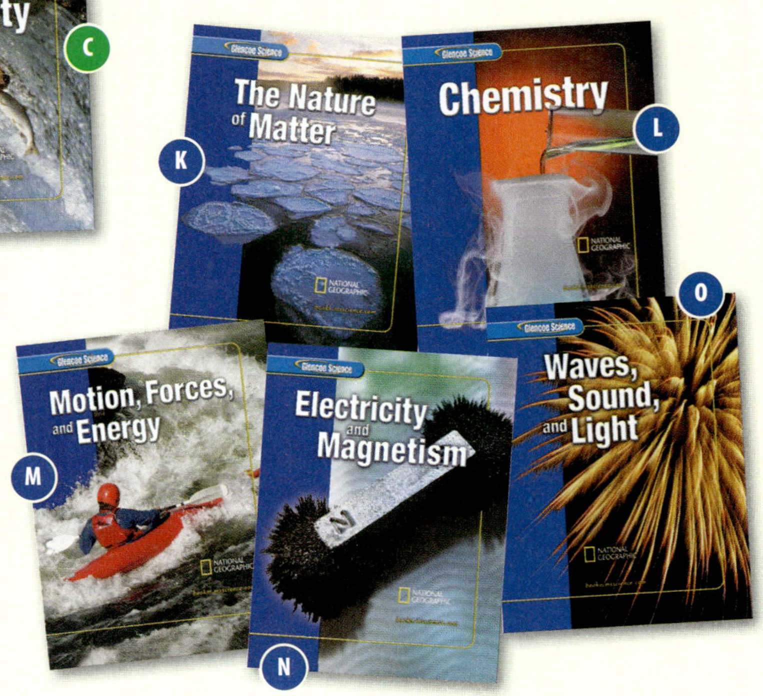

Multi-Level Review and Assessment

Each chapter provides five pages of review and testing to help you evaluate students' knowledge and ability to apply science concepts.

Section Review

- Summary pinpoints important concepts in the section.
- Skill-based questions promote critical thinking skills.

Study Guide

- Main idea summary of each section
- Concept mapping activity to help students visualize the main ideas

Chapter Review

- Using Vocabulary
- Checking Concepts
- Thinking Critically
- Performance Activities
- Applying Math

Standardized Test Practice

- Multiple Choice
- Short Response/ Grid In
- Open-Ended Questions

Dynamic Instruction

The consistent instructional strategies in each chapter strengthen students' learning—from the beginning of each chapter where students see "Chapter Preview," to the end where they have a chance to test the knowledge they have acquired and prepare for the next lesson.

Chapter Opener

- *Chapter Preview* introduces the main concepts.
- *Science Journal* promotes critical-thinking skills
- *Dinah Zike's Foldables*™ let students create interactive study guides.
- *Launch Labs* give students an opportunity to explore new ideas.

Section Opener

- *What You'll Learn* introduces main concepts.
- *Why It's Important* provides an answer to "Why do we have to learn this?"
- *Review Vocabulary* reviews a term that helps students better understand section content.
- *New Vocabulary* highlights new terms students will learn in the section.

Labs

- *Labs* allow students to design their own experiments or follow well-tested procedures, enabling them to learn and practice science processes.
- *MiniLABS* offer students quick and easy-to-do ways to clarify concepts and reinforce skills.

Assessment

Glencoe Science offers the Glencoe Assessment Advantage, a system designed to give you all the tools you need to prepare your students for success in any testing situation.

In the *Student Edition*

Section Review and **Applying Math** questions appear in every chapter.

Chapter Review questions help you evaluate students' knowledge and ability to apply science concepts.

Standardized Test Practice questions at the end of each chapter provide students with additional opportunities to practice their test-taking skills.

In the *Teacher Wraparound Edition*

Assessments located throughout the *Teacher Wraparound Edition* provide methods for assessing students' comprehension with Performance, Process, and Content exercises.

Teacher Classroom Resources

Performance Assessment in the Science Classroom

- Guidelines for assessing the performance of a task
- Reproducible activities for evaluating students
- Sample rubrics and checklists

***Fast File* Chapter Resources** provides six pages of assesement for every chapter including *Testing Concepts, Applying Concepts,* and *Writing Skills.*

Technology Support

MindJogger Videoquizzes are interactive video quizzes set in game show format. Each is designed for the full range of student learning styles.

Exam*View*® Pro Testmaker CD-ROM for Windows® and Macintosh® provides an easy way to create, edit, and customize your tests. Select your own test items by objective from two different levels of difficulty, or write and edit your own. Translate tests from English to Spanish and vice versa.

Rubrics

The following rubrics are sample scoring devices for short response and open-ended questions.

Short Response

Points	Description
2	The student demonstrates a thorough understanding of the science of the task. The response may contain minor flaws that do not detract from the demonstration of a thorough understanding.
1	The student has provided a response that is only partially correct.
0	The student has provided a completely incorrect solution or no response at all.

Open Ended

Points	Description
4	The student demonstrates a thorough understanding of the science of the task. The response may contain minor flaws that do not detract from the demonstration of a thorough understanding.
3	The student demonstrates an understanding of the science of the task. The response is essentially correct and demonstrates an essential but less than thorough understanding of the science.
2	The student demonstrates only a partial understanding of the science of the task. Although the student may have used the correct approach to a solution or may have provided a correct solution, the work lacks an essential understanding of the underlying science concepts.
1	The student demonstrates a very limited understanding of the science of the task. The response is incomplete and exhibits many flaws.
0	The student provides a completely incorrect solution or no response at all.

Time-Saving Teacher Resources

Glencoe Science provides an extensive array of support materials and resources designed to help you create and customize your science course quickly and easily.

FAST FILE Chapter Resources

For each chapter, Chapter Resources contain key reproducible masters along with additional teaching strategies, teacher support, and answer keys.

Teacher Wraparound Edition

The *Teacher Wraparound Edition* is your key to the teaching resources available. In addition to teaching strategies and suggestions, the *Teacher Wraparound Edition* provides a guide for all print and software materials available for each lesson.

Transparencies

Color Transparencies provides three types of transparencies for use while teaching each chapter. The *Section Focus Transparencies* are designed to generate interest and focus students' attention on the topic being presented in the section. The *Teaching Transparency* for each chapter addresses a major concept that will benefit from an extra visual learning aid. The *Assessment Transparency* for each chapter is set up to resemble standardized tests.

Exam*View*® Pro Testmaker CD-ROM

This CD-ROM will help you create, edit, and customize tests. In addition, it will help you create multiple versions of tests, translate tests from English to Spanish and vice versa, and build tests aligned with state standards.

Video Labs

These VHS cassettes contain step-by-step procedures for selected *Student Edition* labs. They also contain lab safety skills, teacher support, and troubleshooting advice.

TeacherWorks™

This CD-ROM is your all-in-one teacher resource center. Personalize a lesson plan, access resources from the *Teacher Wraparound Edition*, connect to the Internet, or make a to-do list. These are only a few of the many features that can assist you in the planning and organizing of your lessons.

StudentWorks™

This CD-ROM is a valuable resource for students to access content online and use online resources to continue learning chapter concepts.

Interactive Chalkboard with Image Bank — PowerPoint® Presentations

This CD-ROM brings Microsoft® PowerPoint® presentations right to your door. With the large number of graphics provided, students can use a visual approach to learning chapter content.

Virtual Labs CD-ROM Program

The Virtual Labs CD-ROM contains a collection of labs that allow students to complete labs that are too expensive, take too long to complete, or might be too dangerous in a classroom laboratory.

Science Online

This website is a portal to hundreds of prescreened Internet sites (Web links) that correlate to content in the text. Visit this link to find interactive activities that review chapter concepts and to access the *Student Edition* online.

McGraw Hill Learning Network mhln.com

mhln.com is an online teaching and learning space for teachers, students, and parents.

Differentiated Instruction

Teaching Strategies

Following each suggested assessment and activity, ability levels are supplied to accommodate all students. For a key to the Teaching Strategies designations, see the C page before each chapter.

Identifying Misconceptions

These short, diagnostic, and perscriptive lessons target common science misconceptions.

Multiple Learning Styles

Look for these italicized designations under various activities to help you target your lessons to each student's preferred learning style.

- ***Kinesthetic*** learners learn through touch, movement, and manipulating objects.
- ***Visual-Spatial*** learners think in terms of images, illustrations, and models.
- ***Interpersonal*** learners understand and work well with other people.
- ***Intrapersonal*** learners can analyze their own strengths and weaknesses and may prefer to work on their own.
- ***Linguistic*** learners write clearly and easily understand the written word.
- ***Logical-Mathematical*** learners understand numbers easily and have highly-developed reasoning skills.

Daily Interventions

Found at the end of each chapter section, this feature is designed to intercept students who are struggling and prescribe a system to help them get back on track. *Reteach* provides reinforcement of the section's concepts through visual activities.

Differentiated Instruction

These activities present various teaching strategies designed to help you meet the special needs of students with learning disabilities, physical challenges, visual impairment, and hearing impairment. *Challenge* activities provide opportunities for students who excel to engage in activities and research projects that extend the chapter's concepts. English-language learners in the classroom will also find exercises that bridge the gap between language barriers and the chapter content.

Cultural Diversity

These readings provide insights into the unique ways in which people of different ethnicities and cultural heritage have approached science. The intent of these features is to build awareness and appreciation for the global community in which we live.

Inquiry-Based Science

The call for more inquiry-based science by the *National Science Education Standards* has been met by Glencoe Science.

Glencoe Science recognizes the importance of conducting inquiry-based science activities in the classroom. The process of doing inquiry models actual science practice, encouraging problem-solving strategies and developing critical thinking skills. Inquiry gets students actively involved in the learning process by allowing them to determine materials, procedures, or the topics and questions they want to investigate.

Inquiry can range from a very structured activity for those students who need more guidance to a more open-ended approach where students lead the investigations. Glencoe Science recognizes that the inquiry activities suggested will not look the same in every classroom. We encourage teachers to modify the suggested activities in a manner that best supports your students.

Glencoe also provides teachers with *Alternative Inquiry Labs,* teaching strategies or suggestions for making existing labs more inquiry-based.

Research-Based Learning Strategies

Glencoe Science incorporates the most current and applicable educational research on science learning and follows recommendations from the American Association for the Advancement of Science and the National Science Teachers Association. The following research-based strategies can be found throughout the text.

Learning Strategies

The following research-based strategies can be found throughout the text:

- **Using Prior Knowledge** Glencoe Science encourages students to use their prior knowledge to learn information because this adds relevance to the material. Students are referred back to other parts of the text or to their own real-life experiences.
- **Practicing Important Tasks** By offering students an opportunity to practice important tasks using a variety of labs and activities in the *Student Edition, Teacher Wraparound Edition,* ancillaries and technology, Glencoe Science makes learning fun and relevant for students.
- **Using Visuals to Communicate, Organize, and Reinforce Learning** High-quality art and photos throughout the text communicate concepts more efficiently and reinforce learning, while allowing students to organize information.
- **Motivating Students to Achieve** Active strategies and real-world experiences motivate students to achieve. Throughout Glencoe's programs, students are encouraged to apply their knowledge in ways that will motivate them to learn.
- **Developing Decoding and Reading Comprehension Strategies** Throughout the text, students are supplied with caption questions, reading checks, and other strategies to aid in comprehension.
- **Using Study Strategies** Through the use of highlighting, outlining, note-taking, summarizing, and other such strategies, students can monitor their own progress and organize information more effectively, thereby increasing their scientific literacy. These strategies are found throughout the text and ancillaries.

The use of these strategies within Glencoe Science will help teachers to achieve the goals set forth by the *National Science Education Standards.*

White Paper

The Glencoe Science White Paper outlines the educational strategies on which this program was based. This document provides specific examples from the *Student Edition, Teacher Wraparound Edition,* ancillary program, and technology resources, highlighting extensive use of educationally sound strategies that help students learn science.

Field Research and Testing

Feedback from students, teachers, curriculum supervisors, department chairpersons, parents, learning specialists, and science content experts was invaluable in the development of this program. The following pre-publication and post-publication research was conducted.

Prior to Publication

- Detailed classroom teacher and curriculum supervisor **surveys** were conducted by independently contracted researchers.
- A **nationwide panel** of science teachers, curriculum supervisors, and department chairpersons provided countless hours of feedback and assistance throughout program development.
- A wide range of **educator and content reviewers** provided in-depth reviews of and suggestions for manuscripts and pre-publication versions of the program.
- **Face-to-face interviews** with science teachers provided insight into teachers' day-to-day challenges.

Field-Test Results

- Field-test research indicates that test scores increased among students using Glencoe Science programs.
- Nine out of ten students earned higher scores after using Glencoe programs.
- Scores improved among both male and female students.
- Scores improved among both minority and non-minority students.
- Overall, the gap between the average pre-test score and a perfect score closed by 33 percent. Stated differently, on average, **scores increased 77 percent after students used the Glencoe program.**

After Publication

- Field tests were conducted in which students and teachers used a pre-publication manuscript in the classroom.
- Follow-up interviews, observations, and surveys of Glencoe Science users provide ongoing opportunities for program development and verification of program success.

KS Studios

National Education Standards

Correlation of *Glencoe Science* to the National Science Education Standards.

Content Standard	Chapter and Section
(UCP) Unifying Concepts and Processes	
1. Systems, order, and organization	F1-1, F1-2, F1-3, F2-1, F2-2, F2-3, F2-4, F3-1, F3-2, F3-3, F4-1, F4-2, F4-3, F5-1, F5-2, F5-3, F6-1, F6-2, F6-3, G1-1, G1-2, G1-3, G2-1, G2-2, G2-3, G3-1, G3-2, G3-3, G4-1, G4-2, G4-3, G5-1, G5-2, G5-3, G6-1, G6-2, G6-3, H2-1, H2-2, H2-3, H2-4, H3-1, H3-2, H3-3, H4-1, H4-2, H4-3, H5-1, H5-2, H5-3, I1-1, I1-2, I1-3, I2-1, I2-2, I2-3, I3-1, I3-2, I3-3, I4-1, I4-2, I4-3, J1-1, J1-2, J1-3, J2-1, J2-2, J2-3, J3-1, J3-2, J3-3, J3-4, J4-1, J4-2, J4-3
2. Evidence, models, and explanation	F1-1, F1-2, F1-3, F2-1, F2-2, F2-3, F2-4, F3-1, F3-2, F3-3, F4-1, F4-2, F4-3, F5-1, F5-2, F5-3, F6-1, F6-2, F6-3, G1-1, G1-2, G1-3, G2-1, G2-2, G2-3, G3-1, G3-2, G3-3, G4-1, G4-2, G4-3, G5-1, G5-2, G5-3, G6-1, G6-2, G6-3, H1-1, H1-2, H1-3, H2-1, H2-2, H2-3, H2-4, H3-1, H3-2, H3-3, H4-1, H4-2, H4-3, H5-1, H5-2, H5-3, I1-1, I1-2, I1-3, I2-1, I2-2, I2-3, I3-1, I3-2, I3-3, I4-1, I4-2, I4-3, J1-1, J1-2, J1-3, J2-1, J2-2, J2-3, J3-1, J3-2, J3-3, J3-4, J4-1, J4-2, J4-3
3. Change, constancy, and measurement	F1-1, F1-2, F1-3, F2-1, F2-2, F2-3, F2-4, F3-1, F3-2, F3-3, F4-1, F4-2, F4-3, F5-1, F5-2, F5-3, F6-1, F6-2, F6-3, G1-1, G1-2, G1-3, G2-1, G2-2, G2-3, G3-1, G3-2, G3-3, G4-1, G4-2, G4-3, G5-1, G5-2, G5-3, G6-1, G6-2, G6-3, H1-1, H1-2, H1-3, H2-1, H2-2, H2-3, H2-4, H3-1, H3-2, H3-3, H4-1, H4-2, H4-3, H5-1, H5-2, H5-3, I1-1, I1-2, I1-3, I2-1, I2-2, I2-3, I3-1, I3-2, I3-3, I4-1, I4-2, I4-3, J1-1, J1-2, J1-3, J2-1, J2-2, J2-3, J3-1, J3-2, J3-3, J3-4, J4-1, J4-2, J4-3
4. Evolution and equilibrium	G6-1, G6-2, J4-3
5. Form and function	F1-1, F1-2, F1-3, F2-1, F2-2, F2-3, F2-4, F3-1, F3-2, F3-3, F4-1, F4-2, F4-3, F5-1, F5-2, F5-3, F6-1, F6-2, F6-3, G1-1, G1-2, G1-3, G2-1, G2-2, G2-3, G3-1, G3-2, G3-3, G4-1, G4-2, G4-3, G5-1, G5-2, G5-3, G6-1, G6-2, G6-3, H1-1, H1-2, H1-3, H2-1, H2-2, H2-3, H2-4, H3-1, H3-2, H3-3, H4-1, H4-2, H4-3, H5-1, H5-2, H5-3, I1-1, I1-2, I1-3, I2-1, I2-2, I2-3, I3-1, I3-2, I3-3, I4-1, I4-2, I4-3, J1-1, J1-2, J1-3, J2-1, J2-2, J2-3, J3-1, J3-2, J3-3, J3-4, J4-1, J4-2, J4-3
(A) Science as Inquiry	
1. Abilities necessary to do scientific inquiry	F1-1, F1-2, F1-3, F2-1, F2-2, F2-3, F2-4, F3-1, F3-2, F3-3, F4-1, F4-2, F4-3, F5-1, F5-2, F5-3, F6-1, F6-2, F6-3, G1-1, G1-2, G1-3, G2-1, G2-2, G2-3, G3-1, G3-2, G3-3, G4-1, G4-2, G4-3, G5-1, G5-2, G5-3, G6-1, G6-2, G6-3, H1-1, H1-2, H1-3, H2-1, H2-2, H2-3, H2-4, H3-1, H3-2, H3-3, H4-1, H4-2, H4-3, H5-1, H5-2, H5-3, I1-1, I1-2, I1-3, I2-1, I2-2, I2-3, I3-1, I3-2, I3-3, I4-1, I4-2, I4-3, J1-1, J1-2, J1-3, J2-1, J2-2, J2-3, J3-1, J3-2, J3-3, J3-4, J4-1, J4-2, J4-3
2. Understandings about scientific theory	F1-1, F1-2, F1-3, F2-1, F2-2, F2-3, F2-4, F3-1, F3-2, F3-3, F4-1, F4-2, F4-3, F5-1, F5-2, F5-3, F6-1, F6-2, F6-3, G1-1, G1-2, G1-3, G2-1, G2-2, G2-3, G3-1, G3-2, G3-3, G4-1, G4-2, G4-3, G5-1, G5-2, G5-3, G6-1, G6-2, G6-3, H1-1, H1-2, H1-3, H2-1, H2-2, H2-3, H2-4, H3-1, H3-2, H3-3, H4-1, H4-2, H4-3, H5-1, H5-2, H5-3, I1-1, I1-2, I1-3, I2-1, I2-2, I2-3, I3-1, I3-2, I3-3, I4-1, I4-2, I4-3, J1-1, J1-2, J1-3, J2-1, J2-2, J2-3, J3-1, J3-2, J3-3, J3-4, J4-1, J4-2, J4-3
(B) Physical Science	
1. Properties and changes of properties in matter	G2-1, G5-3, H1-1, H4-1, H5-2, I4-1, J4-3
2. Motions and forces	F1-2, F4-1, F4-2, F4-3, F5-1, F5-2, F6-1, F6-2, G2-2, G2-3, G3-3, I1-2, I1-3, I2-2, I3-3, J2-1, J2-2, J3-1, J4-3
3. Transfer of energy	H5-2, I1-2, I3-3, J4-3
(C) Life Science	
1. Structure and function in living systems	H2-1, H2-2, H2-3, H5-2
2. Reproduction and heredity	H5-2
3. Regulation and behavior	H5-1, H5-2
4. Populations and ecosystems	H2-1, H2-2, H2-3, H5-1, H5-2, I4-2
5. Diversity and adaptations of organisms	H2-1, H2-2, H2-3, H5-2
(D) Earth and Space Science	
1. Structure of the Earth system	F1-1, F2-1, F2-2, F2-3, F2-4, F3-1, F3-2, F3-3, F4-1, F4-2, F4-3, F5-1, F5-2, F5-3, F6-1, F6-2, G1-1, G1-2, G1-3, G2-1, G2-2, G4-1, G4-2, G4-3, G6-1, G6-2, G6-3, H2-1, H2-2, H3-1, H3-2, H3-3, H4-1, H4-2, H4-3, H5-1, H5-2, I1-1, I1-2, I1-3, I2-1, I2-2, I2-3, I3-1, I3-2, I3-3, J2-1, J2-3
2. Earth's history	F1-1, F1-2, F2-1, F2-2, F2-3, F2-4, F3-1, F4-1, F5-1, F6-1, G1-1, G1-3, G2-1, G2-3, H3-3, I2-3, I3-3
3. Earth and the solar system	F1-3, F2-1, F2-2, F2-3, F2-4, F3-1, F3-3, H4-3, J1-1, J1-2, J1-3, J2-3, J3-2, J4-2

Content Standard	Chapter and Section
(E) Science and Technology	
1. Abilities of technological design	H4-3
2. Understandings about science and technology	H4-3
(F) Science in Personal and Social Perspectives	
1. Personal health	F3-2, G2-2, I4-2
2. Populations, resources, and environments	F3-1, F3-2, F3-3, H1-3, H2-4, H3-2, H5-3, I4-1, I4-2, I4-3
3. Natural hazards	F4-1, F4-2, F4-3, F5-1, F5-2, F5-3, F6-1, F6-2, G3-3, G4-1, G4-3, I2-3, I4-1, I3-3
4. Risks and benefits	G2-1, G3-3, H3-2
5. Science and technology in society	F3-1, G5-1, G5-3, I4-3, J1-3
(G) History and Nature of Science	
1. Science as a human endeavor	F1-3
2. Nature of science	F1-2, F1-3, F4-3, F6-3, I4-3
3. History of science	F1-2, F1-3, F5-2, G5-3, H5-3, I4-3, J2-3, J3-1, J3-4, J4-3

How Glencoe Science Aligns with the National Science Education Standards

The correlations at the left and above show the close alignment between Glencoe Science and the grade-appropriate standards. Glencoe Science allows students to discover concepts within each of the content standards and gives students opportunities to make connections among the science disciplines. Hands-on activities and inquiry-based lessons reinforce the science processes emphasized in the standards.

How Glencoe Science Aligns with the NCTM Standards for Grades 6–8

Throughout Glencoe Science, each Applying Math activity provides students with the opportunity to practice and apply some of the mathematical concepts and applications described in the NCTM Standards. These activities serve to reinforce mathematical skills in real-life situations, thus preparing students to meet their needs in an ever-changing world.

Correlation of *Glencoe Science* to NCTM Standards

Math Standard	Page
1. Number and Operations	F-31, F-54, F-61, F-121, F-143, F-151, F-172, F-181, G-31, G-46, G-59, G-87, G-119, G-149, G-173, H-47, H-63, H-95, H-123, H-130, I-31, I-39, I-61, I-114, I-121, J-35, J-65, J-80, J-99, J-130
2. Algebra	F-54, F-61, F-143, F-151, F-172, F-181, G-31, G-119, G-149, H-47, H-95, H-123, H-130, I-39, I-61, I-114, I-121, J-65
3. Geometry	G-31, H-95, J-16, J-65
4. Measurement	G-31, H-95, I-91, I-121, J-35, J-65
5. Data Analysis and Probability	F-31, F-93, F-151, F-181, G-46, G-149, G-181, H-31, H-63, H-123, H-153, I-31, I-39, I-61, I-121, J-35, J-99, J-130
6. Problem Solving	F-31, G-46, G-87, G-149
7. Reasoning and Proof	F-54, F-143, F-172, G-46, G-105, G-173, H-108, H-130, I-39, J-16, J-80
8. Communication	F-117, F-147, G-83, I-87, J-31
9. Connections	F-54, F-61, F-121, F-143, F-151, F-172, F-181, G-31, G-59, G-119, H-47, H-63, H-95, H-123, H-130, H-153, I-31, I-39, I-61, I-91, I-114, I-121, J-35, J-65, J-80, J-99, J-130
10. Representation	G-31, G-149

Foldables™

Foldables™ are easy-to-make, three-dimensional, interactive graphic organizers that students create out of simple sheets of paper. These unique hands-on tools for studying and reviewing were created exclusively for Glencoe by education specialist Dinah Zike.

Research Behind Foldables™

According to research (Bransford, 1979; Corno, 1994), study strategies help students understand, organize, remember and apply new information presented in science textbooks. Some study strategies include concept mapping, highlighting, outlining, note taking, summarizing, and underlining (Peverly, Brobst, Graham & Shaw, 2003). Glencoe Science offers Dinah Zike's Foldables™ Study Organizers as an organizational tool and study guide for students.

Foldables™

- Build prereading skills
- Encourage active reading and writing
- Summarize content for review

FOLDABLES™ Study Organizer

Primary and Secondary Succession Make the following Foldable to help you illustrate the main ideas about succession.

STEP 1 **Fold** a vertical sheet of paper in half from top to bottom.

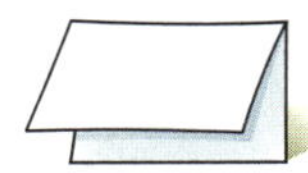

For more ideas on how to incorporate Foldables™ into your lessons consult **Dinah Zike's *Teaching Science with Foldables™***

Educational Partnerships

NATIONAL GEOGRAPHIC

Some topics in the chapter either require or benefit from a larger, more detailed visual explanation. The National Geographic Society has created *Visualizing* features that call out an important concept from the chapter and illustrate it in a way that will inform, excite, and motivate your students.

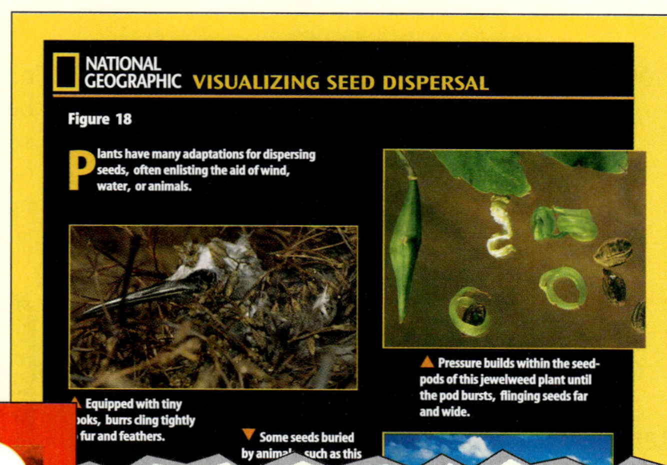

TIME

TIME magazine brings science topics and history together to further explain the chapter's main ideas and show how science relates to real life.

Safety in the Laboratory

All activities are designed to minimize dangers in the laboratory. Careful laboratory planning and management by both the instructor and the student are essential to a safe laboratory experience. **Local, state, and federal laboratory safety laws and regulations must be strictly followed.** The information provided here is one of the many resources to which you can refer for information about laboratory safety.

Classroom and Laboratory Preparation

1. Store equipment properly and securely and other thing.
   a. Clean and dry all equipment before storing.
   b. Protect electronic equipment and microscopes from dust, humidity, and extreme temperatures.
   c. Number, catalog, and organize equipment.
2. Ensure adequate work space for each student.
3. Ensure adequate classroom and storeroom ventilation.
4. Explain and post safety and evacuation guidelines along with expectations of conduct.
5. Ensure that all safety equipment is functioning properly and is clearly visible.
6. Provide hot plates as a heat source whenever possible. If gas burners are used, know where the central gas supply shutoff valve is located.
7. Ensure that each workstation has a GFCI-protected electrical source.
8. Provide safety goggles consistent with ANSI Standard Z87.1 for each student, including students who wear corrective lenses.

Before Each MiniLAB or Lab

1. Arrange the lab in such a way that equipment and supplies are clearly labeled and easily accessible.
2. Have available only equipment and supplies needed to complete the assigned investigation.
3. Review the procedure with students, emphasizing any caution statements or safety symbols that appear.
4. Be sure all students know the proper procedures to follow if an accident should occur.

After the MiniLAB or Lab

1. Be certain that students have returned all equipment and disposed of broken glassware and chemicals properly.
2. Be sure that all hot plates and electrical connections are off.

Storage of Chemicals

Be sure to store all chemicals properly. The following are guidelines commonly used. Your school, city, county, or state may have additional requirements for handling chemicals. It is the responsibility of each teacher to become informed of the rules or guidelines in effect in his or her area.

1. Separate chemicals by reaction type. Strong acids should be stored together. Likewise, strong bases should be stored together and should be separated from acids. Oxidants should be stored away from easily oxidized materials, and so on.
2. Be sure all chemicals are stored in labeled containers indicating contents, concentration, source, date purchased (or prepared), any precautions for handling and storage, and expiration date.
3. Hazardous chemicals require special storage containers and conditions. Be sure to know which chemicals those are and the accepted practices for your area. Some substances must be stored outside the building.

Disposal of Chemicals

Local, state, and federal laws regulate the proper disposal of chemicals. These laws should be consulted before chemical disposal is attempted. Although many substances encountered in the science classroom can be flushed down the drain with plenty of water, it is not safe to assume that this is always true.

DISCLAIMER

Glencoe Publishing Company makes no claims to the completeness of this discussion of laboratory safety and chemical storage. The material presented is not all-inclusive, nor does it address all of the hazards associated with handling, storing, and disposing of chemicals, or with laboratory management.

Preparation of Solutions

It is important to use safe laboratory techniques when handling all chemicals. Always check the MSDS (Material Safety Data Sheet) for each chemical before using it in the classroom. Many substances might appear harmless, but might be toxic, corrosive, or very reactive. Chemicals should never be ingested. Use proper techniques to smell any chemical, wear safety goggles and an apron in the laboratory, and observe the following precautions.

1. **Dilution of Acids and Bases** When diluting acids with water, always add the acids to the water. Never add water to acids. When sulfuric acid and sodium hydroxide are added to water, a large amount of thermal energy is released. Use extra care when handling these substances.

2. **Poisonous and Corrosive Liquids or Vapors** Use a fume hood if possible. Examples include hydrochloric acid, acetic acid, nitric acid, and ammonium hydroxide.

3. **Poisonous and Corrosive to Eyes, Lungs, and Skin** Examples include acids, bases, silver nitrate, iodine, and potassium permanganate.

Bromthymol blue: Add 0.5 g bromthymol blue powder to 500 mL distilled water to make a BTB stock solution. Dilute 40 mL BTB stock solution to 2 L with distilled water. Solution should be bright blue. If not, add one drop of NaOH at a time, swirling to mix. Check color.

Hydrochloric acid (HCL) solution: To make a 5% solution, add 13.6 mL concentrated HCl to 73 mL water while stirring. To make a 0.1*M* solution, add 1 mL concentrated hydrochloric acid to 100 mL water while stirring.

Iodine solution/Iodine stain: Dilute 1 part Lugol's solution with 15 parts water.

Lugol's solution: Dissolve 10 g potassium iodide in 100 mL distilled water. Then add and dissolve 5 g iodine. Store in dark bottle. Keeps indefinitely.

Phenolphthalein indicator: From a drug store, buy a package of any laxative that contains phenolphthalein. To make 1% solution, mash 4 tablets and pour the powder into 10 mL of rubbing alcohol. Let mixture soak for 15 minutes. Pour liquid into and store in a dropper bottle.

Potassium permanganate: For a 0.01*M* solution of potassium permanganate, dissolve 0.15 g $KMnO_4$ in 100 mL water.

Red cabbage concentrate: Put 5 leaves of red cabbage in a pot. Add 1 L of water, bring to a boil, and simmer until water turns a deep purple. Pour liquid through a strainer or piece of cheesecloth into a storage bottle. Keep refrigerated.

Salt solution: For a 3.5% salt (NaCl) solution that simulates the concentration of ocean water, dissolve 35 g of salt (NaCl) in 965 mL of water. For a 1% solution (weak), dissolve 1 g of salt (NaCl) in 99 mL of water. For a 6% solution, dissolve 6 g of salt (NaCl) in 94 mL of water.

Silver nitrate solution: To make a 10% solution, put 5 g of silver nitrate in 50 mL of distilled water.

Sugar solution: Add 1 tablespoon of sugar to 1 cup of warm water in a deep jar or flask. Stir to dissolve.

Sodium hydroxide (dilute): To make a 1% solution, dissolve 1 g NaOH in 99 mL of water.

Equipment and Materials List

Refer to the Chapter Organizer in front of each chapter for a list of equipment and materials used for each laboratory activity in the chapter.

Material	Launch Lab (Chapter)	MiniLAB (Chapter-Section)	Lab (Chapter-Section)
Consumables			
cardboard			3-1
cardboard tube		1-1	
chalk		2-2	
clay, modeling		2-3	
clear-plastic wrap	1		
construction paper, black			3
flour, white	3		
plastic foam cup		2-1	
powdered drink mix, colored or gelatin powder	3		
string		1-2, 2-2	3-1, 3-4
Nonconsumables			
basketball	2		
cake pan, metal	3		
cork		1-2	
flashlight		2-3	
globe			2-2
gooseneck lamp with 75-watt bulb			2-3
light source			2-2
magnet		2-1	
magnifying lenses			1-1
marbles	3		
mirror, concave			1-1
mirror, flat			1-1
pavement		2-2	
plate		2-1	
polystyrene ball			2-2
protractor			2-3
sewing needle		2-1	
thermometer, Celsius			2-3
thumbtacks or pins			3-1
watch			2-3

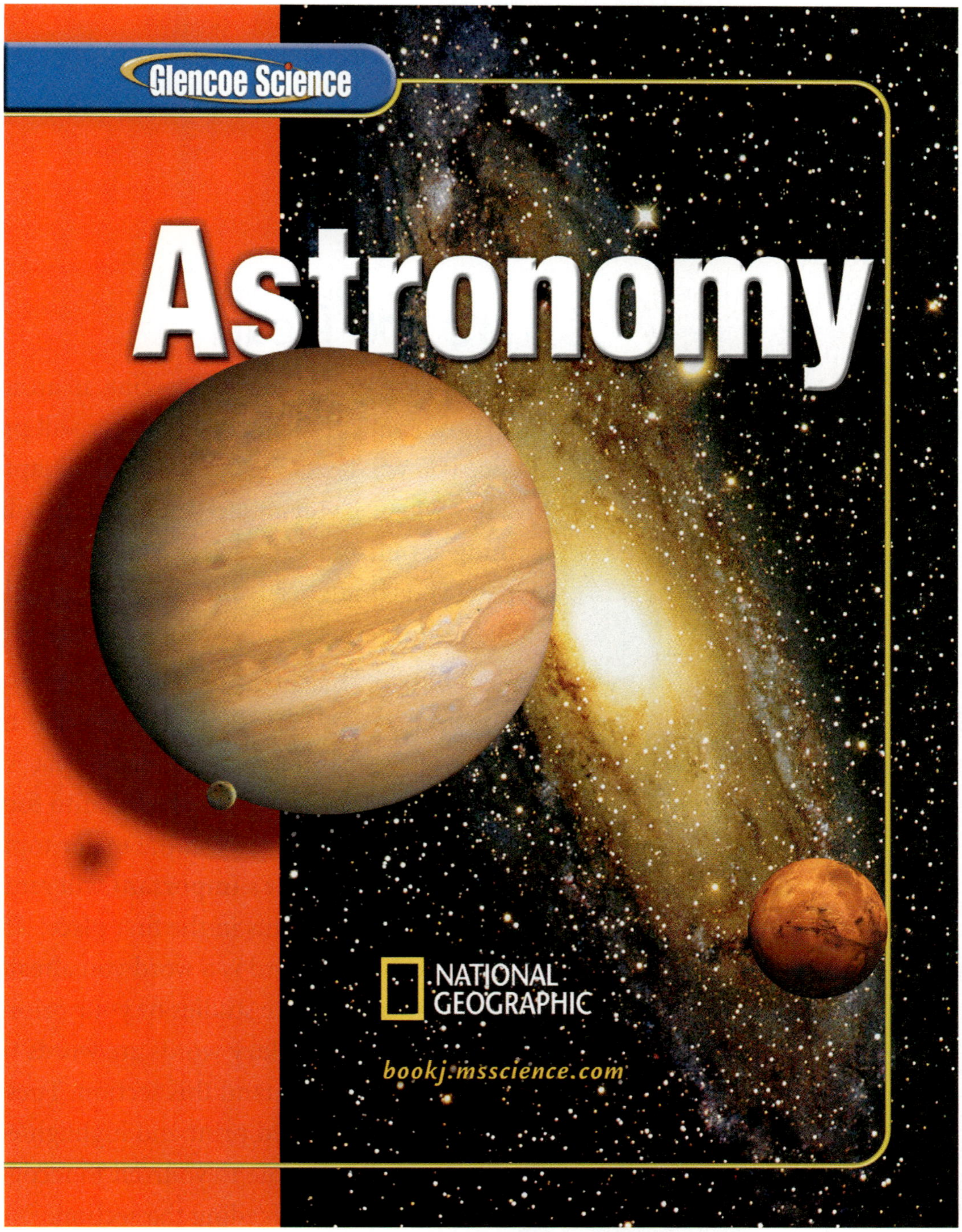

McGraw Hill Glencoe

New York, New York Columbus, Ohio Chicago, Illinois Peoria, Illinois Woodland Hills, California

HOW TO...

Use Your Science Book

Why do I need my science book?

Have you ever been in class and not understood all of what was presented? Or, you understood everything in class, but at home, got stuck on how to answer a question? Maybe you just wondered when you were ever going to use this stuff?

These next few pages are designed to help you understand everything your science book can be used for . . . besides a paperweight!

Before You Read

- **Chapter Opener** Science is occurring all around you, and the opening photo of each chapter will preview the science you will be learning about. The **Chapter Preview** will give you an idea of what you will be learning about, and you can try the **Launch Lab** to help get your brain headed in the right direction. The **Foldables** exercise is a fun way to keep you organized.
- **Section Opener** Chapters are divided into two to four sections. The **As You Read** in the margin of the first page of each section will let you know what is most important in the section. It is divided into four parts. **What You'll Learn** will tell you the major topics you will be covering. **Why It's Important** will remind you why you are studying this in the first place! The **Review Vocabulary** word is a word you already know, either from your science studies or your prior knowledge. The **New Vocabulary** words are words that you need to learn to understand this section. These words will be in **boldfaced** print and highlighted in the section. Make a note to yourself to recognize these words as you are reading the section.

As You Read

- **Headings** Each section has a title in large red letters, and is further divided into blue titles and small red titles at the beginnings of some paragraphs. To help you study, make an outline of the headings and subheadings.
- **Margins** In the margins of your text, you will find many helpful resources. The **Science Online** exercises and **Integrate** activities help you explore the topics you are studying. **MiniLabs** reinforce the science concepts you have learned.
- **Building Skills** You also will find an **Applying Math** or **Applying Science** activity in each chapter. This gives you extra practice using your new knowledge, and helps prepare you for standardized tests.
- **Student Resources** At the end of the book you will find **Student Resources** to help you throughout your studies. These include **Science, Technology,** and **Math Skill Handbooks,** an **English/Spanish Glossary,** and an **Index.** Also, use your **Foldables** as a resource. It will help you organize information, and review before a test.
- **In Class** Remember, you can always ask your teacher to explain anything you don't understand.

FOLDABLES™ Study Organizer

Science Vocabulary Make the following Foldable to help you understand the vocabulary terms in this chapter.

STEP 1 **Fold** a vertical sheet of notebook paper from side to side.

STEP 2 **Cut** along every third line of only the top layer to form tabs.

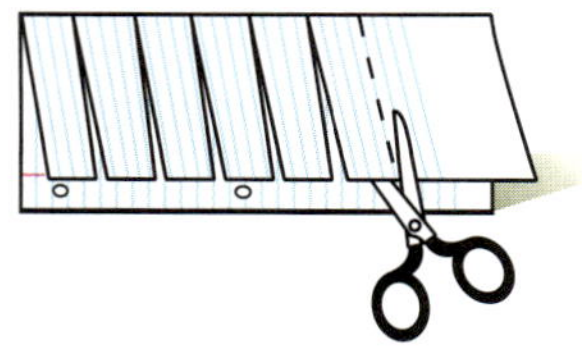

STEP 3 **Label** each tab with a vocabulary word from the chapter.

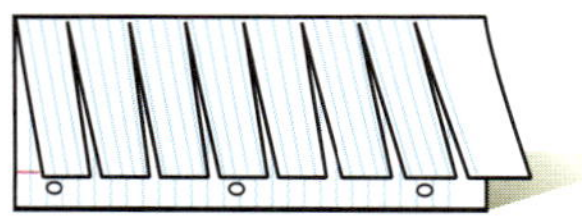

Build Vocabulary As you read the chapter, list the vocabulary words on the tabs. As you learn the definitions, write them under the tab for each vocabulary word.

In Lab

Working in the laboratory is one of the best ways to understand the concepts you are studying. Your book will be your guide through your laboratory experiences, and help you begin to think like a scientist. In it, you not only will find the steps necessary to follow the investigations, but you also will find helpful tips to make the most of your time.

- Each lab provides you with a **Real-World Question** to remind you that science is something you use every day, not just in class. This may lead to many more questions about how things happen in your world.
- Remember, experiments do not always produce the result you expect. Scientists have made many discoveries based on investigations with unexpected results. You can try the experiment again to make sure your results were accurate, or perhaps form a new hypothesis to test.
- Keeping a **Science Journal** is how scientists keep accurate records of observations and data. In your journal, you also can write any questions that may arise during your investigation. This is a great method of reminding yourself to find the answers later.

Before a Test

Admit it! You don't like to take tests! However, there *are* ways to review that make them less painful. Your book will help you be more successful taking tests if you use the resources provided to you.

- Review all of the **New Vocabulary** words and be sure you understand their definitions.
- Review the notes you've taken on your **Foldables,** in class, and in lab. Write down any question that you still need answered.
- Review the **Summaries** and **Self Check questions** at the end of each section.
- Study the concepts presented in the chapter by reading the **Study Guide** and answering the questions in the **Chapter Review.**

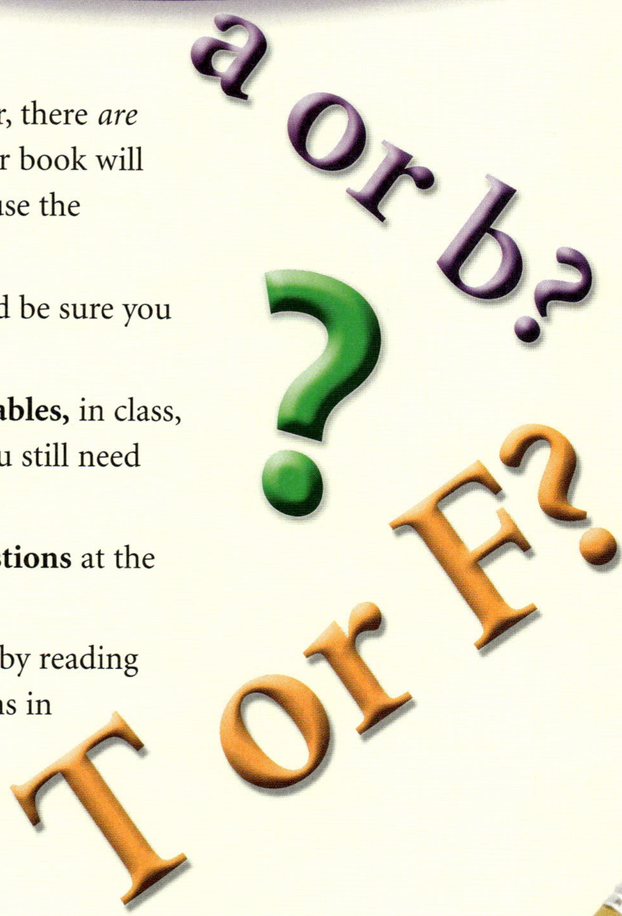

Look For...

- **Reading Checks** and **caption questions** throughout the text.
- the **Summaries** and **Self Check questions** at the end of each section.
- the **Study Guide** and **Review** at the end of each chapter.
- the **Standardized Test Practice** after each chapter.

Scavenger HUNT

Let's Get Started

To help you find the information you need quickly, use the Scavenger Hunt below to learn where things are located in Chapter 1.

1. What is the title of this chapter?
2. What will you learn in Section 1?
3. Sometimes you may ask, "Why am I learning this?" State a reason why the concepts from Section 2 are important.
4. What is the main topic presented in Section 2?
5. How many reading checks are in Section 1?
6. What is the Web address where you can find extra information?
7. What is the main heading above the sixth paragraph in Section 2?
8. There is an integration with another subject mentioned in one of the margins of the chapter. What subject is it?
9. List the new vocabulary words presented in Section 2.
10. List the safety symbols presented in the first Lab.
11. Where would you find a Self Check to be sure you understand the section?
12. Suppose you're doing the Self Check and you have a question about concept mapping. Where could you find help?
13. On what pages are the Chapter Study Guide and Chapter Review?
14. Look in the Table of Contents to find out on which page Section 2 of the chapter begins.
15. You complete the Chapter Review to study for your chapter test. Where could you find another quiz for more practice?

Teacher Advisory Board

The Teacher Advisory Board gave the authors, editorial staff, and design team feedback on the content and design of the Student Edition. They provided valuable input in the development of the 2005 edition of ***Glencoe Science.***

John Gonzales
Challenger Middle School
Tucson, AZ

Rachel Shively
Aptakisic Jr. High School
Buffalo Grove, IL

Roger Pratt
Manistique High School
Manistique, MI

Kirtina Hile
Northmor Jr. High/High School
Galion, OH

Marie Renner
Diley Middle School
Pickerington, OH

Nelson Farrier
Hamlin Middle School
Springfield, OR

Jeff Remington
Palmyra Middle School
Palmyra, PA

Erin Peters
Williamsburg Middle School
Arlington, VA

Rubidel Peoples
Meacham Middle School
Fort Worth, TX

Kristi Ramsey
Navasota Jr. High School
Navasota, TX

Student Advisory Board

The Student Advisory Board gave the authors, editorial staff, and design team feedback on the design of the Student Edition. We thank these students for their hard work and creative suggestions in making the 2005 edition of ***Glencoe Science*** student friendly.

Jack Andrews
Reynoldsburg Jr. High School
Reynoldsburg, OH

Peter Arnold
Hastings Middle School
Upper Arlington, OH

Emily Barbe
Perry Middle School
Worthington, OH

Kirsty Bateman
Hilliard Heritage Middle School
Hilliard, OH

Andre Brown
Spanish Emersion Academy
Columbus, OH

Chris Dundon
Heritage Middle School
Westerville, OH

Ryan Manafee
Monroe Middle School
Columbus, OH

Addison Owen
Davis Middle School
Dublin, OH

Teriana Patrick
Eastmoor Middle School
Columbus, OH

Ashley Ruz
Karrer Middle School
Dublin, OH

The Glencoe middle school science Student Advisory Board taking a timeout at COSI, a science museum in Columbus, Ohio.

Contents

Nature of Science: Life on Mars—2

In each chapter, look for these opportunities for review and assessment:

- Reading Checks
- Caption Questions
- Section Review
- Chapter Study Guide
- Chapter Review
- Standardized Test Practice
- Online practice at bookj.msscience.com

chapter 4 Stars and Galaxies—102

Student Resources

Cross-Curricular Readings/Labs

[video lab icon] available as a video lab

NATIONAL GEOGRAPHIC VISUALIZING

TIME SCIENCE AND Society

TIME SCIENCE AND HISTORY

Oops! Accidents in SCIENCE

SCIENCE Stats

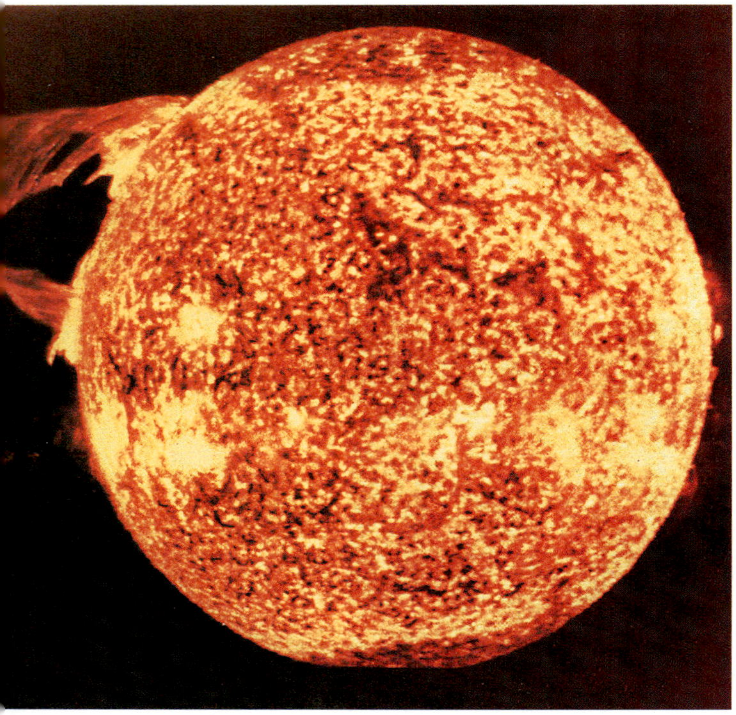

Launch LAB

Mini LAB

Mini LAB Try at Home

One-Page Labs

Two-Page Labs

Labs/Activities

Design Your Own Labs

Model and Invent

Use the Internet Labs

Applying Math

Applying Science

INTEGRATE

Science Online

Standardized Test Practice

The NATURE OF SCIENCE

Life on Mars

Introduction

This feature introduces students to scientific inquiry by capturing their interest and then guiding them through some steps scientists are using to try to determine whether life existed on Mars. Scientists who are examining this question use scientific methods by:

1. Asking the question "Has life ever existed on Mars?"
2. Gathering and using information to form hypotheses;
3. Testing hypotheses by making observations;
4. Describing and analyzing what they observe; and
5. Drawing conclusions about their data.

1 Motivate

Write responses on the board as students brainstorm what they know about Mars. Then ask: **If life did exist on Mars, what do you think it was like?** Have students use the list to help them answer. Possible answer: Organisms might have to get nutrients from sources other than plants.

The NATURE OF SCIENCE

Scientific Methods

Life on Mars

Is there life on Mars? Ever since the 1600s, when scientists first looked at the sky with telescopes and determined that Mars is the most Earthlike planet in the solar system, they have asked this question.

In 1877, Italian scientist Giovanni Schiaparelli saw a network of lines on the surface of Mars and believed they were channels. Later, the American scientist Percival Lowell saw the same lines and claimed they were canals dug by martians.

Today scientists know that flowing water created many of the martian surface features. But scientists still wonder whether simple life-forms existed on Mars or might even exist today. To answer this question, they began undertaking space missions. One objective of the missions is to gather information on whether Mars has or ever had the conditions necessary for life, such as the presence of flowing water.

In 1964, scientists sent a space probe to take photographs of Mars. Examining the photos, they decided the planet is too cold and dry for life. Later probes showed that Mars might have been warm and wet billions of years ago. However, scientists still thought it had been cold and dry since those times.

Figure 1 This martian rock fell to Earth (Antarctica) as a meteorite.

Figure 2 Scientists now know that the lines on Mars's surface were created by flowing water.

Visual Learning

Figure 1 Ask students to study the photograph and think of places on Earth that look similar. Possible answers: Southwest U.S., dried-up riverbeds. **How have these features on Earth formed?** Some have formed by running water that is now gone, and some were formed by plate tectonics.

Then, in June 2000, scientists made an astounding discovery. Photographs taken by a new space probe showed evidence of recent erosion by running water. But if Mars is so cold, how could liquid water exist?

Further study of the new photographs convinced scientists that lava had flowed on Mars in the recent past. This means that Mars's interior must be warmer than previously thought. This heat could melt underground ice and allow it to flow to the surface as liquid water. Liquid water could help support life.

In 1984, scientists in Antarctica found a martian meteorite—a small piece of rock that was blasted into space when Mars was hit by a much larger meteorite. When scientists examined the meteorite with microscopes, they discovered strange shapes inside.

Similar shapes have been found in Earth's rocks and are thought to be the fossilized remains of bacteria that lived billions of years ago. Some scientists thought that the shapes in the martian meteorite were fossils of tiny forms of martian life.

Others thought the shapes were only globules of minerals formed when water changed the rocks on Mars. To test this idea, scientists tried to reproduce the shapes in a laboratory. When their experiment was completed, they saw globules of minerals like those in the martian meteorite. They concluded that the shapes probably were not fossils of martian life-forms.

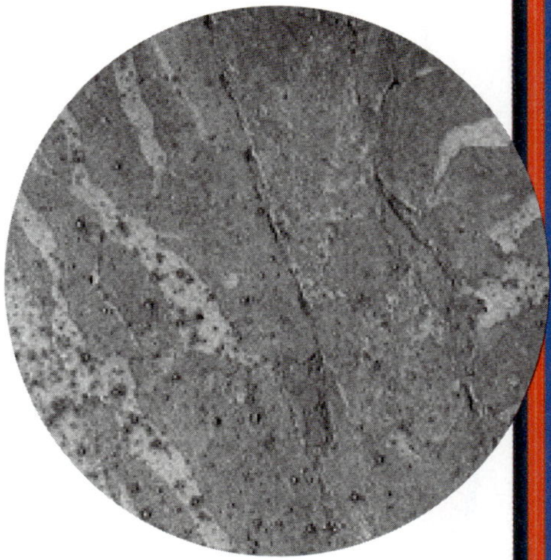

Figure 3 The darker areas in this photograph are newer lava flows that broke up along their edges.

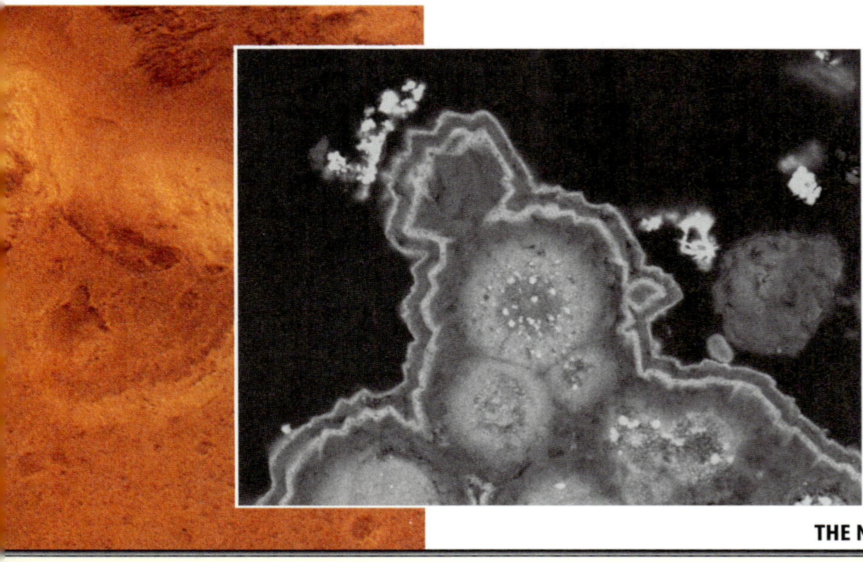

Figure 4 These globules of minerals made in a lab look like the shapes inside the martian meteorite.

2 Teach

Content Background

Mars may not have life on it now, but early in its history it may have been a warm and wet planet with a much thicker atmosphere than it has now. Mars explorers will be looking for evidence of present or past life. If evidence of life is found, it would show that life elsewhere in the universe is also possible.

To study just one geologic formation would require field trips by humans to many different sites. Researchers would like to develop robot technology for use on Mars. One problem with using robots to do in-depth work is distance from Earth to Mars. A robot with artificial intelligence would need to be programmed to do a task and would need some ability to "think" on its own.

Discussion

Why would robots on Mars need to be able to "think" on their own? Signals from Earth to Mars would take so long that a robot could not react immediately to a situation. **What are advantages of a piloted mission to Mars?** Humans can make judgments about what is being observed. They can change the mission if there is a need. Humans might also be able to find clues about the formation of the solar system or life on other planets.

Curriculum Connection

Math Have students calculate how many years it would take to travel from Earth's orbit to Mars' orbit, if they were traveling at 88km/hr, a common highway speed. The average distance between the orbits is about 78 million km. The trip would take about 886,000 years. LS **Logical-Mathematical**

Make a Model

Provide encyclopedias and other books or magazines that contain information about and photographs of the surface of Mars. Using descriptions and photos have students make clay models of the planet Mars. They can contour the clay to match surface features on Mars including mountains, craters, and channels. If different colors of clay are available, students can indicate specific features of the planet, such as the reddish soil. L2 LS **Kinesthetic**

Use Science Words

Word Meaning Have students find the derivation of the word *Mars.* The word *Mars* comes from the Latin root *Mart-* or *Mars.* Mars is the Roman god of war. The planet Mars was named by the Romans after the war god because of its red color.

Discussion

What is a controlled experiment? A control is used and only one factor is changed during the experiment. **Do the actions described in the paragraph on Testing Hypotheses involve a controlled experiment?** No. Point out that scientific methods do not always involve an experiment. Scientists can answer some questions by making observations. Students can work in pairs or small groups to make a list of steps they might use to test their hypotheses. LS **Interpersonal**

The NATURE OF SCIENCE

Science

Trying to find out whether life ever existed on Mars is just one example of doing science. Science is the process of observing, experimenting, and thinking about the universe to create knowledge. In fact, the word *science* comes from the Latin word *scientia,* which means knowledge. Every time you answer a question by observing the world or testing an idea, you are doing science.

The Earth sciences study Earth—its land, oceans, and atmosphere—as well as other objects in the universe. In this book, you will learn about astronomy, the study of outer space.

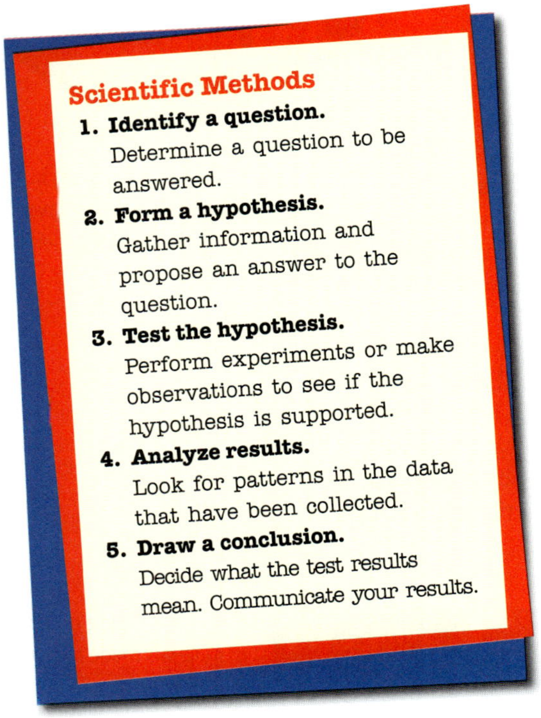

Figure 5 Scientists use scientific methods to answer questions about life on Mars.

Scientific Methods

Many scientists are working to answer the question of whether life ever existed on Mars. These scientists use a variety of methods to try to answer the question. These methods are commonly called scientific methods. **Scientific methods** are procedures used to investigate a question scientifically.

Identifying a Question

The first step in doing science is identifying a question. One such question is *Did life ever exist on Mars?* Answering this question could lead to many others. Scientists might want to know under what kinds of conditions life can survive. They also might want to ask whether the surface of Mars could have met such conditions. If you have ever participated in a science fair, you had to identify a question before you began your project.

Forming a Hypothesis

The next step is to gather information about the question and form a hypothesis. You can find information by going to the library and reading books or magazines, by using the Internet, or by talking to other people about the question. A **hypothesis** is a possible answer to a question. One hypothesis about the shapes in the martian meteorite is that *the shapes are fossils of tiny life-forms that lived on Mars long ago.* Another hypothesis is that *the shapes are globules of minerals formed inside martian rocks.*

Differentiated Instruction

Learning Disabled Reinforce that each step in a scientific method must be completed in a logical order. Caution students to not jump ahead to analyze their data while they are observing and recording. They should write logical steps in their Science Journals.

Testing the Hypothesis

To find out whether a hypothesis is correct, scientists must test it. They do this by performing experiments or making observations. When scientists tried to produce globules of minerals that looked like the shapes in the martian meteorite, they were testing their hypothesis.

Analyzing Results

As scientists perform tests, they collect lots of information, or **data,** that must be analyzed. Data about the martian meteorite include measurements, microscope photographs, and chemical studies of the strange shapes. The test data must be organized and studied. Many times scientists make graphs so they can see patterns in the data. They also use computers to check the data.

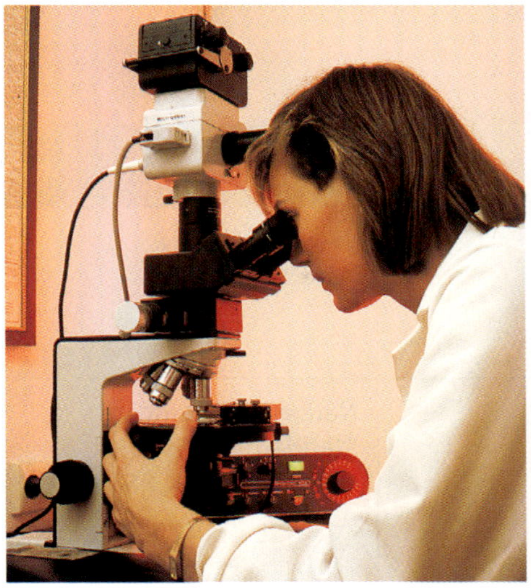

Figure 6 Scientists often use microscopes and other equipment to test hypotheses.

Drawing a Conclusion

Often, the last step in a scientific method is to draw a conclusion. In this step, scientists decide what the results of their tests and observations mean. Sometimes the original hypothesis is not supported by the data. When this happens, the scientists begin again with a new hypothesis. Other times, though, the original hypothesis is supported. If a hypothesis is supported by repeatable experiments and many observations over time, it could become a theory. In science, a **theory** is an idea that has been tested and can explain a large set of observations. For instance, the claim that liquid water has, at some time, flowed over the martian surface is a theory. It might be many years before scientists know whether any of the hypotheses about the martian meteorite and life on Mars are correct.

You Do It

In recent years, scientists have discovered microscopic organisms living kilometers beneath the surface of Earth. Some scientists have hypothesized that simple life-forms might exist deep below the surface of Mars, too. Describe one way that scientists could test this hypothesis.

Activity

Ask students to list possible ways that data can be organized. Possible answers: line, bar, and circle graphs, a spreadsheet, tables. Have students consider the question "What is the composition of soil on Mars?" Provide these data on Martian soil to students: Silicon oxide—60%, aluminum oxide—20%, iron oxide—5%, calcium oxide—5%, other—10%. Have students make circle or bar graphs of these data.

Extension

Have students consider what other factors would need to be examined to determine if the soil on Mars is similar to that of Earth. Possible answer: the composition of soil on Earth **How could the different factors be compared?** Possible answer: Draw circle graphs of each. Have students make a graph that shows the average composition of soil on Earth and draw conclusions about whether the soil on Mars is similar to that of Earth.

3 Assess

Use the You Do It to assess understanding. Answers should relate to what is known about Mars and should be as specific as possible.

You Do It

Lead students through this exercise as a class to reinforce the steps in a scientific method. Emphasize that there are no right or wrong answers. Scientists may ask the same question in different ways and use many different methods to test hypotheses. Scientists also may interpret data differently. **How are hypotheses supported?** Hypotheses are supported when experiments and observations provide data that can be explained logically by the hypothesis. Hypotheses are not supported when data demand an alternate explanation.

chapter 1 Organizer

Section/Objectives	Standards		Labs/Features
	National	**State/Local**	
Chapter Opener	See pp. 9T–10T for a Key to Standards.		**Launch Lab:** An Astronomer's View, p. 7 **Foldables,** p. 7
Section 1 Radiation from Space 2 sessions 1 block 1. **Explain** the electromagnetic spectrum. 2. **Identify** the differences between refracting and reflecting telescopes. 3. **Recognize** the differences between optical and radio telescopes.	National Content Standards: UCP.1, UCP.2, UCP.3, UCP.5, A.1, A.2, D.3		**Integrate Health,** p. 9 **MiniLAB:** Observing Effects of Light Pollution, p. 12 **Lab:** Building a Reflecting Telescope, p. 14
Section 2 Early Space Missions 1 session 0.5 block 4. **Compare and contrast** natural and artificial satellites. 5. **Identify** the differences between artificial satellites and space probes. 6. **Explain** the history of the race to the Moon.	National Content Standards: UCP.1, UCP.2, UCP.3, UCP.5, A.1, A.2, D.3		**Applying Math:** Drawing by Numbers, p. 16 **Integrate Career,** p. 18 **Visualizing Space Probes,** p. 19 **Science Online,** p. 20 **MiniLAB:** Modeling a Satellite, p. 21
Section 3 Current and Future Space Missions 4 sessions 2 blocks 7. **Explain** the benefits of the space shuttle. 8. **Identify** the usefulness of orbital space stations. 9. **Explore** future space missions. 10. **Identify** the applications of space technology to everyday life.	National Content Standards: UCP.1, UCP.2, UCP.3, UCP.5, A.1, A.2, D.3, F.5		**Science Online,** p. 25 **Science Online,** p. 27 **Lab:** Star Sightings, pp. 30–31 **Science and Society:** Cities in Space, p. 32

Glencoe Exclusive! TeacherWorks™ All-In-One Planner and Resource Center

Lab Materials	Reproducible Resources	Section Assessment	Technology
Launch Lab: clear-plastic wrap (15 cm), textbook	**Chapter *FAST FILE* Resources** Foldables Worksheet, p. 13 Directed Reading Overview, p. 15 Note-taking Worksheets, pp. 29–31	Glencoe's Assessment Advantage	TeacherWorks includes: • Interactive Teacher Edition • Lesson Planner with calendar • Access to all program blacklines • Correlations to standards • Web links
MiniLAB: cardboard tube **Lab:** flat mirror, curved concave mirror, magnifying lenses of different magnifications (3–4)	**Chapter *FAST FILE* Resources** Transparency Activity, p. 638 MiniLAB, p. 3 Enrichment, p. 26 Reinforcement, p. 23 Directed Reading, p. 16 Lab Worksheets, pp. 5–6 Transparency Activity, pp. 43–44 Lab Activity, pp. 9–12 **Cultural Diversity,** p. 51	**Portfolio** Science Journal, p. 11 **Performance** MiniLAB, p. 12 Applying Math, p. 13 **Content** Section Review, p. 13	Section Focus Transparency Teaching Transparency Guided Reading Audio Program Interactive Chalkboard CD-ROM
MiniLAB: string, cork *Need materials?* Contact Science Kit at 1-800-828-7777 or www.sciencekit.com on the Internet.	**Chapter *FAST FILE* Resources** Transparency Activity, p. 41 MiniLAB, p. 4 Enrichment, p. 27 Reinforcement, p. 24 Directed Reading, p. 17 **Mathematics Skill Activities,** p. 15 **Reading and Writing Skill Activities,** p. 11	**Portfolio** Activity, p. 19 **Performance** Applying Math, p. 16 MiniLAB, p. 21 Applying Math, p. 22 **Content** Section Review, p. 22	Section Focus Transparency Virtual Labs CD-ROM Guided Reading Audio Program Interactive Chalkboard CD-ROM
Lab: computer, Internet and other resources on space technology	**Chapter *FAST FILE* Resources** Transparency Activity, p. 42 Enrichment, p. 28 Reinforcement, p. 25 Directed Reading, pp. 17, 18 Lab Worksheets, pp. 7–8 **Home and Community Involvement,** p. 31 **Physical Science Critical Thinking/Problem Solving,** p. 5 **Cultural Diversity,** p. 25	**Portfolio** Curriculum Connection, p. 25 **Performance** Applying Math, p. 29 **Content** Section Review, p. 29	Section Focus Transparency Guided Reading Audio Program Interactive Chalkboard CD-ROM

End of Chapter Assessment

Blackline Masters	Technology	Professional Series
Chapter *FAST FILE* Resources Chapter Review, pp. 33–34 Chapter Tests, pp. 35–38 **Standardized Test Practice,** pp. 7–10	MindJogger Videoquiz Virtual Labs CD-ROM Exam*View*® Pro Testmaker TeacherWorks CD-ROM Interactive Chalkboard CD-ROM	**Performance Assessment in the Science Classroom (PASC)**

chapter 1 Exploring Space

Transparencies

Section Focus

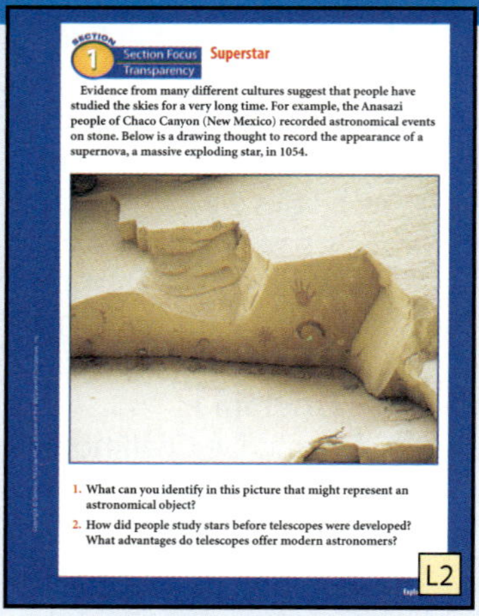

Section 1 Section Focus Transparency **Superstar**

Evidence from many different cultures suggest that people have studied the skies for a very long time. For example, the Anasazi people of Chaco Canyon (New Mexico) recorded astronomical events on stone. Below is a drawing thought to record the appearance of a supernova, a massive exploding star, in 1054.

1. What can you identify in this picture that might represent an astronomical object?
2. How did people study stars before telescopes were developed? What advantages do telescopes offer modern astronomers?

L2

Section 2 Section Focus Transparency **The Outer Limits of the Solar System**

Where does the solar system end? The planets end at Pluto, but the effects of the Sun's gravitational pull extend much farther. The last items held by the Sun's gravitational pull are in Oort's Cloud.

1. How can satellites and space probes help us get a better understanding of the planets in our solar system?
2. How can space missions help us study Earth?

L2

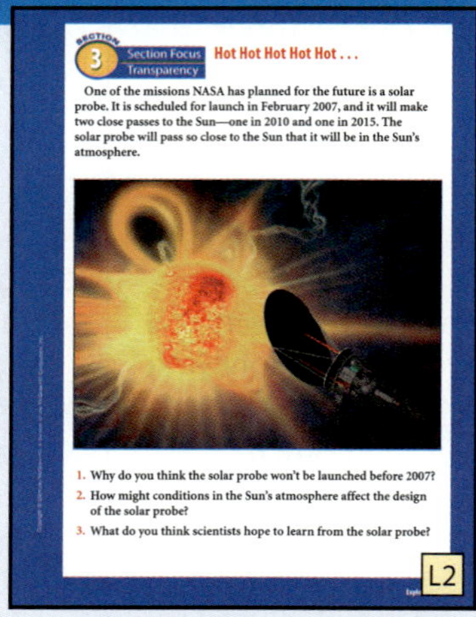

Section 3 Section Focus Transparency **Hot Hot Hot Hot . . .**

One of the missions NASA has planned for the future is a solar probe. It is scheduled for launch in February 2007, and it will make two close passes to the Sun—one in 2010 and one in 2015. The solar probe will pass so close to the Sun that it will be in the Sun's atmosphere.

1. Why do you think the solar probe won't be launched before 2007?
2. How might conditions in the Sun's atmosphere affect the design of the solar probe?
3. What do you think scientists hope to learn from the solar probe?

L2

This is a representation of key blackline masters available in the Teacher Classroom Resources. See Resource Manager boxes within the chapter for additional information.

Assessment

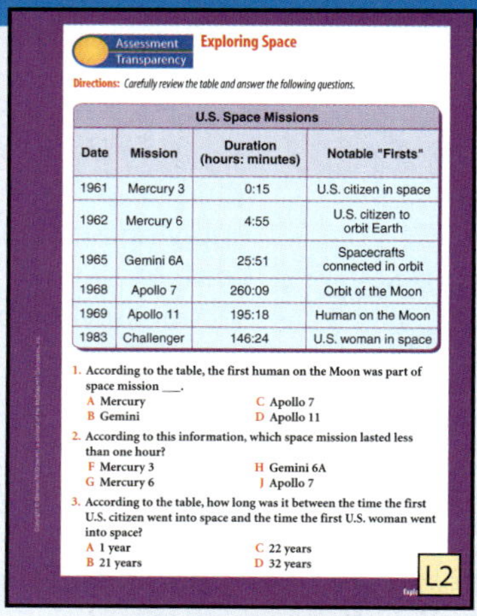

Assessment Transparency **Exploring Space**

Directions: *Carefully review the table and answer the following questions.*

U.S. Space Missions

Date	Mission	Duration (hours: minutes)	Notable "Firsts"
1961	Mercury 3	0:15	U.S. citizen in space
1962	Mercury 6	4:55	U.S. citizen to orbit Earth
1965	Gemini 6A	25:51	Spacecrafts connected in orbit
1968	Apollo 7	260:09	Orbit of the Moon
1969	Apollo 11	195:18	Human on the Moon
1983	Challenger	146:24	U.S. woman in space

1. According to the table, the first human on the Moon was part of space mission ___.
   A Mercury C Apollo 7
   B Gemini D Apollo 11
2. According to this information, which space mission lasted less than one hour?
   F Mercury 3 H Gemini 6A
   G Mercury 6 J Apollo 7
3. According to the table, how long was it between the time the first U.S. citizen went into space and the time the first U.S. woman went into space?
   A 1 year C 22 years
   B 21 years D 32 years

L2

Teaching

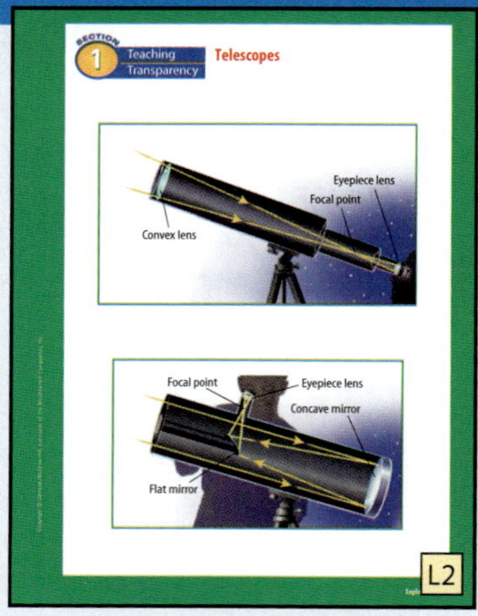

Section 1 Teaching Transparency **Telescopes**

L2

Key to Teaching Strategies

The following designations will help you decide which activities are appropriate for your students.

L1 Level 1 activities should be appropriate for students with learning difficulties.

L2 Level 2 activities should be within the ability range of all students.

L3 Level 3 activities are designed for above-average students.

ELL ELL activities should be within the ability range of English Language Learners.

COOP LEARN Cooperative Learning activities are designed for small group work.

LS Multiple Learning Styles logos, as described on page 6T, are used throughout to indicate strategies that address different learning styles.

P These strategies represent student products that can be placed into a best-work portfolio.

PBL Problem-Based Learning activities apply real-world situations to learning.

Hands-on Activities

Student Text Lab Worksheet

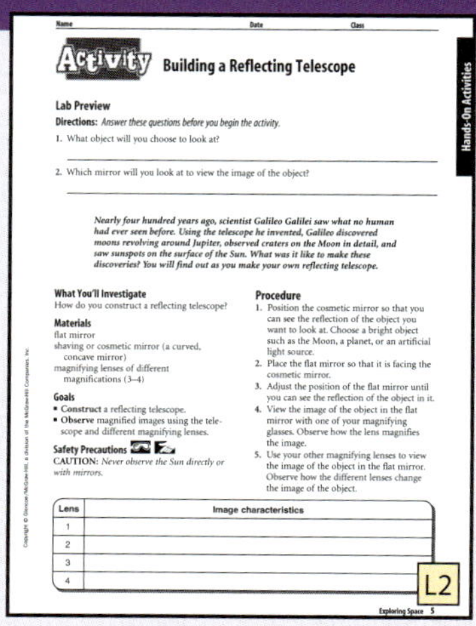

Activity **Building a Reflecting Telescope**

L2

Laboratory Activities

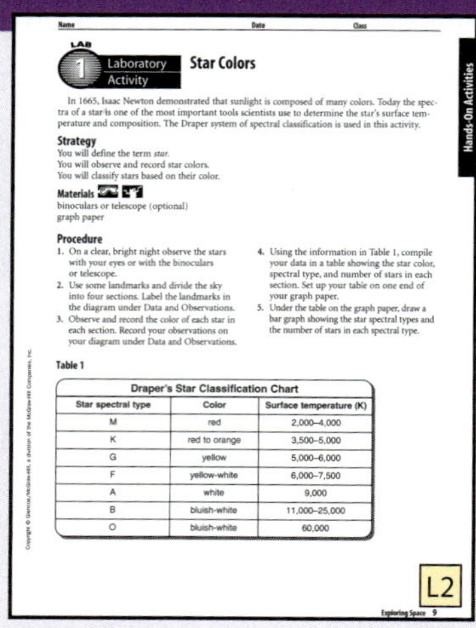

Laboratory Activity 1 **Star Colors**

Table 1

Draper's Star Classification Chart

Star spectral type	Color	Surface temperature (K)
M	red	2,000–4,000
K	red to orange	3,500–5,000
G	yellow	5,000–6,000
F	yellow-white	6,000–7,500
A	white	9,000
B	bluish-white	11,000–25,000
O	bluish-white	60,000

L2

Meeting Different Ability Levels

Content Outline

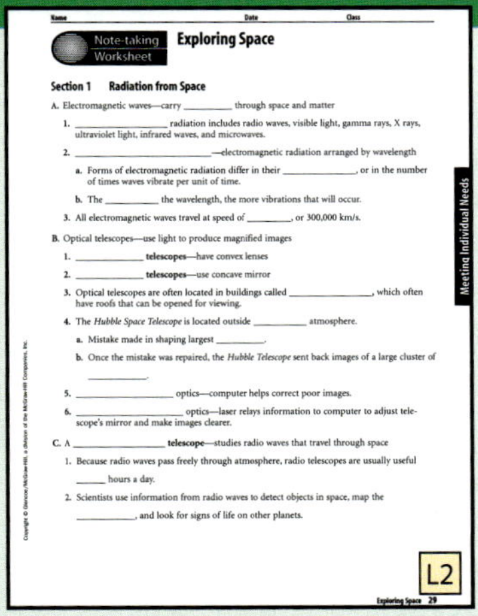

Reinforcement

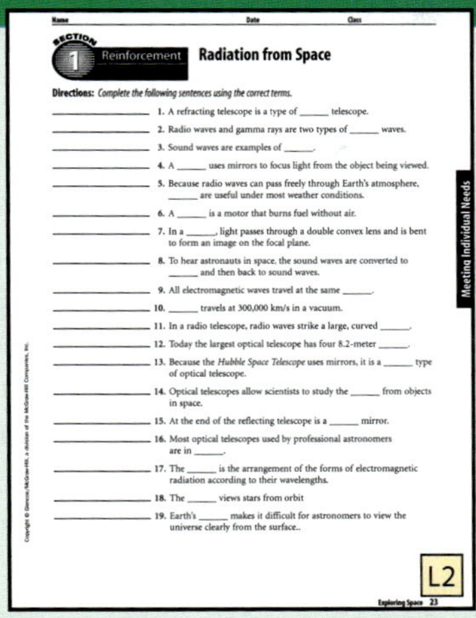

Enrichment

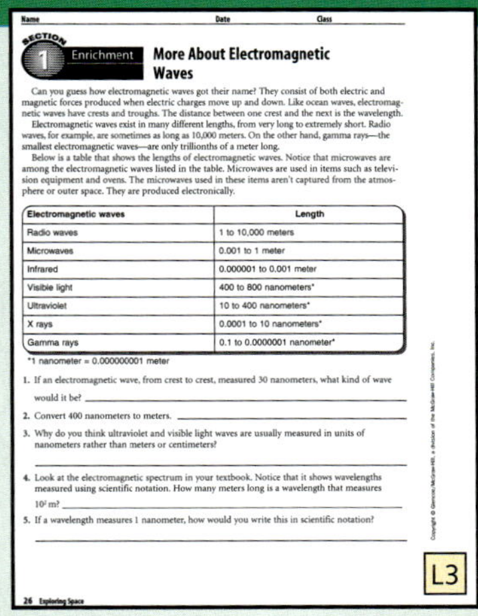

Directed Reading (English/Spanish)

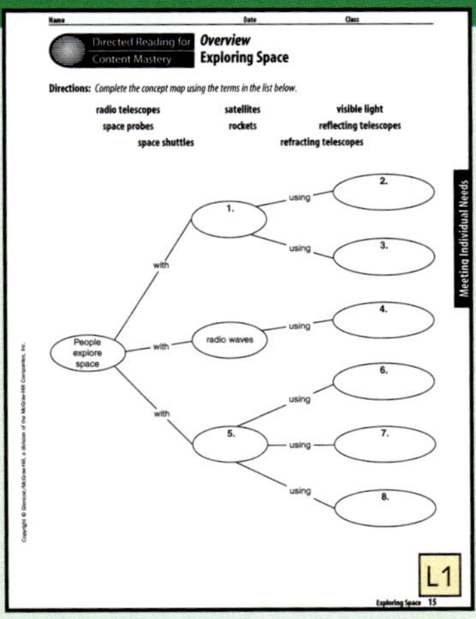

Study Guide

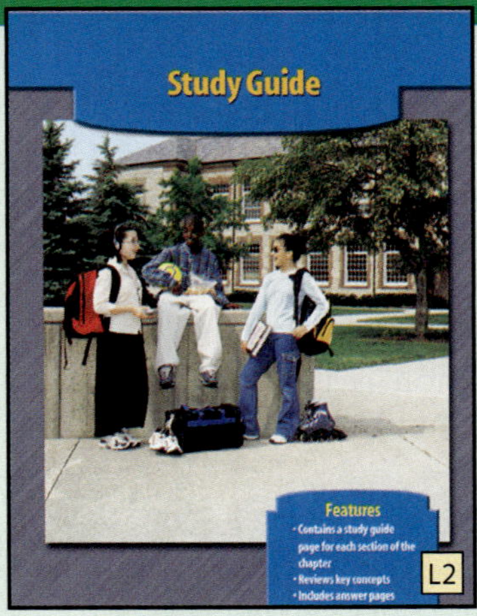

Reading Essentials

Assessment

Test Practice Workbook

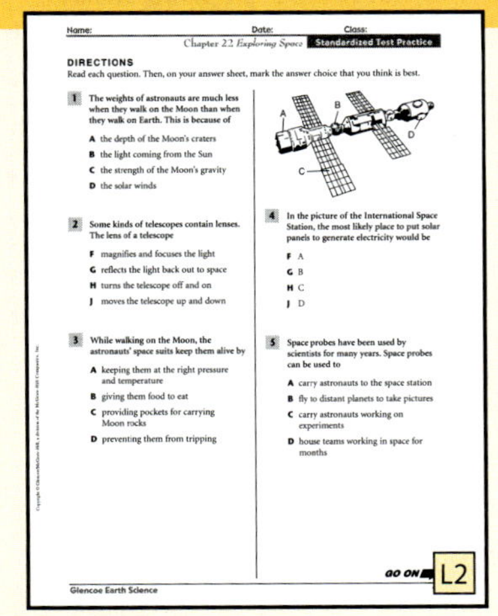

Chapter Review

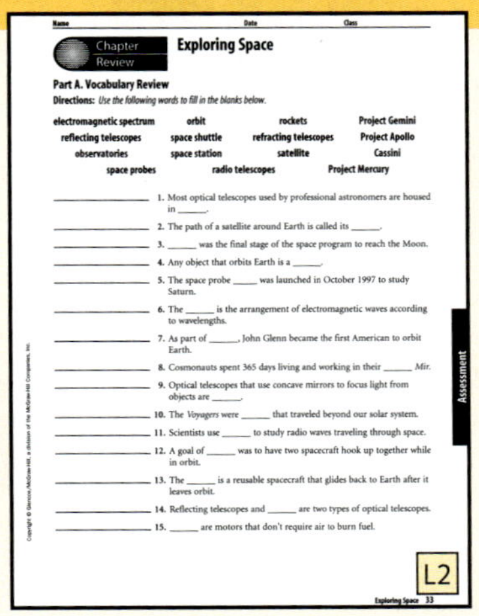

Chapter Tests

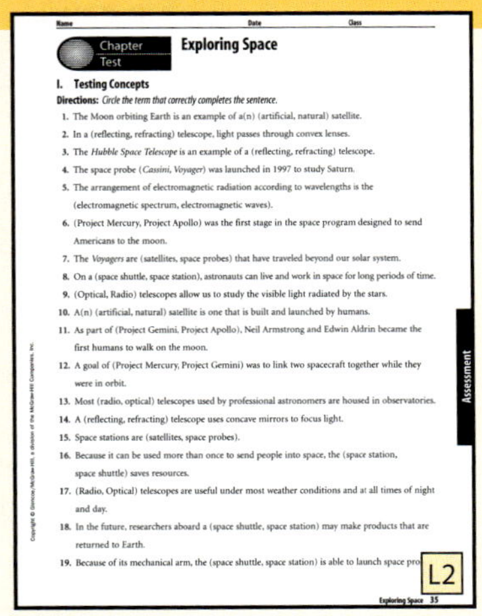

Science Content Background

section 1 Radiation from Space

Large Telescopes

On April 21, 1998, the largest telescope on Earth captured light from its first star. The European Southern Observatory's Very Large Telescope (VLT) is composed of four identical 8.2-m individual telescopes and several movable 1.8-m telescopes.

Other large telescopes include the twin 10-m Keck reflecting telescopes on Mauna Kea in Hawaii. The primary mirror of each telescope has been constructed from 36 hexagonal glass segments, each 1.8 m in diameter.

The NASA Orbital Debris Observatory (NODO) uses a liquid mirror telescope. The objective mirror is made of a shallow container of the liquid element mercury. It is 3 m in diameter and is constantly spinning at one revolution every 6.02315 seconds. This spinning forces the liquid mercury to form a parabolic reflector. The telescope is kept level; thus, it observes only the sky's zenith. Its main mission is to locate and count orbital debris (space junk) as small as 1 cm in diameter.

The Very Large Array (VLA) of telescopes near Socorro, New Mexico, is composed of 27 radio telescopes. All 27 telescopes can be used at once to operate as one large telescope.

Light Pollution

Light pollution is a glow in the sky caused by city lights. The glow makes it difficult to see dim stars.

In several United States cities, work has begun to reduce light pollution. Tucson, Arizona, located only 80 km from the Kitt Peak National Observatory, has replaced its street lighting with low-pressure sodium lamps. These lamps shine at wavelengths that can be filtered out by astronomers. Other cities have put hoods on billboards, parking-lot lights, and floodlights so they illuminate the object or the ground rather than the sky.

Teacher to Teacher
Ralph M. Feather, Jr.
Derry Area High School
Derry, PA

"To help my students understand that the Moon's phases are caused by the position of the Moon in relation to Earth and Sun, I have them make observations of the Moon at approximately the same time each day for three to four days."

Ralph Feather

section 2 Early Space Missions

The First Missions into Space

The first mission into space was conducted by the Soviet Union, which launched the uncrewed *Sputnik 1*. It was designed to prove that human-made satellites could orbit Earth in the manner of a natural satellite, such as the moon. Within a year of that launch, the United States had built its own satellite and sent it into space. Known as *Explorer 1*, it was followed by three more attempts, all unsuccessful, to send probes into an orbit similar to that of Earth's Moon.

In 1959 the Soviet Union did what the United States hadn't been able to do. Their probe, *Luna 1*, became the first human-made satellite to travel past the Moon. Two months later, American attempts at a lunar flyby, or voyage past the Moon, became successful with *Pioneer 4*. The United States followed this successful effort with a series of probes known as *Ranger 1* and *2*. Both were failures. Finally, with *7*, *8*, and *9*, the United States was able to capture detailed pictures of the Moon's surface. This led into the Apollo program, which placed astronauts on the Moon itself.

section 3 Current and Future Space Missions

Water on Other Planets

Jupiter's moon Europa may contain water. Recent preliminary studies of Europa by the *Galileo* spacecraft indicate an ocean of water or ice under the moon's crustal layer of ice.

The *Cassini* mission to Saturn will have similar goals to the Galileo mission to Jupiter. The science goals of the Cassini mission involve studying and collecting data on Saturn, its magnetosphere, rings, icy moons, and Titan.

The International Space Station

The *International Space Station* will be a permanently crewed platform in which teams of astronauts from many nations will work cooperatively in space. Launch of materials and modules that are part of the *International Space Station* assembly began with the launch of the space shuttle *Endeavor*. Work continues on this international undertaking.

Everyday Space Technology

Much of the technology developed by NASA to achieve its goals in space is now being used by people throughout the world. These technologies are called spin-offs. For example, NASA developed lightweight, compact breathing systems for astronauts to carry as they ventured out of their spacecraft and onto the Moon. Today, firefighters use these breathing systems as well as fire-resistant uniforms originally designed as flight suits for NASA pilots. The lightweight material in the suits won't burn or crack. A material designed for boots worn by astronauts on the Moon is now found in some athletic shoes. Other materials have been incorporated into ski goggles, blankets, and bicycle seats.

People who are visually impaired also have benefited from spin-offs. One device vibrates ink on a printed page, enabling them to read materials that are not printed in Braille. Another device determines the denomination of currency and generates an audible signal. Other spin-offs include pens that write without the help of gravity and sunglasses that adjust to various light levels.

David Ducros/Photo Researchers, Inc.

chapter content resources

Internet Resources

For additional content background, visit **bookj.msscience.com** to:

- access your book online
- find references to related articles in popular science magazines
- access Web links with related content background
- access current events with science journal topics

Print Resources

Celestial Delights, Francis Reddy and Greg Walz-Chojnocki, Celestial Arts, 2002

A+ Projects in Astronomy, Janice Van Cleave, John Wiley & Sons, 2002

The Stars—A New Way to See Them, H.A. Rey, Houghton Mifflen Co., 1997

How to Identify the Night Sky, Storm Dunlop and Wil Tircon, Barnes & Noble Inc., 2003

chapter 1

Chapter Vocabulary

electromagnetic spectrum, p. 9
refracting telescope, p. 10
reflecting telescope, p. 10
observatory, p. 10
radio telescope, p. 13
rocket, p. 15
satellite, p. 17
orbit, p. 17
space probe, p. 18
Project Mercury, p. 21
Project Gemini, p. 21
Project Apollo, p. 22
space shuttle, p. 23
space station, p. 24

Science Journal Student answers will vary, depending on their value on space exploration.

INTERACTIVE CHALKBOARD with Image Bank

PowerPoint® Presentations

This CD-ROM is an editable Microsoft® PowerPoint® presentation that includes:

- a pre-made presentation for every chapter
- interactive graphics
- animations
- audio clips
- image bank
- all new section and chapter questions
- Standardized Test Practice
- transparencies
- pre-lab questions for all labs
- Foldables directions
- links to bookj.msscience.com

chapter 1

Exploring Space

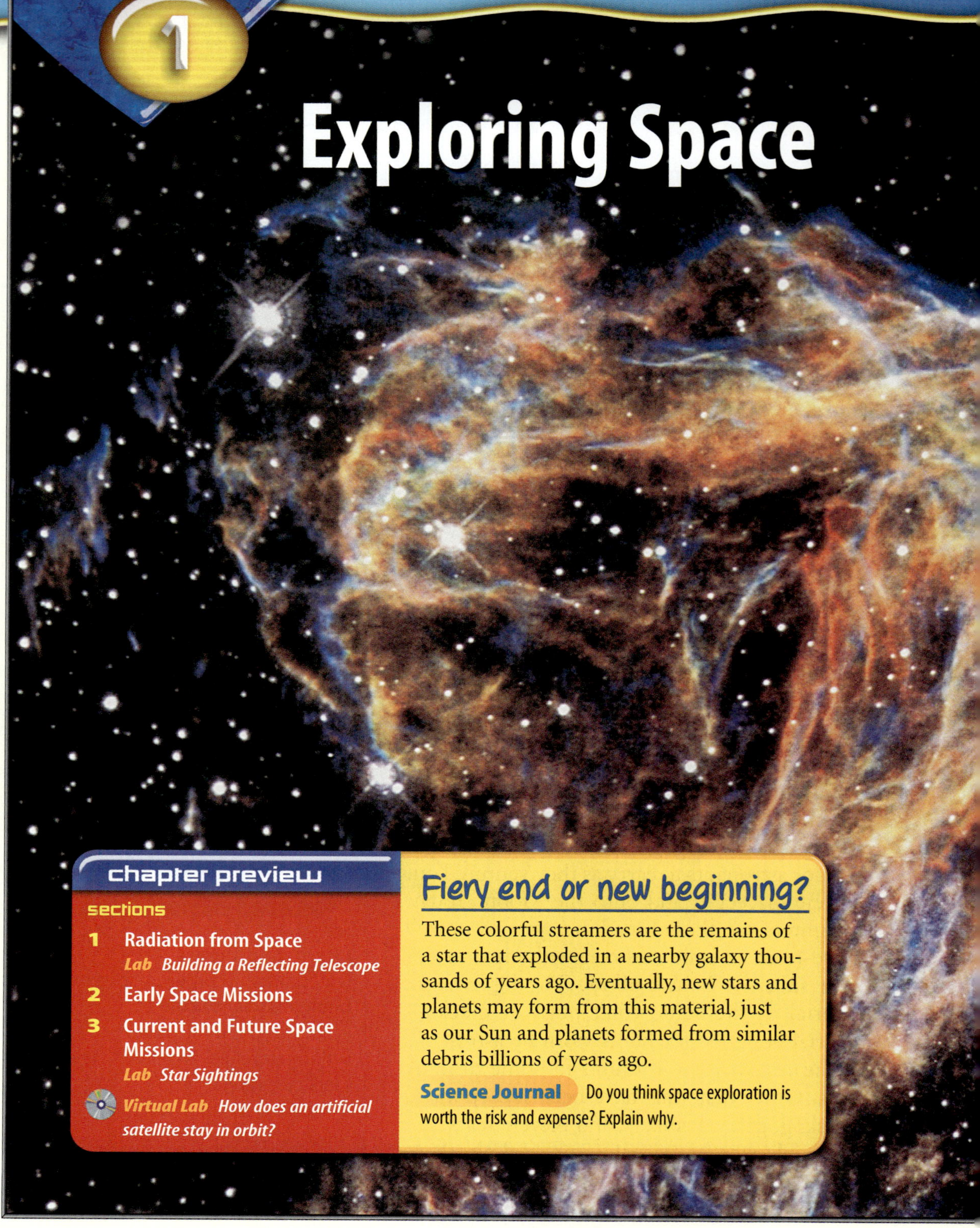

chapter preview

sections

1. **Radiation from Space**
   Lab Building a Reflecting Telescope
2. **Early Space Missions**
3. **Current and Future Space Missions**
   Lab Star Sightings

Virtual Lab How does an artificial satellite stay in orbit?

Fiery end or new beginning?

These colorful streamers are the remains of a star that exploded in a nearby galaxy thousands of years ago. Eventually, new stars and planets may form from this material, just as our Sun and planets formed from similar debris billions of years ago.

Science Journal Do you think space exploration is worth the risk and expense? Explain why.

Theme Connection

Energy Students are introduced to the way astronomers use various types of electromagnetic radiation (energy) to explore the solar system and the rest of the universe.

About the Photo

Supernova The remnants of the supernova explosion shown here are located in the Large Magellanic Cloud, a small companion galaxy to our own Milky Way. Somewhere in this swirling mass of gas and dust is a spinning neutron star or pulsar. Pulsars emit no visible radiation but can be detected in other electromagnetic spectra, such as X rays and gamma rays. Gamma ray emissions from this pulsar were detected in 1979.

Start-Up Activities

Launch LAB

An Astronomer's View

You might think exploring space with a telescope is easy because the stars seem so bright and space is dark. But starlight passing through Earth's atmosphere, and differences in temperature and density of the atmosphere can distort images.

1. Cut off a piece of clear plastic wrap about 15 cm long.
2. Place an opened book in front of you and observe the clarity of the text.
3. Hold the piece of plastic wrap close to your eyes, keeping it taut using both hands.
4. Look at the same text through the plastic wrap.
5. Fold the plastic wrap in half and look at the text again through both layers.
6. **Think Critically** Write a paragraph in your Science Journal comparing reading text through plastic wrap to an astronomer viewing stars through Earth's atmosphere. Predict what might occur if you increased the number of layers.

Science Online: Preview this chapter's content and activities at bookj.msscience.com

FOLDABLES Study Organizer

Exploring Space Make the following Foldable to help identify what you already know, what you want to know, and what you learned about exploring space.

STEP 1 **Fold** a vertical sheet of paper from side to side with the front edge about 1.25 cm shorter than the back.

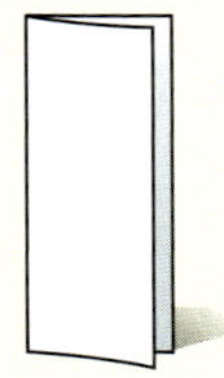

STEP 2 **Turn** lengthwise and **fold** into thirds.

STEP 3 **Unfold and cut** only the top layer along both folds to make three tabs. **Label** each tab.

Identify Questions Before you read the chapter, write what you already know about exploring space under the left tab of your Foldable, and write questions about what you'd like to know under the center tab. After you read the chapter, list what you learned under the right tab.

Launch LAB

Purpose Use this lab to introduce students to how the atmosphere distorts viewing.

Preparation Have available a large roll of clear-plastic wrap. No other advance preparation is needed.

Materials Give each student a piece of plastic wrap about 15 cm long.

Teaching Strategy Allow students to follow directions. Check to see that they are keeping the wrap taut without wrinkles. Wrap should be held about 3 cm from their eyes or glasses if they are wearing them.

Think Critically

Some students should notice a slight difference looking through one layer. Encourage them to compare their views with and without the wrap. Adding the second layer makes the difference more obvious and a third layer causes more distortion. Point out that even when the air seems clear, slight distortions may be present. The layers of plastic mimic the effect of layers of atmosphere through which starlight must pass.

Assessment

Process Have students infer from their results why astronomical observatories are built on high mountains. The atmosphere affects starlight. On high mountains, starlight passes through less air and celestial objects can be seen more clearly. Also, they are above layers of smog and other pollution. Use **Performance Assessment in the Science Classroom**, p. 89.

FOLDABLES Study Organizer — Dinah Zike Study Fold

Student preparation materials for this Foldable are available in the Chapter *FAST FILE* Resources.

1 Motivate

INTERACTIVE CHALKBOARD
PowerPoint® Presentations

Bellringer

Section Focus Transparencies also are available on the Interactive Chalkboard CD-ROM. L2 ELL

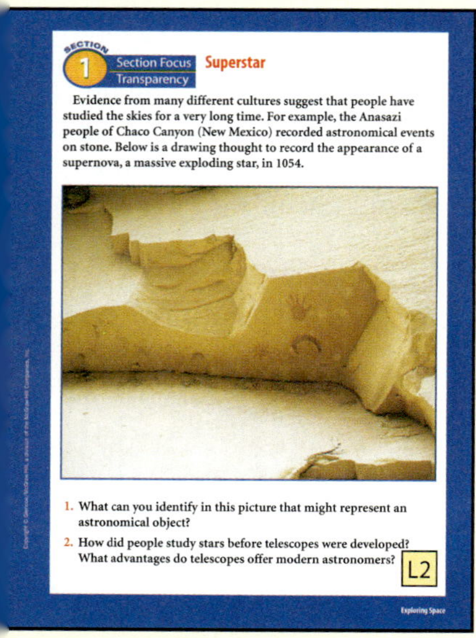
SECTION 1 Section Focus Transparency **Superstar**

Evidence from many different cultures suggest that people have studied the skies for a very long time. For example, the Anasazi people of Chaco Canyon (New Mexico) recorded astronomical events on stone. Below is a drawing thought to record the appearance of a supernova, a massive exploding star, in 1054.

1. What can you identify in this picture that might represent an astronomical object?
2. How did people study stars before telescopes were developed? What advantages do telescopes offer modern astronomers? L2

Exploring Space

Tie to Prior Knowledge

Electromagnetic Radiation Ask for a show of hands for each of these questions: Have you ever used electromagnetic radiation? How many have had an X ray? Used a microwave? Listened to the radio? Seen sunshine or felt the Sun's heat? Explain that anyone who has raised his or her hand has experienced electromagnetic radiation.

section 1

Radiation from Space

as you read

What You'll Learn

- **Explain** the electromagnetic spectrum.
- **Identify** the differences between refracting and reflecting telescopes.
- **Recognize** the differences between optical and radio telescopes.

Why It's Important

Learning about space can help us better understand our own world.

Review Vocabulary

telescope: an instrument that can magnify the size of distant objects

New Vocabulary

- electromagnetic spectrum
- refracting telescope
- reflecting telescope
- observatory
- radio telescope

Electromagnetic Waves

As you have read, we have begun to explore our solar system and beyond. With the help of telescopes like the *Hubble,* we can see far into space, but if you've ever thought of racing toward distant parts of the universe, think again. Even at the speed of light it would take many years to reach even the nearest stars.

Light from the Past When you look at a star, the light that you see left the star many years ago. Although light travels fast, distances between objects in space are so great that it sometimes takes millions of years for the light to reach Earth.

The light and other energy leaving a star are forms of radiation. Radiation is energy that is transmitted from one place to another by electromagnetic waves. Because of the electric and magnetic properties of this radiation, it's called electromagnetic radiation. Electromagnetic waves carry energy through empty space and through matter.

Electromagnetic radiation is everywhere around you. When you turn on the radio, peer down a microscope, or have an X ray taken—you're using various forms of electromagnetic radiation.

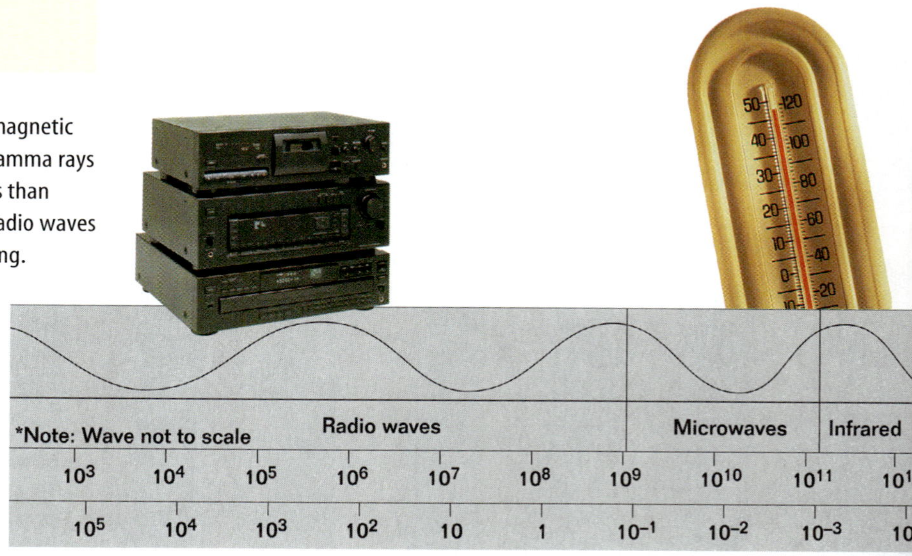

Figure 1 The electromagnetic spectrum ranges from gamma rays with wavelengths of less than 0.000 000 000 01 m to radio waves more than 100,000 m long. **Observe** *how frequency changes as wavelength shortens.*

Section 1 Resource Manager

Chapter *FAST FILE* Resources
Transparency Activity, pp. 40, 43–44
Directed Reading for Content Mastery, pp. 15–16
Note-taking Worksheets, pp. 29–31
Enrichment, p. 26
Reinforcement, p. 23
MiniLab, p. 3
Lab Worksheets, pp. 5–6
Lab Activities, pp. 9–12
Cultural Diversity, p. 3

Electromagnetic Radiation Sound waves, which are a type of mechanical wave, can't travel through empty space. How, then, do we hear the voices of the astronauts while they're in space? When astronauts speak into a microphone, the sound waves are converted into electromagnetic waves called radio waves. The radio waves travel through space and through Earth's atmosphere. They're then converted back into sound waves by electronic equipment and audio speakers.

Radio waves and visible light from the Sun are just two types of electromagnetic radiation. Other types include gamma rays, X rays, ultraviolet waves, infrared waves, and microwaves. **Figure 1** shows these forms of electromagnetic radiation arranged according to their wavelengths. This arrangement of electromagnetic radiation is called the **electromagnetic spectrum.** Forms of electromagnetic radiation also differ in their frequencies. Frequency is the number of wave crests that pass a given point per unit of time. The shorter the wavelength is, the higher the frequency, as shown in **Figure 1.**

Speed of Light Although the various electromagnetic waves differ in their wavelengths, they all travel at 300,000 km/s in a vacuum. This is called the speed of light. Visible light and other forms of electromagnetic radiation travel at this incredible speed, but the universe is so large that it takes millions of years for the light from some stars to reach Earth.

When electromagnetic radiation from stars and other objects reaches Earth, scientists use it to learn about its source. One tool for studying such electromagnetic radiation is a telescope.

INTEGRATE Health

Ultraviolet Light Many newspapers include an ultraviolet (UV) index to urge people to minimize their exposure to the Sun. Compare the wavelengths and frequencies of red and violet light, shown below in **Figure 1.** Infer what properties of UV light cause damage to tissues of organisms.

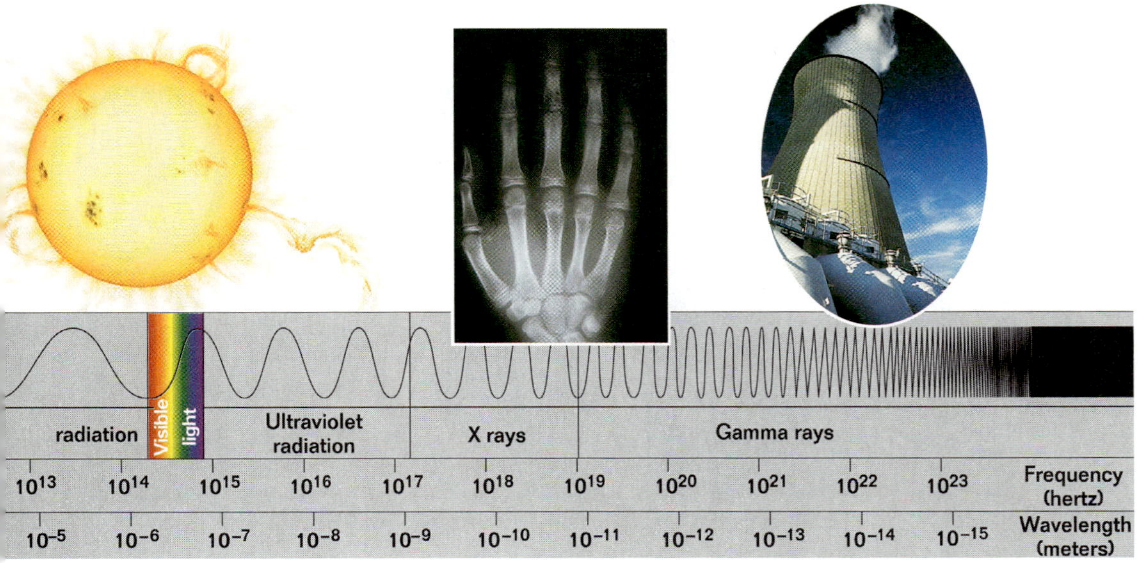

2 Teach

Caption Answer

Figure 1 Frequency increases.

INTEGRATE Health

Ultraviolet Light Violet light has a shorter wavelength and therefore a higher frequency than red light. Since UV light has an even shorter wavelength, it also has an even higher frequency and more energy. The added energy in UV light damages living tissues.

Discussion

Waves in a Vacuum Do you think electromagnetic waves can travel through a vacuum? Explain. Yes; solar energy is electromagnetic radiation. It travels to Earth through the vacuum of space.

Inquiry Lab

Electromagnetic Spectrum

Purpose to help students to better visualize the wavelengths of the electromagnetic spectrum

Possible Materials meterstick, metric ruler, chalkboard, paper

Estimated Time 30 minutes

Teaching Strategies

- Students can use a meterstick to explore the dimensions of some radio waves by drawing on a chalkboard. Waves of 0.1 m (10^{-1} m) and 1 m should be doable.
- Students can explore the waves in the microwave range on paper using a metric ruler as a guide. They could draw 1-cm (10^{-2} m) and 1-mm (10^{-3} m) waves.
- Then have them compare the waves and notice the differences in size, that each is 10 times the length of the previous wave.
- Encourage them to visualize the wavelengths that are too small or too large to reproduce easily.

Differentiated Instruction

Learning Disabled Have each student investigate a different type of electromagnetic radiation, noting its characteristics, what natural or artificial objects generate it, and how (if this is the case) it is used by people. Have students work together to make an informative bulletin board that includes diagrams and labels explaining their findings. L2 ELL COOP LEARN LS **Interpersonal**

SECTION 1 Radiation from Space J ◆ 9

Fun Fact

The *Hubble Space Telescope* is so powerful that if Earth's surface were not curved and you were in Washington, D.C., you could use it to read a billboard in Missouri.

Quick Demo

Telescope Lens

Materials double convex lens with a 15-cm focal length, and a sheet of white paper

Estimated Time 10 minutes

Procedure Hold the lens in one hand and the paper in the other. Turn off room lights and hold the lens so that light coming in a window passes through it and falls on the sheet of paper. Move the lens until students can see an inverted image of the window and what can be seen through it.

Visual Learning

Figure 2 Have students compare the two types of light telescopes. Which one seems as though it would be easier to construct? Why? The refracting telescope is easier; it has a simpler design.

Activity

Reflecting Telescopes Hand out some polished metal spoons and have students observe their inverted reflections in the bowls. Explain that this resembles the concave mirror in a reflecting telescope. In fact, the first working reflecting telescope built by Isaac Newton used a metal mirror made of tin and copper.

Optical Telescopes

Optical telescopes use light, which is a form of electromagnetic radiation, to produce magnified images of objects. Light is collected by an objective lens or mirror, which then forms an image at the focal point of the telescope. The focal point is where light that is bent by the lens or reflected by the mirror comes together to form an image. The eyepiece lens then magnifies the image. The two types of optical telescopes are shown in **Figure 2.**

A **refracting telescope** uses convex lenses, which are curved outward like the surface of a ball. Light from an object passes through a convex objective lens and is bent to form an image at the focal point. The eyepiece magnifies the image.

A **reflecting telescope** uses a curved mirror to direct light. Light from the object being viewed passes through the open end of a reflecting telescope. This light strikes a concave mirror, which is curved inward like a bowl and located at the base of the telescope. The light is reflected off the interior surface of the bowl to the focal point where it forms an image. Sometimes, a smaller mirror is used to reflect light into the eyepiece lens, where it is magnified for viewing.

Figure 2 These diagrams show how each type of optical telescope collects light and forms an image.

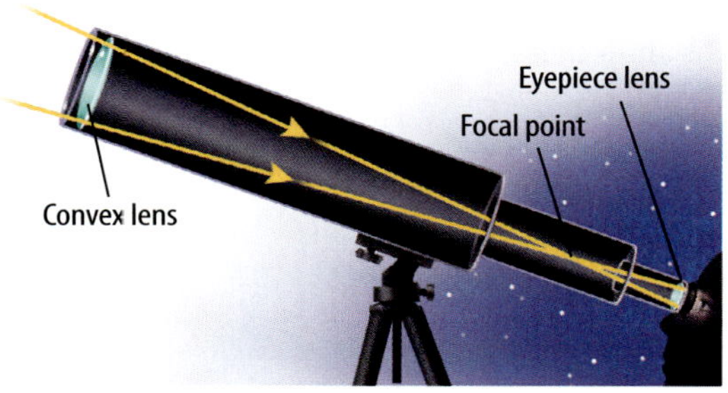

In a refracting telescope, a convex lens focuses light to form an image at the focal point.

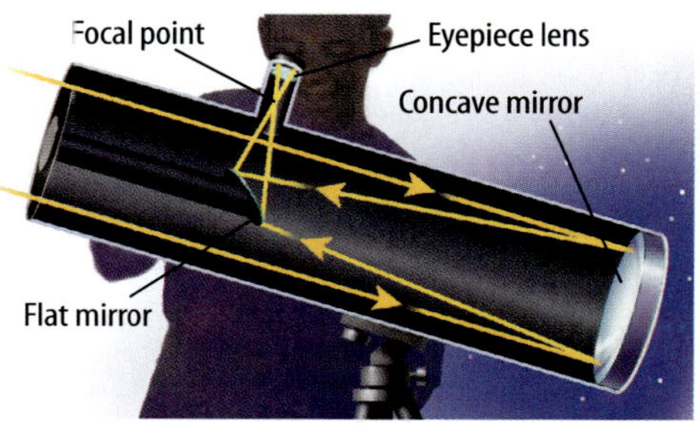

In a reflecting telescope, a concave mirror focuses light to form an image at the focal point.

Using Optical Telescopes Most optical telescopes used by professional astronomers are housed in buildings called **observatories.** Observatories often have dome-shaped roofs that can be opened up for viewing. However, not all telescopes are located in observatories. The *Hubble Space Telescope* is an example.

Optical telescopes are widely available for use by individuals.

Differentiated Instruction

Challenge Have students research the index of refraction and relate what they learn to the use of lenses in a telescope. The index of refraction of a substance is equal to the sine of the angle of incidence of light from a vacuum divided by the sine of the angle of refraction. The index of refraction of glass is used when determining how to shape the objective lens of a telescope. L3

Science Journal

Telescopes Ask students to write a brief paragraph in their Science Journals that answers this question: What advantages does a radio telescope have over an optical telescope? Radio telescopes can be used in all kinds of weather and at any time of day or night. Radio telescopes can detect objects that are invisible to optical telescopes. L2 LS **Logical-Mathematical and Linguistic**

Hubble Space Telescope The *Hubble Space Telescope* was launched in 1990 by the space shuttle *Discovery.* Because *Hubble* is located outside Earth's atmosphere, which absorbs and distorts some of the energy received from space, it should have produced clear images. However, when the largest mirror of this reflecting telescope was shaped, a mistake was made. As a result, images obtained by the telescope were not as clear as expected. In December 1993, a team of astronauts repaired the *Hubble Space Telescope* by installing a set of small mirrors designed to correct images obtained by the faulty mirror. Two more missions to service *Hubble* were carried out in 1997 and 1999, shown in **Figure 3.** Among the objects viewed by *Hubble* after it was repaired in 1999 was a large cluster of galaxies known as Abell 2218.

Reading Check ***Why is* Hubble *located outside Earth's atmosphere?***

Figure 3 The *Hubble Space Telescope* was serviced at the end of 1999. Astronauts replaced devices on *Hubble* that are used to stabilize the telescope.

Teacher FYI

Hubble Space Telescope The *Hubble Space Telescope* is a reflecting telescope with a light-gathering mirror 240 cm in diameter. It is in orbit 610 km above Earth's surface. NASA controls the telescope with radio commands beamed to it from the ground. The images obtained by *Hubble* are sent to the ground as radio signals as well.

Reading Check

Answer to avoid the interference caused by Earth's atmosphere

Discussion

Sharing Science Tell students that NASA built the *Hubble Space Telescope* and put it in orbit. Now astronomers from all over the world share the use of the telescope and the images it provides. Do you think the United States should share resources like the *Hubble* with other nations? Explain. Possible answers: Yes; everyone should share any tool that advances scientific discovery. No; the United States built it and U.S. scientists should have the benefit of sole use of the telescope.

Science Journal

Hubble Space Telescope Have students research and write about some of the discoveries made by astronomers using the *Hubble Space Telescope.* Discoveries should include galaxies at the edge of the known universe and galaxies colliding and tearing each other apart.

L2 LS **Linguistic** P

Reading Check

Answer about 400 years

Mini LAB

Purpose Students contrast the number of stars visible from each of their homes and determine whether light pollution accounts for the difference. L2

LS **Visual-Spatial**

Materials cardboard tube

Teaching Strategy Explain that light pollution is a glow in the sky caused by city lights. The glow makes it difficult to see dim stars. Remind them to record light conditions at their viewing spot.

Analysis

1. Students in areas away from street lights will see more stars than students in urban areas or on main streets.
2. More stars are visible in areas with less background light.

Assessment

Process Have students observe and record common sources of light pollution. Possible answers: streetlights, lights from homes, apartment buildings, and shopping centers. Use **Performance Assessment in the Science Classroom,** p. 97.

Try at Home

Mini LAB

Observing Effects of Light Pollution

Procedure

1. Obtain a **cardboard tube** from an empty roll of paper towels.
2. Go outside on a clear night about two hours after sunset. Look through the cardboard tube at a specific constellation decided upon ahead of time.
3. Count the number of stars you can see without moving the observing tube. Repeat this three times.
4. Calculate the average number of observable stars at your location.

Analysis

1. Compare and contrast the number of stars visible from other students' homes.
2. Explain the causes and effects of your observations.

Try at Home

Large Reflecting Telescopes Since the early 1600s, when the Italian scientist Galileo Galilei first turned a telescope toward the stars, people have been searching for better ways to study what lies beyond Earth's atmosphere. For example, the twin Keck reflecting telescopes, shown in **Figure 4,** have segmented mirrors 10 m wide. Until 2000, these mirrors were the largest reflectors ever used. To cope with the difficulty of building such huge mirrors, the Keck telescope mirrors are built out of many small mirrors that are pieced together. In 2000, the European Southern Observatory's telescope, in Chile, consisted of four 8.2-m reflectors, making it the largest optical telescope in use.

Reading Check *About how long have people been using telescopes?*

Active and Adaptive Optics The most recent innovations in optical telescopes involve active and adaptive optics. With active optics, a computer corrects for changes in temperature, mirror distortions, and bad viewing conditions. Adaptive optics is even more ambitious. Adaptive optics uses a laser to probe the atmosphere and relay information to a computer about air turbulence. The computer then adjusts the telescope's mirror thousands of times per second, which lessens the effects of atmospheric turbulence. Telescope images are clearer when corrections for air turbulence, temperature changes, and mirror-shape changes are made.

Figure 4 The twin Keck telescopes on Mauna Kea in Hawaii can be used together, more than doubling their ability to distinguish objects. A Keck reflector is shown in the inset photo. Currently, plans include using these telescopes, along with four others to obtain images that will help answer questions about the origin of planetary systems.

Teacher FYI

Keck's Mirrors Each of the Keck's mirrors actually consists of 36 hexagonal mirrors mounted extremely close together to form one reflecting surface 10 m across. An electronic sensing system keeps each of the mirrors in place and adjusts each automatically if it becomes misaligned.

Radio Telescopes

As shown in the spectrum illustrated in **Figure 1,** stars and other objects radiate electromagnetic energy of various types. Radio waves are an example of long-wavelength energy in the electromagnetic spectrum. A **radio telescope,** such as the one shown in **Figure 5,** is used to study radio waves traveling through space. Unlike visible light, radio waves pass freely through Earth's atmosphere. Because of this, radio telescopes are useful 24 hours per day under most weather conditions.

Radio waves reaching Earth's surface strike the large, concave dish of a radio telescope. This dish reflects the waves to a focal point where a receiver is located. The information allows scientists to detect objects in space, to map the universe, and to search for signs of intelligent life on other planets.

Figure 5 This radio telescope is used to study radio waves traveling through space.

section 1 review

Summary

Electromagnetic Waves

- Light is a form of electromagnetic radiation.
- Electromagnetic radiation includes radio waves, microwaves, X rays, gamma rays, and infrared and ultraviolet radiation.
- Light travels at 300,000 km/s in a vacuum.

Optical Telescopes

- A refracting telescope uses lenses to collect, focus, and view light.
- A reflecting telescope uses a mirror to collect and focus light and a lens to view the image.
- Computers and lasers are used to reduce problems caused by looking through Earth's atmosphere.
- These telescopes are housed in domed buildings called observatories.
- Placing a telescope in space avoids problems caused by Earth's atmosphere.

Radio Telescopes

- Radio telescopes collect and measure radio waves coming from stars and other objects.

Self Check

1. **Identify** one advantage of radio telescopes over optical telescopes.
2. **Infer** If red light has a longer wavelength than blue light, which has a greater frequency?
3. **Explain** the difference between sound waves and radio waves.
4. **Describe** how adaptive optics in a telescope help solve problems caused by atmospheric turbulence.
5. **Think Critically** It takes light from the closest star to Earth (other than the Sun) about four years to reach Earth. If intelligent life were on a planet circling that star, how long would it take for scientists on Earth to send them a radio transmission and for the scientists to receive their reply?

Applying Math

6. **Calculate** how long it takes for a radio signal to reach the Moon, which is about 380,000 km away.
7. **Use Numbers** If an X ray has a frequency of 10^{18} hertz and a gamma ray has a frequency of 10^{21} hertz, how many times greater is the frequency of the gamma ray?

Science Online bookj.msscience.com/self_check_quiz

Fun Fact

The Very Large Array (VLA) in New Mexico consists of 27 movable dishes, each 25 m wide. The dishes operate as one telescope, making the VLA one of the world's most powerful radio telescopes.

3 Assess

DAILY INTERVENTION

Check for Understanding

Visual-Spatial Give students diagrams containing labeled cross sections of both refracting and reflecting telescopes. Ask them to identify the telescopes as to type and to indicate how each type focuses light.

Reteach

Radio Telescopes Remind students of the wavelength of radio waves and point out that radio telescopes are often very large. Show them some pictures of radio telescopes, such as the one at Arecibo, Puerto Rico, which is suspended between mountain peaks and the Very Large Array in New Mexico, which consists of 27 linked, dish antennas.

Assessment

Process Have students make a table describing the types of telescopes covered in this section and the type of electromagnetic radiation examined by each. Use **Performance Assessment in the Science Classroom,** p. 109.

section 1 review

1. Radio telescopes use a dish antenna to collect and focus radio waves, whereas optical telescopes use lenses or mirrors to collect and focus visible light. Radio telescopes can be used day and night, even when weather conditions make optical viewing impossible.
2. blue light
3. Radio waves are electromagnetic waves and can travel through a vacuum. Sound waves are not electromagnetic and need a medium, such as air or water. Sound waves cannot travel in a vacuum.
4. Adaptive optics uses a computer to adjust the telescope's objective mirror to decrease the effects of air turbulence.
5. about 8 years
6. If light travels 300,000 km/s, you must divide the distance (380,000 km) by this speed. This gives the answer 1.3 s.
7. This is practice with exponents. If you subtract the exponent 18 from 21 you get 3 and 10^3 equals 1,000.

SECTION 1 Radiation from Space J ◆ 13

BENCH TESTED

Real-World Question

Purpose Students will construct a working, reflecting telescope. L2 COOP LEARN LS **Visual-Spatial**

Process Skills observe, construct, design

Time Required 25 minutes

Procedure

Safety Precautions Caution students never to touch broken glass.

Teaching Strategies Collect photographs of the Moon or other celestial objects for students to view.

Troubleshooting Be certain that each pair of students has both a flat and curved mirror. Label the two types of mirrors for easy reference.

Analyze Your Data

1. The object's image will appear larger or smaller depending upon the level of magnification of each lens.
2. The cosmetic mirror reflects the light onto the flat mirror. The flat mirror reflects the light onto the magnifying lens.
3. Magnifying lenses.

Conclude and Apply

1. The curved mirror gathered the light reflecting from the object and focused the light on the flat mirror. The flat mirror reflected the light onto the magnifying lens, and each lens magnified the image.
2. Convex lenses would have been used instead of mirrors.

LAB

Building a Reflecting Telescope

Nearly four hundred years ago, Galileo Galilei saw what no human had ever seen. Using the telescope he built, he saw moons around Jupiter, details of lunar craters, and sunspots. What was it like to make these discoveries? Find out as you make your own reflecting telescope.

Real-World Question

How do you construct a reflecting telescope?

Goals

- **Construct** a reflecting telescope.
- **Observe** magnified images using the telescope and different magnifying lenses.

Materials

flat mirror
shaving or cosmetic mirror (a curved, concave mirror)
magnifying lenses of different magnifications (3–4)

Safety Precautions

WARNING: *Never observe the Sun directly or with mirrors.*

Procedure

1. Position the cosmetic mirror so that you can see the reflection of the object you want to look at. Choose an object such as the Moon, a planet, or an artificial light source.
2. Place the flat mirror so that it is facing the cosmetic mirror.
3. Adjust the position of the flat mirror until you can see the reflection of the object in it.
4. View the image of the object in the flat mirror with one of your magnifying lenses. Observe how the lens magnifies the image.
5. Use your other magnifying lenses to view the image of the object in the flat mirror. Observe how the different lenses change the image of the object.

Analyze Your Data

1. **Describe** how the image changed when you used different magnifying lenses.
2. **Identify** the part or parts of your telescope that reflected the light of the image.
3. **Identify** the parts of your telescope that magnified the image.

Conclude and Apply

1. **Explain** how the three parts of your telescope worked to reflect and magnify the light of the object.
2. **Infer** how the materials you used would have differed if you had constructed a refracting instead of a reflecting telescope.

Communicating Your Data

Write an instructional pamphlet for amateur astronomers about how to construct a reflecting telescope.

14 ◆ J CHAPTER 1 Exploring Space

Communicating Your Data

Students may want to include a labeled diagram with their pamphlets.

Assessment

Portfolio Ask students to construct a permanent reflecting telescope with a housing of their own design. Use **Performance Assessment in the Science Classroom,** p. 117.

section 2

Early Space Missions

The First Missions into Space

You're offered a choice—front-row-center seats for this weekend's rock concert, or a copy of the video when it's released. Wouldn't you rather be right next to the action? Astronomers feel the same way about space. Even though telescopes have taught them a great deal about the Moon and planets, they want to learn more by going to those places or by sending spacecraft where humans can't go.

Rockets The space program would not have gotten far off the ground using ordinary airplane engines. To break free of gravity and enter Earth's orbit, spacecraft must travel at speeds greater than 11 km/s. The space shuttle and several other spacecrafts are equipped with special engines that carry their own fuel. **Rockets,** like the one in **Figure 6,** are engines that have everything they need for the burning of fuel. They don't even require air to carry out the process. Therefore, they can work in space, which has no air. The simplest rocket engine is made of a burning chamber and a nozzle. More complex rockets have more than one burning chamber.

Rocket Types Two types of rockets are distinguished by the type of fuel they use. One type is the liquid-propellant rocket and the other is the solid-propellant rocket. Solid-propellant rockets are generally simpler but they can't be shut down after they are ignited. Liquid-propellant rockets can be shut down after they are ignited and can be restarted. For this reason, liquid-propellant rockets are preferred for use in long-term space missions. Scientists on Earth can send signals that start and stop the spacecraft's engines whenever they want to modify its course or adjust its orbit. Liquid propellants successfully powered many space probes, including the two *Voyagers* and *Galileo.*

Figure 6 Rockets differ according to the types of fuel used to launch them. Liquid oxygen is used often to support combustion.

as you read

What You'll Learn

- **Compare and contrast** natural and artificial satellites.
- **Identify** the differences between artificial satellites and space probes.
- **Explain** the history of the race to the Moon.

Why It's Important

Early missions that sent objects and people into space began a new era of human exploration.

Review Vocabulary

thrust: the force that propels an aircraft or missile

New Vocabulary

- rocket
- satellite
- orbit
- space probe
- Project Mercury
- Project Gemini
- Project Apollo

1 Motivate

Bellringer

INTERACTIVE CHALKBOARD PowerPoint® Presentations

Section Focus Transparencies also are available on the Interactive Chalkboard CD-ROM. L2 ELL

Tie to Prior Knowledge

Moon Landing Ask students if astronauts have ever traveled to the Moon. Write responses on the board as students brainstorm facts they know about the Moon landings. Possible answers: First one occurred in 1969; first person to step on the Moon was Neil Armstrong. Have students review this list after reading the section to see how many "facts" were correct.

Section 2 Resource Manager

Chapter *FAST FILE* Resources
- Transparency Activity, p. 41
- Directed Reading for Content Mastery, p. 17
- Enrichment, p. 27
- MiniLAB, p. 4
- Reinforcement, p. 24

Mathematics Skill Activities, p. 15

Reading and Writing Skill Activities, p. 11

2 Teach

Discussion

Rocket Propulsion Remind students that solid propellant rockets are simpler than liquid propellant rockets. What advantage might a liquid propellant rocket have over a solid propellant rocket? Possible answer: the ability to start and stop the rocket makes it easier to control the rocket's velocity.

Quick Demo

How Rockets Move

Materials long balloon, drinking straw, string, tape

Estimated Time 10 minutes

Procedure Have students thread a long string through a drinking straw. Inflate the balloon and hold the end closed. Have students tape the straw to the side of the balloon. Then have two students hold the ends of the string and pull it taut. Release the balloon. Students should see that gases escaping from behind the balloon push it forward along the string.

Applying Math

National Math Standards

Correlation to Mathematics Objectives

3

Answers to Practice Problems

1. Students' graphs should look like a line drawing of an astronaut.
2. Have students compare drawings made by classmates to the originals.

Figure 7 The space shuttle uses both liquid and solid fuels. Here the red liquid fuel tank is visible behind a white, solid rocket booster.

Rocket Launching Solid-propellant rockets use a rubberlike fuel that contains its own oxidizer. The burning chamber of a rocket is a tube that has a nozzle at one end. As the solid propellant burns, hot gases exert pressure on all inner surfaces of the tube. The tube pushes back on the gas except at the nozzle where hot gases escape. Thrust builds up and pushes the rocket forward.

Liquid-propellant rockets use a liquid fuel and an oxidizer, such as liquid oxygen, stored in separate tanks. To ignite the rocket, the oxidizer is mixed with the liquid fuel in the burning chamber. As the mixture burns, forces are exerted and the rocket is propelled forward. **Figure 7** shows the space shuttle, with both types of rockets, being launched.

Applying Math — Make and Use Graphs

DRAWING BY NUMBERS Points are defined by two coordinates, called an ordered pair. To plot an ordered pair, find the first number on the horizontal *x*-axis and the second on the vertical *y*-axis. The point is placed where these two coordinates intersect. Line segments are drawn to connect points. Draw a symmetrical house by using an *x*-*y* grid and these coordinates: (1,1), (5,1), (5,4), (3,6), (1,4)

Solution

1. *On a piece of graph paper, label and number the* x*-axis 0 to 6 and the* y*-axis 0 to 6, as shown here.*
2. *Plot the above points and connect them with straight line segments, as shown here.*

Section	Points
1	(1, –8) (3,–13) (6, –21) (9,–21) (9,–17) (8,–15) (8,–12) (6,–8) (5,–4) (4,–3) (4,–1) (5,1) (6,3) (8,3) (9,4) (9,7) (7,11) (4,14) (4,22) (–9,22) (–9,10) (–10,5) (–11,–1) (–11,–7) (–9,–8) (–8,–7) (–8,–1) (–6,3) (–6,–3) (–6,–9) (–7,–20) (–8,–21) (–4,–21) (–4,–18) (–3,–14) (–1,–8)
2	(0,11) (2,13) (2,17) (0,19) (–4,19) (–6,17) (–6,13) (–4,11)
3	(–4,9) (1,9) (1,5) (–1,5) (–2,6) (–4,6)

Practice Problems

1. Label and number the *x*-axis –12 to 10 and the *y*-axis –22 to 23. Draw an astronaut by plotting and connecting the points in each section. Do not draw segments to connect points in different sections.
2. Make your own drawing on graph paper and write its coordinates as ordered pairs. Then give it to a classmate to solve.

Science online — For more practice, visit bookj.msscience.com/math_practice

Curriculum Connection

History In 1926, American scientist Robert Goddard tested the first liquid-fueled rocket. The small rocket rose more than 12 m before crashing. It was the forerunner of the rockets we send into space today. Have students research that first rocket launch and do a report on videotape or audiotape as if they were journalists reporting from the scene. Have students play the tapes in class. L2 ELL LS **Musical-Auditory**

Cultural Diversity

Chinese Rockets Historians and scientists generally agree that the Chinese invented the rocket centuries ago. During the 1200s, Chinese soldiers fired small rockets (called "fire arrows" by some) as weapons in battle. Have students research these early rockets and write reports on what they were like as well as how they contrast with rockets today. L2 LS **Logical-Mathematical**

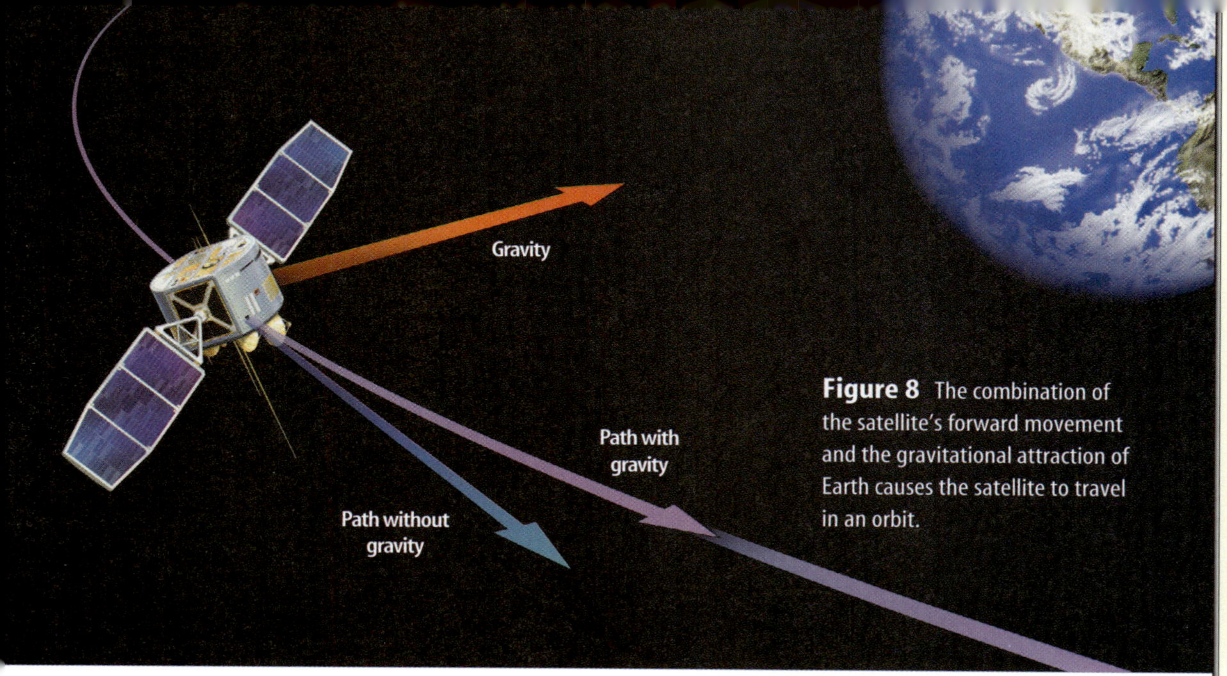

Figure 8 The combination of the satellite's forward movement and the gravitational attraction of Earth causes the satellite to travel in an orbit.

Satellites The space age began in 1957 when the former Soviet Union used a rocket to send *Sputnik I* into space. *Sputnik I* was the first artificial satellite. A **satellite** is any object that revolves around another object. When an object enters space, it travels in a straight line unless a force, such as gravity, makes it turn. Earth's gravity pulls a satellite toward Earth. The result of the satellite traveling forward while at the same time being pulled toward Earth is a curved path, called an **orbit,** around Earth. This is shown in **Figure 8.** *Sputnik I* orbited Earth for 57 days before gravity pulled it back into the atmosphere, where it burned up.

Satellite Uses *Sputnik I* was an experiment to show that artificial satellites could be made and placed into orbit around Earth.

Today, thousands of artificial satellites orbit Earth. Communication satellites transmit radio and television programs to locations around the world. Other satellites gather scientific data, like those shown in **Figure 9,** which can't be obtained from Earth, and weather satellites constantly monitor Earth's global weather patterns.

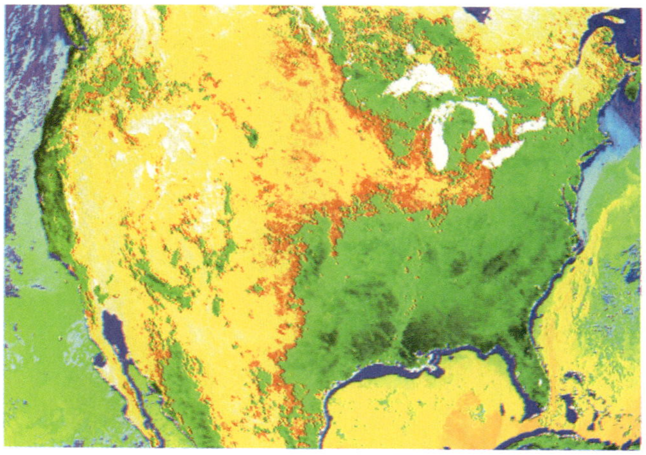

Figure 9 Data obtained from the satellite *Terra,* launched in 1999, illustrates the use of space technology to study Earth. This false-color image includes data on spring growth, sea-surface temperature, carbon monoxide concentrations, and reflected sunlight, among others.

Teacher FYI

Soviet Space Missions Although not as successful as the American effort, the Soviet Union (today's Russia) also had a space program that sent several probes to other planets. The Soviet Union's Venera probes made several voyages to Venus between the late 1960s and the 1980s. The probes sent back data on Venus's atmosphere, beamed back pictures from Venus's surface, and analyzed rocks from the planet's surface.

Discussion

Satellites How has life on Earth changed since man-made satellites were placed in orbit? Discuss some of the uses of satellites and the types of information they carry. Satellites have improved weather forecasting and communications, especially satellite telephone links. Satellite data aid in resource conservation. GPS navigation may also be mentioned.

Virtual Labs

Satellites *How does an artificial satellite stay in orbit?*

Active Reading

Think-Pair-Share Ask students to respond to a question by writing a response. After thinking for a few minutes, have partners share responses to the question. Finally, ask the students to share responses with the class. Have students become involved in a Think-Pair-Share about the importance of exploring space. L1 COOP LEARN IN **Interpersonal**

SECTION 2 Early Space Missions J ◆ 17

INTEGRATE Career

Astronomy The American Astronomical Society advises those interested in a career in astronomy to study as much science and math as they can in high school, including pre-calculus if possible. In college, students may lean towards either math or physics, but computer science is of great benefit also.

Use Science Words

Word Usage Have students use the term *space probe* in a sentence. Have pairs take turns reading their sentences and critiquing one another on whether the term was used correctly.

Visual Learning

Table 1 Direct students to look carefully at the names of the space probes in the table. Ask them to determine what the names have in common. They all concern adventure and exploration. Mariners are seafarers. Pioneers are people who are among the first to travel to a place. Vikings were ancient Scandinavian explorers. Magellan commanded the first expedition to sail around the world in the early 1500s. LS **Linguistic**

INTEGRATE Career

Astronomy Astronomers today have more choices than ever before. Some still use optical telescopes to study stars and galaxies. Others explore the universe using the radio, X-ray, infrared, or even gamma-ray regions of the electromagnetic spectrum. Still others deal with theory and work with physicists to understand the big bang and the nature of matter in the universe. Government, universities, and private industry offer jobs for astronomers.

Space Probes

Not all objects carried into space by rockets become satellites. Rockets also can be used to send instruments into space to collect data. A **space probe** is an instrument that gathers information and sends it back to Earth. Unlike satellites that orbit Earth, space probes travel into the solar system as illustrated in **Figure 10.** Some even have traveled to the edge of the solar system. Among these is *Pioneer 10,* launched in 1972. Although its transmitter failed in 2003, it continues on through space. Also, both *Voyager* spacecrafts should continue to return data on the outer reaches of the solar system until about 2020.

Space probes, like many satellites, carry cameras and other data-gathering equipment, as well as radio transmitters and receivers that allow them to communicate with scientists on Earth. **Table 1** shows some of the early space probes launched by the National Aeronautics and Space Administration (NASA).

Table 1 Some Early Space Missions

Mission Name		Date Launched	Destination	Date Obtained
Mariner 2		August 1962	Venus	verified high temperatures in Venus's atmosphere
Pioneer 10	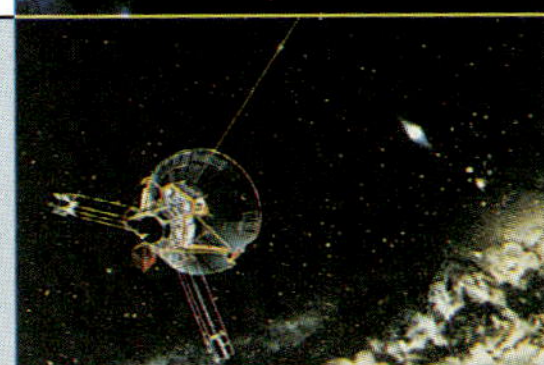	March 1972	Jupiter	sent back photos of Jupiter—first probe to encounter an outer planet
Viking 1		August 1975	Mars	orbiter mapped the surface of Mars; lander searched for life on Mars
Magellan		May 1989	Venus	mapped Venus's surface and returned data on the composition of Venus's atmosphere

Curriculum Connection

Social Studies Have students speculate why so many developments occurred in the space program during the 1960s. Possible answer: Students should realize that the United States and the former Soviet Union were participating in a "race for space." L2 LS **Logical-Mathematical**

Differentiated Instruction

Challenge Have students research space junk, the unused and nonfunctional materials orbiting Earth. Such junk ranges from objects the size of grains of sand, such as paint chips, to large, nonfunctioning satellites. Students can research and write a brief report summarizing the growing threat that space junk might pose to the safety of future space flights. Have them give specific examples, if possible. L3

NATIONAL GEOGRAPHIC VISUALIZING SPACE PROBES

Figure 10

Probes have taught us much about the solar system. As they travel through space, these car-size craft gather data with their onboard instruments and send results back to Earth via radio waves. Some data collected during these missions are made into pictures, a selection of which is shown here.

A In 1974, *Mariner 10* obtained the first good images of the surface of Mercury.

B A Soviet *Venera* probe took this picture of the surface of Venus on March 1, 1982. Parts of the spacecraft's landing gear are visible at the bottom of the photograph.

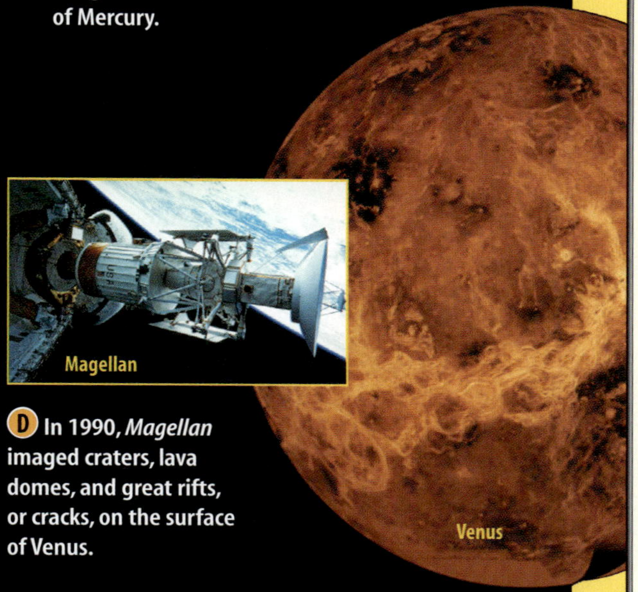

D In 1990, *Magellan* imaged craters, lava domes, and great rifts, or cracks, on the surface of Venus.

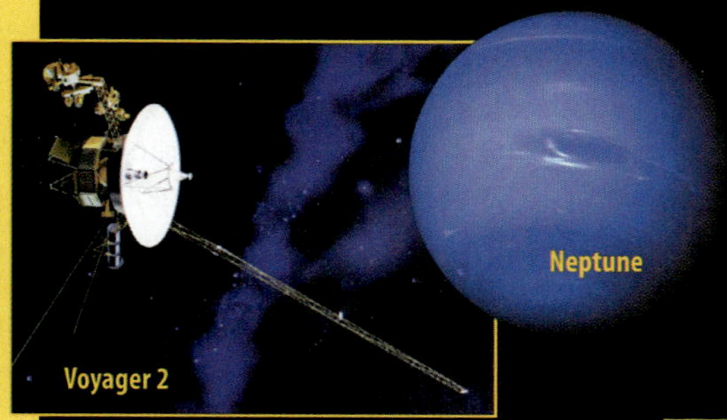

C The *Voyager 2* mission included flybys of the outer planets Jupiter, Saturn, Uranus, and Neptune. *Voyager* took this photograph of Neptune in 1989 as the craft sped toward the edge of the solar system.

E NASA's veteran space traveler *Galileo* nears Jupiter in this artist's drawing. The craft arrived at Jupiter in 1995 and sent back data, including images of Europa, one of Jupiter's 61 moons, seen below in a color-enhanced view.

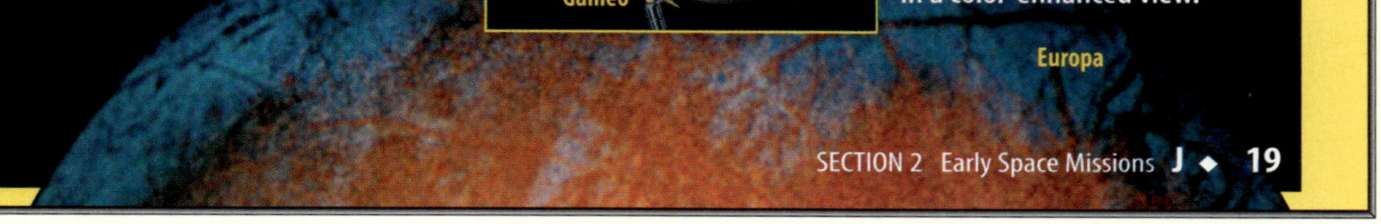

NATIONAL GEOGRAPHIC

Visualizing Space Probes

Have students examine the pictures and read the captions. Then ask the following questions.

What features on the pictured space probes might have helped them communicate with Earth? All have some type of antennae. *Mariner*, *Magellan*, *Voyager 2*, and *Galileo*, for example, have "dish" antennae for receiving signals from Earth. Solar panels on *Mariner* and *Magellan* might have helped power communication equipment.

In what manner is the scale of the planets and moon exaggerated? Possible answer: The smallest planet pictured, Mercury, appears larger than Jupiter, the largest. Venus, while larger than Mercury, would be dwarfed by Neptune and Jupiter and appears far larger than true scale. Europa is the smallest body shown here and is vastly inflated in scale.

Activity

Model Space Probes Have students brainstorm to create their own space probes. They can draw their models using colored pencils or pens. Beneath each model, have students explain the purpose of each feature and why they included it on the probe. P

Differentiated Instruction

English-Language Learners Supply students with a diagram of the solar system with the planets labeled. Then have them cut the outlines of the space probes from photocopies of this figure or elsewhere and match them with the planets they visited. They can follow the paths of the *Voyager* probes from the text on the next page. ELL

Activity

Planet Photos Have students research to find photographs of Saturn, Jupiter, Uranus, or Neptune sent back to Earth by the *Voyager* probes. Have each student print out a photo and describe what it shows about the planet. Make a bulletin board with the photos and student captions.

Reading Check

Answer Cracks in Europa's outer layer of ice might indicate geological activity that could heat Europa's ice, forming liquid water under the surface.

Science Online

Topic: Galileo Mission

Go to bookj.msscience.com for Web links to informaiton about the Galileo mission.

Activity Make a list of the places the *Galileo* spacecraft visited on its mission.

Voyager and Pioneer Probes Space probes *Voyager 1* and *Voyager 2* were launched in 1977 and now are heading toward deep space. *Voyager 1* flew past Jupiter and Saturn. *Voyager 2* flew past Jupiter, Saturn, Uranus, and Neptune. These probes will explore beyond the solar system as part of the Voyager Interstellar Mission. Scientists expect these probes to continue to transmit data to Earth for at least 20 more years.

Pioneer 10, launched in 1972, was the first probe to survive a trip through the asteroid belt and encounter an outer planet, Jupiter. As of 2003, *Pioneer 10* was more than 12 billion km from Earth, and will continue beyond the solar system. The probe carries a gold medallion with an engraving of a man, a woman, and Earth's position in the galaxy.

Galileo Launched in 1989, *Galileo* reached Jupiter in 1995. In July 1995, *Galileo* released a smaller probe that began a five-month approach to Jupiter. The small probe took a parachute ride through Jupiter's violent atmosphere in December 1995.

Before being crushed by the atmospheric pressure, it transmitted information about Jupiter's composition, temperature, and pressure to the satellite orbiting above. *Galileo* studied Jupiter's moons, rings, and magnetic fields and then relayed this information to scientists who were waiting eagerly for it on Earth.

INTEGRATE Life Science Studies of Jupiter's moon Europa by *Galileo* indicate that an ocean of water may exist under the surface of Europa. A cracked outer layer of ice makes up Europa's surface, shown in **Figure 11.** The cracks in the surface may be caused by geologic activity that heats the ocean underneath the surface. Sunlight penetrates these cracks, further heating the ocean and setting the stage for the possible existence of life on Europa. *Galileo* ended its study of Europa in 2000. More advanced probes will be needed to determine whether life exists on this icy moon.

Reading Check *What features on Europa suggest the possibility of life existing on this moon?*

In October and November of 1999, *Galileo* approached Io, another one of Jupiter's moons. It came within 300 km and took photographs of a volcanic vent named Loki, which emits more energy than all of Earth's volcanoes combined. *Galileo* also discovered eruption plumes that shoot gas made of sulfur and oxygen.

Figure 11 Future missions will be needed to determine whether life exists on Europa.

Cultural Diversity

History Early standards for astronauts limited which Americans could participate. For example, the earliest astronauts had to be test pilots, but women were not allowed to train as test pilots until 1970. Astronaut candidate selection broadened in the late 1970s. The 1978 astronaut group included six women, three African Americans, and an Asian American.

Teacher FYI

Voyager The most distant human-made object as of April 2003 was *Voyager 1* space probe. At that time, it was 13 billion km (87 AU) from Earth. *Voyager 1* is expected to send data back to Earth until it runs out of electrical power, around 2020. *Voyager 1*, along with *Voyager 2*, is now part of the Voyager Interstellar Mission (VIM).

Moon Quest

Throughout the world, people were shocked when they turned on their radios and television sets in 1957 and heard the radio transmissions from *Sputnik I* as it orbited Earth. All that *Sputnik I* transmitted was a sort of beeping sound, but people quickly realized that launching a human into space wasn't far off.

In 1961, Soviet cosmonaut Yuri A. Gagarin became the first human in space. He orbited Earth and returned safely. Soon, President John F. Kennedy called for the United States to send humans to the Moon and return them safely to Earth. His goal was to achieve this by the end of the 1960s. The race for space was underway.

The U.S. program to reach the Moon began with **Project Mercury.** The goals of Project Mercury were to orbit a piloted spacecraft around Earth and to bring it back safely. The program provided data and experience in the basics of space flight. On May 5, 1961, Alan B. Shepard became the first U.S. citizen in space. In 1962, *Mercury* astronaut John Glenn became the first U.S. citizen to orbit Earth. **Figure 12** shows Glenn preparing for liftoff.

Figure 12 An important step in the attempt to reach the Moon was John Glenn's first orbit around Earth.

Reading Check *What were the goals of Project Mercury?*

Project Gemini

The next step in reaching the Moon was called **Project Gemini.** Teams of two astronauts in the same *Gemini* spacecraft orbited Earth. One *Gemini* team met and connected with another spacecraft in orbit—a skill that would be needed on a voyage to the Moon.

The *Gemini* spacecraft was much like the *Mercury* spacecraft, except it was larger and easier for the astronauts to maintain. It was launched by a rocket known as a *Titan II,* which was a liquid fuel rocket.

In addition to connecting spacecraft in orbit, another goal of Project Gemini was to investigate the effects of space travel on the human body.

Along with the *Mercury* and *Gemini* programs, a series of robotic probes was sent to the Moon. *Ranger* proved that a spacecraft could be sent to the Moon. In 1966, *Surveyor* landed gently on the Moon's surface, indicating that the Moon's surface could support spacecraft and humans. The mission of *Lunar Orbiter* was to take pictures of the Moon's surface that would help determine the best future lunar landing sites.

Mini LAB

Modeling a Satellite

WARNING: *Stand a safe distance away from classmates.*

Procedure

1. Tie one end of a strong, 50-cm-long **string** to a small **cork.**
2. Hold the other end of the string tightly with your arm fully extended.
3. Move your hand back and forth so that the cork swings in a circular motion.
4. Gradually decrease the speed of the cork.

Analysis

1. What happened as the cork's motion slowed?
2. How does the motion of the cork resemble that of a satellite in orbit?

Reading Check

Answer to orbit a piloted spacecraft around Earth and bring it back safely

Mini LAB

Purpose Students will model how satellites stay in orbit. LS **Kinesthetic**

Materials string (2 m), cork

Teaching Strategy Discuss gravity—the attractive force between objects.

Safety Precautions Make sure students stand a safe distance from one another. Suggest students wear safety glasses.

Analysis

1. It dropped out of orbit.
2. The force of gravity pulls the satellite toward Earth; the satellite's inertia resists this pull. Together, these forces keep a satellite in orbit.

Assessment

Process Predict what would happen to a satellite whose forward motion was very slow. It would fall to Earth.

Differentiated Instruction

Challenge Have different pairs of students use the school library to research one of the seven original Project Mercury astronauts. Pairs should collaborate on written reports about the missions each astronaut performed in the *Mercury* and other space missions, and what these astronauts are doing now. L2 COOP LEARN LS **Linguistic**

Science Journal

Project Apollo Have students write a paragraph in their Science Journals why they do or do not think it was important for the United States to win the race to the Moon. Encourage students to fully explain the reasons for their point of view. Use these paragraphs as a springboard for a class debate. L2 LS **Linguistic**

SECTION 2 Early Space Missions J ◆ 21

3 Assess

DAILY INTERVENTION

Check for Understanding

Interpersonal Have students work in pairs and each make a list of ways that the space program has directly affected their lives. Then they should compare their lists, noting which things they agree or disagree with. If time allows, have them research these and see, which choices are correct. Possibilities are hand-held calculators, computers, and materials for athletic shoes.

Reteach

Flash Cards Have pairs of students make flash cards of the artificial satellites and space probes presented in this section. Have them write the spacecraft name on one side and its mission and discoveries on the other. Have partners quiz each other to review the material. L1 COOP LEARN IS **Interpersonal**

Assessment

Portfolio Ask each student to write a poem about one of the American missions to space or a famous figure from the United States space program. Encourage students to read their poems in class. Use **Performance Assessment in the Science Classroom,** p. 157.

Figure 13 The Lunar Rover vehicle was first used during the *Apollo 15* mission. Riding in the moon buggy, *Apollo 15, 16,* and *17* astronauts explored the lunar surface.

Project Apollo The final stage of the U.S. program to reach the Moon was **Project Apollo.** On July 20, 1969, *Apollo 11* landed on the Moon's surface. Neil Armstrong was the first human to set foot on the Moon. His first words as he stepped onto its surface were, "That's one small step for man, one giant leap for mankind." Edwin Aldrin, the second of the three *Apollo 11* astronauts, joined Armstrong on the Moon, and they explored its surface for two hours. While they were exploring, Michael Collins remained in the Command Module; Armstrong and Aldrin then returned to the Command Module before beginning the journey home. A total of six lunar landings brought back more than 2,000 samples of moon rock and soil for study before the program ended in 1972. **Figure 13** shows an astronaut exploring the Moon's surface from the Lunar Rover vehicle.

section 2 review

Summary

First Missions into Space

- Rockets are engines that have everything they need to burn fuel.
- Rockets may be fueled with liquid or solid propellants.
- A satellite is any object that revolves around another object.

Space Probes

- A space probe is an instrument that gathers information and sends it back to Earth.
- *Voyager* and *Pioneer* are probes designed to explore the solar system and beyond.
- *Galileo* is a space probe that explored Jupiter and its moons.

Moon Quest

- Project Mercury sent the first piloted spacecraft around Earth.
- *Ranger* and *Surveyor* probes explored the Moon's surface.
- *Gemini* orbited teams of two astronauts.
- Project Apollo completed six lunar landings.

Self Check

1. **Explain** why Neptune has eleven satellites even though it is not orbited by human-made objects.
2. **Explain** why *Galileo* was considered a space probe as it traveled to Jupiter. However, once there, it became an artificial satellite.
3. **List** several discoveries made by the *Voyager 1* and *Voyager 2* space probes.
4. **Sequence** Draw a time line beginning with *Sputnik* and ending with Project Apollo. Include descriptions of important missions.
5. **Think Critically** Is Earth a satellite of any other body in space? Explain.

Applying Math

6. **Solve Simple Equations** A standard unit of measurement in astronomy is the astronomical unit, or AU. It equals is about 150,000,000,000 (1.5×10^{11}) m. In 2000, *Pioneer 10* was more than 11 million km from Earth. How many AUs is this?
7. **Convert Units** A spacecraft is launched at a velocity of 40,200 km/h. Express this speed in kilometers per second. Show your work.

Science Online bookj.msscience.com/self_check_quiz

section 2 review

1. They are all natural satellites of Neptune.
2. Once *Galileo* arrived at Jupiter, it went into orbit around that planet.
3. Discoveries include: ring around Jupiter; additional moons of Jupiter, Saturn, Uranus, and Neptune; erupting volcanoes on Io; complexity of Saturn's rings; geysers on Triton; storms on Neptune.
4. Time line should begin in 1957 with *Sputnik*. Time line should include Project Mercury, 1961–1962; Project Gemini, 1965–1966; Project Apollo, 1967–1972, especially *Apollo 11*, 1969.
5. Yes; Earth is a natural satellite of the Sun because Earth revolves around the Sun.
6. Because 1 AU equals 150 million km, you must divide the distance 11 billion by this to get 73.3 AU.
7. Because there are 60 seconds in a minute and 60 minutes in an hour you must divide the 40,200 km/hr by 3,600 to get 11.2 km/s.

section 3

Current and Future Space Missions

The Space Shuttle

Imagine spending millions of dollars to build a machine, sending it off into space, and watching its 3,000 metric tons of metal and other materials burn up after only a few minutes of work. That's exactly what NASA did with the rocket portions of spacecraft for many years. The early rockets were used only to launch a small capsule holding astronauts into orbit. Then sections of the rocket separated from the rest and burned when reentering the atmosphere.

A Reusable Spacecraft NASA administrators, like many others, realized that it would be less expensive and less wasteful to reuse resources. The reusable spacecraft that transports astronauts, satellites, and other materials to and from space is called the **space shuttle**, shown in **Figure 14,** as it is landing.

At launch, the space shuttle stands on end and is connected to an external liquid-fuel tank and two solid-fuel booster rockets. When the shuttle reaches an altitude of about 45 km, the emptied, solid-fuel booster rockets drop off and parachute back to Earth. These are recovered and used again. The external liquid-fuel tank separates and falls back to Earth, but it isn't recovered.

as you read

What You'll Learn

- **Explain** the benefits of the space shuttle.
- **Identify** the usefulness of orbital space stations.
- **Explore** future space missions.
- **Identify** the applications of space technology to everyday life.

Why It's Important

Experiments performed on future space missions may benefit you.

Review Vocabulary
cosmonaut: astronaut of the former Soviet Union or present-day Russian space program

New Vocabulary
- space shuttle
- space station

Work on the Shuttle After the space shuttle reaches space, it begins to orbit Earth. There, astronauts perform many different tasks. In the cargo bay, astronauts can conduct scientific experiments and determine the effects of spaceflight on the human body. When the cargo bay isn't used as a laboratory, the shuttle can launch, repair, and retrieve satellites. Then the satellites can be returned to Earth or repaired onboard and returned to space. After a mission, the shuttle glides back to Earth and lands like an airplane. A large landing field is needed as the gliding speed of the shuttle is 335 km/h.

Figure 14 The space shuttle is designed to make many trips into space.

1 Motivate

INTERACTIVE CHALKBOARD PowerPoint® Presentations

Bellringer

Section Focus Transparencies also are available on the Interactive Chalkboard CD-ROM. L2 ELL

SECTION 3 Section Focus Transparency **Hot Hot Hot Hot Hot . . .**

One of the missions NASA has planned for the future is a solar probe. It is scheduled for launch in February 2007, and it will make two close passes to the Sun—one in 2010 and one in 2015. The solar probe will pass so close to the Sun that it will be in the Sun's atmosphere.

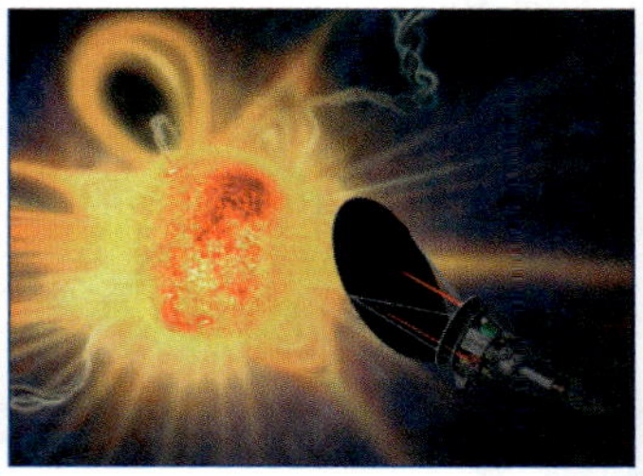

1. Why do you think the solar probe won't be launched before 2007?
2. How might conditions in the Sun's atmosphere affect the design of the solar probe?
3. What do you think scientists hope to learn from the solar probe?

L2

Exploring Space

Tie to Prior Knowledge

Space Shuttle Have students recall watching the launch or landing of a space shuttle. Have each student try to contribute one fact about the shuttle. Make a list of the facts on the board. Review the list with students after reading this section.

Section 3 Resource Manager

Chapter *FAST FILE* Resources

Transparency Activity, p. 42

Directed Reading for Content Mastery, pp. 17, 18

Lab Worksheets, pp. 7-8

Reinforcement, p. 25

Cultural Diversity, p. 25

Home and Community Involvement, p. 31

Physical Science Critical Thinking/Problem Solving, p. 5

Lab Management and Safety, p. 73

2 Teach

Quick Demo

Shuttle Layout

Materials model of the space shuttle, photos or video of the space shuttle

Estimated Time 20 minutes

Procedure Use the model to demonstrate how the cargo bay is used to carry artificial satellites from Earth and how the mobile arm is used to place them into orbit or bring them back into the cargo bay for repairs. If possible, follow up using photos or a video showing the mobile arm and cargo bay in use. L2 **Visual-Kinesthetic**

IDENTIFYING Misconceptions

People in Space Students may think that people on the *International Space Station* will walk around, work, and play normally, much as they see in science fiction TV shows. Remind students that astronauts in orbit around Earth experience microgravity, or weightlessness, which will affect their movements.

Discussion

Space Cooperation Should the United States cooperate with other nations to build a space station, or should the United States build it alone? Accept all answers as a basis for discussion. Some students might think cooperation would cut costs and bring nations together. Other students might think the United States should not share space technology with other nations.

Caption Answer

Figure 16 It was an important first step in building and operating the *International Space Station.*

Figure 15 Astronauts performed a variety of tasks while living and working in space onboard *Skylab.*

Space Stations

Astronauts can spend only a short time living in the space shuttle. Its living area is small, and the crew needs more room to live, exercise, and work. A **space station** has living quarters, work and exercise areas, and all the equipment and support systems needed for humans to live and work in space.

In 1973, the United States launched the space station *Skylab,* shown in **Figure 15.** Crews of astronauts spent up to 84 days there, performing experiments and collecting data on the effects on humans of living in space. In 1979, the abandoned *Skylab* fell out of orbit and burned up as it entered Earth's atmosphere.

Crews from the former Soviet Union have spent more time in space, onboard the space station *Mir,* than crews from any other country. Cosmonaut Dr. Valery Polyakov returned to Earth after 438 days in space studying the long-term effects of weightlessness.

Cooperation in Space

In 1995, the United States and Russia began an era of cooperation and trust in exploring space. Early in the year, American Dr. Norman Thagard was launched into orbit aboard the Russian *Soyuz* spacecraft, along with two Russian cosmonaut crewmates. Dr. Thagard was the first U.S. astronaut launched into space by a Russian booster and the first American resident of the Russian space station *Mir.*

Figure 16 Russian and American scientists have worked together to further space exploration. **Explain** *why the docking of the space shuttle with* Mir *was so important.*

In June 1995, Russian cosmonauts rode into orbit onboard the space shuttle *Atlantis,* America's 100th crewed launch. The mission of *Atlantis* involved, among other studies, a rendezvous and docking with the space station *Mir.* The cooperation that existed on this mission, as shown in **Figure 16,** continued through eight more space shuttle-*Mir* docking missions. Each of the eight missions was an important step toward building and operating the *International Space Station.* In 2001, the abandoned *Mir* space station fell out of orbit and burned up upon reentering the atmosphere. Cooperation continued as the *International Space Station* began to take form.

Differentiated Instruction

Hearing Impaired and Learning Disabled Provide illustrations and photos of the spacecraft discussed. Have students do research to write extended captions for the images. Students can also work on the Internet to find and print out photos of various probes and satellites to make a bulletin board of United States space travel. LS **Visual-Spatial**

Teacher FYI

Skylab When first launched in 1973, *Skylab* lost part of its protective covering and one of its solar panels. The other panel didn't fully open and malfunctioned. At first, the station had limited power and temperatures inside rose. Astronauts heading to the station first had to repair it. Once repaired, *Skylab* went on to complete its mission.

The International Space Station The *International Space Station (ISS)* will be a permanent laboratory designed for long-term research projects. Diverse topics will be studied, including research on the growth of protein crystals. This particular project will help scientists determine protein structure and function, which is expected to enhance work on drug design and the treatment of many diseases.

The *ISS* will draw on the resources of 16 nations. These nations will build units for the space station, which then will be transported into space onboard the space shuttle and Russian launch rockets. The station will be constructed in space. **Figure 17** shows what the completed station will look like.

Figure 17 This is a picture of what the proposed *International Space Station* will look like when it is completed in 2006.

Reading Check *What is the purpose of the* **International Space Station?**

Phases of *ISS* NASA is planning the *ISS* program in phases. Phase One, now concluded, involved the space shuttle-*Mir* docking missions. Phase Two began in 1998 with the launch of the Russian-built *Zarya Module,* also known as the Functional Cargo Block. In December 1998, the first assembly of *ISS* occurred when a space shuttle mission attached the Unity module to *Zarya.* During this phase, crews of three people were delivered to the space station. Phase Two ended in 2001 with the addition of a U.S. laboratory.

Living in Space The project will continue with Phase Three when the Japanese Experiment Module, the European Columbus Orbiting Facility, and another Russian lab will be delivered.

It is hoped that the *International Space Station* will be completed in 2006. Eventually, a seven-person crew should be able to work comfortably onboard the station. A total of 47 separate launches will be required to take all the components of the *ISS* into space and prepare it for permanent habitation. NASA plans for crews of astronauts to stay onboard the station for several months at a time. NASA already has conducted numerous tests to prepare crews of astronauts for extended space missions. One day, the station could be a construction site for ships that will travel to the Moon and Mars.

Science Online

Topic: *International Space Station*

Visit bookj.msscience.com for Web links to information about the *International Space Station.*

Activity You can see the station travel across the sky with an unaided eye. Find out the schedule and try to observe it.

Visual Learning

Figure 17 What do you think the flat panels do? They are solar panels that provide electricity for the space station.

Reading Check

Answer It is a laboratory that will be used for long-term research projects.

Make a Model

Space Station Have pairs of students construct models of the *International Space Station*, using the photos in the text or images from other sources. Have each pair design and add one new feature to the station. Each group should make a presentation, explaining what the new feature would do. Display the models in class. L2 COOP LEARN IS **Interpersonal**

Curriculum Connection

Health The space program has been used to benefit humans in many fields, including health and medicine. Have students write a report in their Science Journals about advancements in health and medicine that have occurred as a direct result of the space program. P

Visual Learning

Figure 18 Have students contrast the surface of Earth and the surface of Mars. Possible answer: Earth's surface is mostly water while Mars's surface is mostly land.

Reading Check

Answer high-quality photographs of the planet's surface

Discussion

Mars Exploration Do you think that successful completion of missions such as the Mars Odyssey make the public more or less supportive of future Mars probes? Explain. Possible answers: Less supportive because of the waste of millions of dollars that could be spent elsewhere; more supportive because of renewed determination to make the next mission successful

Figure 18 Gulleys, channels, and aprons of sediment imaged by the *Mars Global Surveyor* are similar to features on Earth known to be caused by flowing water. This water is thought to seep out from beneath the surface of Mars.

Exploring Mars

Two of the most successful missions in recent years were the 1996 launchings of the *Mars Global Surveyor* and the *Mars Pathfinder. Surveyor* orbited Mars, taking high-quality photos of the planet's surface as shown in **Figure 18.** *Pathfinder* descended to the Martian surface, using rockets and a parachute system to slow its descent. Large balloons absorbed the shock of landing. *Pathfinder* carried technology to study the surface of the planet, including a remote-controlled robot rover called Sojourner. Using information gathered by studying photographs taken by *Surveyor,* scientists determined that water recently had seeped to the surface of Mars in some areas.

Reading Check ***What type of data were obtained by the* Mars Global Surveyor?**

Another orbiting spacecraft, the *Mars Odyssey* began mapping the surface of Mars in 2002. Soon after, its data confirmed the findings of *Surveyor*—that Martian soil contains frozen water in the southern polar area. The next step was to send robots to explore the surface of Mars. Twin rovers named *Spirit* and *Opportunity* were launched in 2003 with schedules to reach their separate destinations on Mars in January 2004. Their primary goals are to analyze Martian rocks and soils to tell scientists more about Martian geology and provide clues about the role of water on Mars. Future plans include *Phoenix* in 2008, a robot lander capable of digging over a meter into the surface.

LAB DEMONSTRATION

Purpose to demonstrate simulated gravity

Materials turntable, LP record, scissors, construction paper, tape, marbles

Preparation Fold 8-cm wide strips of construction paper in half lengthwise. Unfold.

Procedure Wrap the strips along the fold around the circumference of the record so there is a 4-cm wall. Securely tape the remainder underneath. Place the record on the turntable, place three marbles at its center, and switch it on.

Expected Outcome The marbles are accelerated outward; the paper exerts a force and stops them.

Assessment

How could this procedure generate artificial gravity in a space station? If the space station were spinning, inertia would cause objects to accelerate toward the outer edge, producing a feeling of gravity.

New Millennium Program

To continue space missions into the future, NASA has created the New Millennium Program (NMP). The goal of the NMP is to develop advanced technology that will let NASA send smart spacecraft into the solar system. This will reduce the amount of ground control needed. They also hope to reduce the size of future spacecraft to keep the cost of launching them under control. NASA's challenge is to prove that certain cutting-edge technologies, as well as mission concepts, work in space.

Exploring the Moon

Does water exist in the craters of the Moon's poles? This is one question NASA intends to explore with data gathered from the *Lunar Prospector* spacecraft shown in **Figure 19.** Launched in 1998, the *Lunar Prospector's* one-year mission was to orbit the Moon, mapping its structure and composition. Data obtained from the spacecraft indicate that water ice might be present in the craters at the Moon's poles. Scientists first estimated up to 300 million metric tons of water may be trapped as ice, and later estimates are much higher. In the permanently shadowed areas of some craters, the temperature never exceeds −230°C. Therefore water delivered to the Moon by comets or meteorites early in its history could remain frozen indefinitely.

At the end of its mission, *Lunar Prospector* was deliberately crashed into a lunar crater. Using special telescopes, scientists hoped to see evidence of water vapor thrown up by the collision. None was seen, however scientists still believe that much water ice is there. If so, this water would be useful if a colony is ever built on the Moon.

Science Online

Topic: New Millennium Program

Visit bookj.msscience.com for Web links to information about NASA's New Millennium Program.

Activity Prepare a table listing proposed missions, projected launch dates, and what they will study.

Figure 19 The *Lunar Prospector* analyzed the Moon's composition during its one-year mission.
Explain *why* Lunar Prospector *was deliberately crashed on the Moon.*

Activity

Polling Students Ask students: What do other students in your school think about the future of the space program? Do they want it to continue? If so, what do they think NASA should explore? Should uncrewed probes be sent or should there be many more crewed missions to space? Have groups of students design a poll to find the answers to these questions. They should finalize the questions, decide how many students they will survey, carry out the poll, and report the results. Have different student groups share their results in class.

Discussion

Moon Water Why is it important for scientists to determine if there is water on the moon? Possible answer: Having water on the Moon would make it easier to set up a moon base. Water would be available for drinking, and also could be used to produce oxygen.

Caption Answer

Figure 19 Scientists hoped that the collision would throw up water vapor. However, none was seen.

Differentiated Instruction

Challenge NASA has planned a number of Mars missions. Have students do research to find out about these missions and write a short report on one of them. Have students include the design of the probe, when it was launched, or is scheduled to be launched, how long it will take to reach Mars, and what NASA scientists want it to accomplish. L2

Use an Analogy

Walking the Equator To help students understand the vast distances space probes travel, tell them that the *Cassini* probe, headed toward Saturn, must travel roughly 1.3 billion km one way. That's nearly equivalent to walking around Earth along the equator more than 100,000 times.

Figure 20 *Cassini* is currently on its way to Saturn. After it arrives, it will spend four years studying Saturn and its surrounding area.

Cassini

In October 1997, NASA launched the space probe *Cassini.* This probe's destination is Saturn. *Cassini,* shown in **Figure 20,** will not reach its goal until 2004. At that time, the space probe will explore Saturn and surrounding areas for four years. One part of its mission is to deliver the European Space Agency's *Huygens* probe to Saturn's largest moon, Titan. Some scientists theorize that Titan's atmosphere may be similar to the atmosphere of early Earth.

The Next Generation Space Telescope Not all space missions involve sending astronauts or probes into space. Plans are being made to launch a new space telescope that is capable of observing the first stars and galaxies in the universe. The *James Webb Space Telescope,* shown in **Figure 21,** will be the successor to the *Hubble Space Telescope.* As part of the Origins project, it will provide scientists with the opportunity to study the evolution of galaxies, the production of elements by stars, and the process of star and planet formation. To accomplish these tasks, the telescope will have to be able to see objects 400 times fainter than those currently studied with ground-based telescopes such as the twin Keck telescopes. NASA hopes to launch the *James Webb Space Telescope* as early as 2010.

Figure 21 The *James Webb Space Telescope* honors the NASA administrator who contributed greatly to the Apollo Program. It will help scientists learn more about how galaxies form.

Differentiated Instruction

Challenge Have students research NASA's Origins Program. Ask them to find out when the program began, what stages are involved, and where the program is heading. L3

Learning Disabled Help your students better understand the *Cassini* mission by showing them photos of the Saturn system with its many rings and moons. These will be the objects that *Cassini* will study. Point out Titan and explain to students that *Cassini* will deliver the probe *Huygens* to study this large moon. LS **Visual-Spatial**

Everyday Space Technology Many people have benefited from research done for space programs. Medicine especially has gained much from space technology. Space medicine led to better ways to diagnose and treat heart disease here on Earth and to better heart pacemakers. A screening system that works on infants is helping eye doctors spot vision problems early. Cochlear implants that help thousands of deaf people hear were developed using knowledge gained during the space shuttle program.

Space technology can even help catch criminals and prevent accidents. For example, a method to sharpen images that was devised for space studies is being used by police to read numbers on blurry photos of license plates. Equipment using space technology can be placed on emergency vehicles. This equipment automatically changes traffic signals as an emergency vehicle approaches intersections, so that crossing vehicles have time to stop safely. A hand-held device used for travel directions is shown in **Figure 22.**

Figure 22 Global Positioning System (GPS) technology uses satellites to determine location on Earth's surface.

Reading Check *How have research and technology developed for space benefited people here on Earth?*

section 3 review

Summary

The Space Station

- A space station is an orbiting laboratory.
- The new *International Space Station (ISS)* is being built with the aid of 16 nations.
- The space shuttle transports astronauts, satellites, and other materials to and from the *ISS*.

Exploring Mars and the Moon

- The *Mars Global Surveyor* orbited Mars and the *Mars Pathfinder* studied its surface.
- *Lunar Prospector* orbited the Moon, mapping its structure and composition.
- Recent data indicate that water ice crystals may exist in shadows of lunar craters.

Future Missions

- The *Cassini* probe is scheduled to explore Saturn and its moons.
- The successor to the *Hubble* will be the *James Webb Space Telescope*.

Self Check

1. **Identify** the main advantage of the space shuttle.
2. **Describe** the importance of space shuttle-*Mir* docking missions.
3. **Explain** how *International Space Station* is used.
4. **Identify** three ways that space technology is a benefit to everyday life.
5. **Think Critically** What makes the space shuttle more versatile than earlier spacecraft?

Applying Math

6. **Solve One-Step Equations** *Voyager 1* had about 30 kg of hydrazine fuel left in 2003. If it uses about 500 g per year, how long will this fuel last?
7. **Use Percentages** Suppose you're in charge of assembling a crew of 50 people. Decide how many to assign each task, such as farming, maintenance, scientific experimentation, and so on. Calculate the percent of the crew assigned to each task. Justify your decisions.

3 Assess

Reading Check

Answer It can provide better medical diagnoses, implants, and assist in law enforcement.

DAILY INTERVENTION

Check for Understanding

Linguistic Challenge students to make a chart that summarizes the NASA missions on the drawing board for the next ten years, and the goals of each. Have the class pool the information they find to make one large chart on poster board. If the next mission takes place in the near future, have students compare the goals of the mission to what it actually accomplishes. L2

Reteach

Compare Have students compare and contrast the space shuttle and the *International Space Station*. Possible answers: Both orbit Earth and are used for research; astronauts can live longer on the *ISS*, and the *ISS* has more room; shuttle returns to Earth while *ISS* does not; shuttle is an American ship while *ISS* is international. L1

Assessment

Oral Have each student make up a question and answer on information in this section for a brief oral quiz. Arrange students in a circle, then have each student address his or her question to the person on his or her left. Use **Performance Assessment in the Science Classroom,** p. 91.

section 3 review

1. It is a reusable spacecraft.
2. They allowed astronauts to practice a maneuver that would be needed to build the ISS.
3. It will be used for long-term research projects. It could one day be a construction site for ships that go to the moon or Mars.
4. Benefits include better heart pacemakers, vision screening, cochlear implants and enhanced imaging systems.
5. It's reusable and can carry cargo into orbit, including satellites. It can also retrieve objects from orbit and take astronauts into orbit to repair satellites.
6. If the spacecraft uses 500 g/y and has 30 kg, you must first convert kilograms to grams which gives 30,000 g and then divide by 500 g/y to get 60 years.
7. Lists could include people to grow food, make repairs, treat medical emergencies, and pilot spacecraft. Accept reasonable answers.

BENCH TESTED

Real-World Question

Purpose Students will use information and sightings of the star Polaris to estimate the circumference of Earth.

Process Skills observe, plan, use tables, record observations, communicate, interpret data

Time Required 2 hours

Possible Materials astrolabe or cardboard, string, metal bolts for weights to make astrolabe

Safety Precautions Make sure students observe warning about not looking at the Sun directly.

Internet Have students visit bookj.msscience.com/internet_lab to obtain directions for making an astrolabe. The astrolabe will be used for star sightings that can help students determine Earth's circumference.

Non-Internet Sources Students can check their calculations by using an encyclopedia to look up Earth's circumference.

Make a Plan

Preparation

Internet To run through the steps that students will follow, visit bookj.msscience.com/internet_lab.

Non-Internet Have books on astronomy and the history of astronomy as well as an encyclopedia available for students to use. Also provide atlases to use in order to determine the distance between cities.

LAB Use the Internet

Star Sightings

Goals

- **Record** your sightings of Polaris.
- **Share** the data with other students to calculate the circumference of Earth.

Data Source

Science Online

Go to bookj.msscience.com/internet_lab to obtain instructions on how to make an astrolabe. Also visit the Web site for more information about the location of Polaris, and for data from other students.

Safety Precautions

WARNING: *Do not use the astrolabe during the daytime to observe the Sun.*

Real-World Question

For thousands of years, people have measured their position on Earth using the position of Polaris, the North Star. At any given observation point, it always appears at the same angle above the horizon. For example, at the north pole, Polaris appears directly overhead, and at the equator it is just above the northern horizon. Other locations can be determined by measuring the height of Polaris above the horizon using an instrument called an astrolabe. Could you use Polaris to determine the size of Earth? You know that Earth is round. Knowing this, do you think you can estimate the circumference of Earth based on star sightings?

Make a Plan

1. Obtain an astrolabe or construct one using the instructions posted by visiting the link above.
2. **Design** a data table in your Science Journal similar to the one below.

Polaris Observations

Your Location:

Date	Time	Astrolabe Reading
Readings will vary with school's location.		

Alternative Inquiry Lab

Paths of Planets Ask students how they might use the astrolabe to identify the ecliptic path that the Sun and planets follow. Students could follow one easily visible planet, such as Jupiter, Saturn, Mars, or Venus at different hours. Suggest they devise a graph to show the path of the planet. Encourage them to research when planets are visible. Another possibility is to follow the path of the *International Space Station.*

Using Scientific Methods

3. Decide as a group how you will make your observations. Does it take more than one person to make each observation? When will it be easiest to see Polaris?

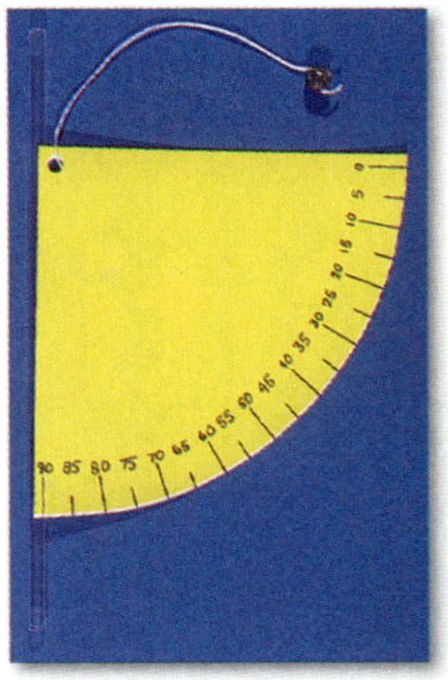

Follow Your Plan

1. Make sure your teacher approves your plan before you start.
2. Carry out your observations.
3. **Record** your observations in your data table.
4. Average your readings and post them in the table provided at the link shown below.

Analyze Your Data

1. **Research** the names of cities that are at approximately the same longitude as your hometown. Gather astrolabe readings from students in one of those cities at the link shown below.
2. **Compare** your astrolabe readings. Subtract the smaller reading from the larger one.
3. **Determine** the distance between your star sighting location and the other city.
4. **Calculate** the circumference of Earth using the following relationship.

   Circumference $= (360°) \times$ (distance between locations)/difference between readings

Conclude and Apply

1. **Analyze** how the circumference of Earth that you calculated compares with the accepted value of 40,079 km.
2. **Determine** some possible sources of error in this method of establishing the size of Earth. What improvements would you suggest?

Communicating Your Data

Find this lab using the link below. **Create** a poster that includes a table of your data and data from students in other cities. **Perform** a sample circumference calculation for your class.

Science Online

bookj.msscience.com/internet_lab

Follow Your Plan

Teaching Strategies

- Have students make repeated star sightings of Polaris and other stars over a few nights and at different times.
- After students have made a few observations, explain the circumference equation. Have them practice explaining the logic behind the circumference equation to a partner.

Analyze Your Data

1. Students can use an atlas to locate cities at approximately the same longitude as your hometown.
2. Answers will vary depending on readings.
3. Answers will vary on cities chosen.
4. Earth's circumference at the equator is 24,901 mi (40,079 km).

Conclude and Apply

1. Values should be close.
2. Possible answers: making errors in calculations, choosing a city not on your longitude, misreading the astrolabe. Students might suggest being more careful or repeating calculations several times.

Assessment

Process Have students draw an events chain concept map that shows the steps used in completing this activity. Use **Performance Assessment in the Science Classroom,** p. 163.

Communicating Your Data

Have students use a computer spreadsheet program to record data on star sightings at their location and at various other cities on the same line of longitude.

TIME

Content Background

The critics of space travel claim that the effort and money spent on space colonization would be better spent on programs to solve current problems. The advocates see space as a new frontier equal or greater in potential to the Americas in the 1600s. Advocates also point out that, if the materials and systems required for the construction of self-sufficient space colonies can be developed, these same technologies can be applied to similar problems on planet Earth.

Activity

Essentials Organize the class into teams to look into these aspects of space habitation: breathable air, energy (for all purposes), water, building materials, sustainable food supplies, protection from cosmic radiation, gravity, and transportation. Instruct the teams to prepare charts indicating the technology required compared with available technology. They can then pool their research and try to determine when they feel space colonization will be feasible.

Investigate the Issue

The settling of Mars and the Moon has been compared to inhabiting harsh climates on the Earth. Have students compare the conditions on Mars with those found in a desert, the Arctic, and underwater. Which are most similar? Most different? What problems are unique to each place?

TIME SCIENCE AND Society

SCIENCE ISSUES THAT AFFECT YOU!

Cities in Space

Should the U.S. spend money to colonize space?

Humans have landed on the Moon, and spacecrafts have landed on Mars. But these space missions are just small steps that may lead to a giant new space program. As technology improves, humans may be able to visit and even live on other planets. But is it worth the time and money involved?

Those in favor of living in space point to the International Space Station that already is orbiting Earth. It's an early step toward establishing floating cities where astronauts can live and work. As Earth's population continues to increase and there is less room on this planet, why not expand to other planets or build a floating city in space? Also, the fact that there is little pollution in space makes the idea appealing to many.

Critics of colonizing space think we should spend the hundreds of billions of dollars that it would cost to colonize space on projects to help improve people's lives here on Earth. Building better housing, developing ways to feed the hungry, finding cures for diseases, and increasing funds for education should come first, these people say. And, critics continue, if people want to explore, why not explore right here on Earth, for example, the ocean floor.

Moon or Mars? If humans were to move permanently to space, the two most likely destinations would be Mars and the Moon, both bleak places. But those in favor of moving to these places say humans could find a way to make them livable as they have made homes in harsh climates and in many rugged areas here on Earth.

Water may be locked in lunar craters, and photos suggest that Mars once had liquid water on its surface. If that water is frozen underground, humans may be able to access it. NASA is studying whether it makes sense to send astronauts and scientists to explore Mars.

Transforming Mars into an Earthlike place with breathable air and usable water will take much longer, but some small steps are being taken. Experimental plants are being developed that could absorb Mars's excess carbon dioxide and release oxygen. Solar mirrors that could warm Mars's surface are available.

Those for and against colonizing space agree on one thing—it will take large amounts of money, research, and planning. It also will take the same spirit of adventure that has led history's pioneers into so many bold frontiers—deserts, the poles, and the sky.

Debate with your class the pros and cons of colonizing space. Do you think the United States should spend money to create space cities or use the money now to improve lives of people on Earth?

Science Online
For more information, visit bookj.msscience.com/time

Debate Each side should be able to produce valid reasons. Those in favor may feel that the money spent will help more people on the long term through new knowledge and scientific developments. Those opposed can cite many of the pressing problems that need attention here on Earth, such as famine and poverty, and diseases such as AIDS or Alzheimers.

Resources for Teachers and Students

On to Mars: Colonizing a New World, by Frank Crossman and Robert Zubrin, Apogee Books, 2002

Making Space Happen: Private Space Ventures and the Visionaries Behind Them, by Paula Berinstein and Fiorella Terenzi, 2002

Entering Space: Creating a Spacefaring Civilization by Robert Zubrin, 2000

Reviewing Main Ideas

Section 1 Radiation from Space

1. The arrangement of electromagnetic waves according to their wavelengths is the electromagnetic spectrum.
2. Optical telescopes produce magnified images of objects.
3. Radio telescopes collect and record radio waves given off by some space objects.

Section 2 Early Space Missions

1. A satellite is an object that revolves around another object. The moons of planets are natural satellites. Artificial satellites are those made by people.
2. A space probe travels into the solar system, gathers data, and sends them back to Earth.
3. American piloted space programs included the Gemini, Mercury, and Apollo Projects.

Section 3 Current and Future Space Missions

1. Space stations provide the opportunity to conduct research not possible on Earth. The *International Space Station* is being constructed in space with the cooperation of more than a dozen nations.
2. The space shuttle is a reusable spacecraft that carries astronauts, satellites, and other cargo to and from space.
3. Space technology is used to solve problems on Earth, too. Advances in engineering related to space travel have aided medicine, environmental sciences, and other fields.

Visualizing Main Ideas

Copy and complete the following concept map about the race to the Moon. Use the phrases: first satellite, Project Gemini, Project Mercury, team of two astronauts orbits Earth, Project Apollo.

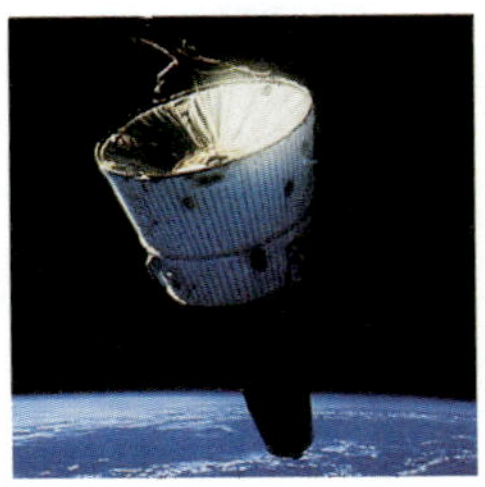

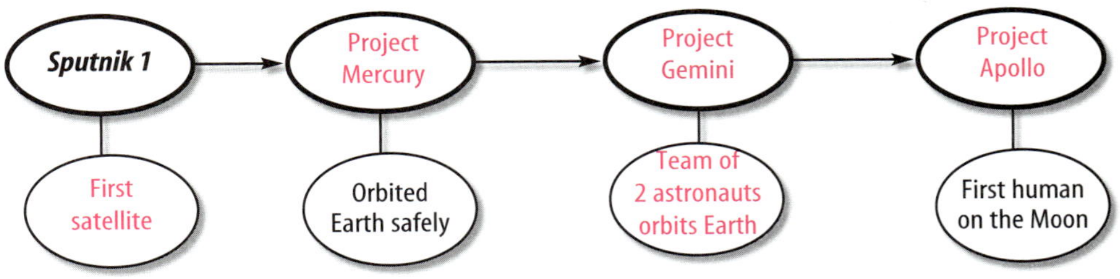

Science Online bookj.msscience.com/interactive_tutor

chapter 1 Study Guide

Reviewing Main Ideas

Summary statements can be used by students to review the major concepts of the chapter.

Visualizing Main Ideas

See student page.

Science Online

Visit bookj.msscience.com
/self_check_quiz
/interactive_tutor
/vocabulary_puzzlemaker
/chapter_review
/standardized_test

Assessment Transparency

For additional assessment questions, use the *Assessment Transparency* located in the transparency book.

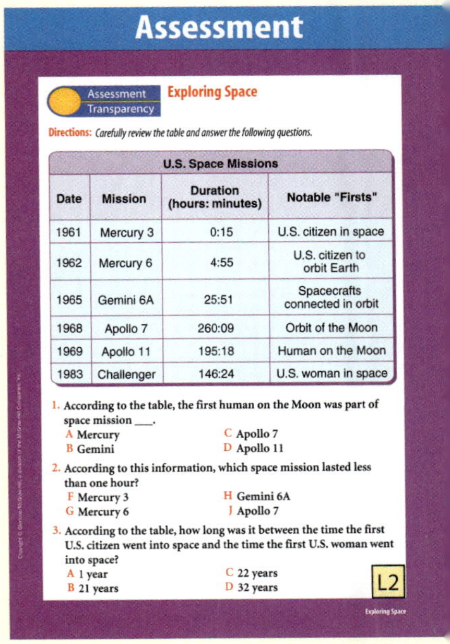

Assessment

Assessment Transparency **Exploring Space**

Directions: *Carefully review the table and answer the following questions.*

U.S. Space Missions			
Date	Mission	Duration (hours: minutes)	Notable "Firsts"
1961	Mercury 3	0:15	U.S. citizen in space
1962	Mercury 6	4:55	U.S. citizen to orbit Earth
1965	Gemini 6A	25:51	Spacecrafts connected in orbit
1968	Apollo 7	260:09	Orbit of the Moon
1969	Apollo 11	195:18	Human on the Moon
1983	Challenger	146:24	U.S. woman in space

1. According to the table, the first human on the Moon was part of space mission ___.
   A Mercury
   B Gemini
   C Apollo 7
   D Apollo 11
2. According to this information, which space mission lasted less than one hour?
   F Mercury 3
   G Mercury 6
   H Gemini 6A
   J Apollo 7
3. According to the table, how long was it between the time the first U.S. citizen went into space and the time the first U.S. woman went into space?
   A 1 year
   B 21 years
   C 22 years
   D 32 years

L2

Exploring Space

Using Vocabulary

1. refracting telescope
2. satellite
3. Project Mercury
4. space shuttle
5. electromagnetic spectrum

Checking Concepts

6. D
7. D
8. B
9. A
10. B
11. A
12. D
13. C
14. C

Thinking Critically

15. Because the Moon has no atmosphere, light and other forms of energy can reach its surface without distortion.
16. The same kind of sensing system could detect dangerous fumes in a factory.
17. Possible answers: Robotic space probes are cheaper and can go on one-way deep space missions that people cannot. Crewed flights provide information about living in space.

chapter 1 Review

Using Vocabulary

electromagnetic spectrum p. 9	reflecting telescope p. 10
observatory p. 10	refracting telescope p. 10
orbit p. 17	rocket p. 15
Project Apollo p. 22	satellite p. 17
Project Gemini p. 21	space probe p. 18
Project Mercury p. 21	space shuttle p. 23
radio telescope p. 13	space station p. 24

Fill in the blanks with the correct vocabulary word(s).

1. A(n) ________ telescope uses lenses to bend light.
2. A(n) ________ is an object that revolves around another object in space.
3. ________ was the first piloted U.S. space program.
4. A(n) ________ carries people and tools to and from space.
5. In the ________, electromagnetic waves are arranged, in order, according to their wavelengths.

Checking Concepts

Choose the word or phrase that best answers the question.

6. Which spacecraft has sent images of Venus to scientists on Earth?
   A) *Voyager* **C)** *Apollo 11*
   B) *Viking* **D)** *Magellan*

7. Which kind of telescope uses mirrors to collect light?
   A) radio
   B) electromagnetic
   C) refracting
   D) reflecting

8. What was *Sputnik I?*
   A) the first telescope
   B) the first artificial satellite
   C) the first observatory
   D) the first U.S. space probe

9. Which kind of telescope can be used during the day or night and during bad weather?
   A) radio
   B) electromagnetic
   C) refracting
   D) reflecting

10. When fully operational, what is the maximum number of people who will crew the *International Space Station?*
    A) 3 **C)** 15
    B) 7 **D)** 50

11. Which space mission's goal was to put a spacecraft into orbit and bring it back safely?
    A) Project Mercury
    B) Project Apollo
    C) Project Gemini
    D) *Viking I*

12. Which of the following is a natural satellite of Earth?
    A) *Skylab*
    B) the space shuttle
    C) the Sun
    D) the Moon

13. What does the space shuttle use to place a satellite into space?
    A) liquid-fuel tank
    B) booster rocket
    C) mechanical arm
    D) cargo bay

14. What part of the space shuttle is reused?
    A) liquid-fuel tanks
    B) *Gemini* rockets
    C) booster engines
    D) Saturn rockets

Science Online bookj.msscience.com/vocabulary_puzzlemaker

Use the ExamView® Pro Testmaker CD-ROM to:
- create multiple versions of tests
- create modified tests with one mouse click for inclusion students
- edit existing questions and add your own questions
- build tests aligned with state standards using built-in State Curriculum Tags
- change English tests to Spanish with one mouse click and vice versa

Thinking Critically

15. **Compare and contrast** the advantages of a moon-based telescope with an Earth-based telescope.

16. **Infer** how sensors used to detect toxic chemicals in the space shuttle could be beneficial to a factory worker.

17. **Drawing Conclusions** Which do you think is a wiser method of exploration—space missions with people onboard or robotic space probes? Why?

18. **Explain** Suppose two astronauts are outside the space shuttle orbiting Earth. The audio speaker in the helmet of one astronaut quits working. The other astronaut is 1 m away and shouts a message. Can the first astronaut hear the message? Support your reasoning.

19. **Make and Use Tables** Copy and complete the table below. Use information from several resources.

United States Space Probes

Probe	Launch Date(s)	Planets or Objects Visited
Vikings 1 and *2*		
Galileo	Answers will vary.	
Lunar Prospector		
Pathfinder		

20. **Classify** the following as a satellite or a space probe: *Cassini, Sputnik I, Hubble Space Telescope*, space shuttle, and *Voyager 2*.

21. **Compare and contrast** space probes and artificial satellites.

Performance Activities

22. **Display** Make a display showing some of the images obtained from the *Hubble Space Telescope*. Include samples of three types of galaxies, nebulae, and star clusters.

Applying Math

23. **Space Communications** In May 2003 *Voyager 1* was 13 billion km from the Sun. Calculate how long it takes for a signal to travel this far assuming it travels at 3×10^8 m/s.

Use the graph below to answer question 24.

Average Velocities of Satellites A & B

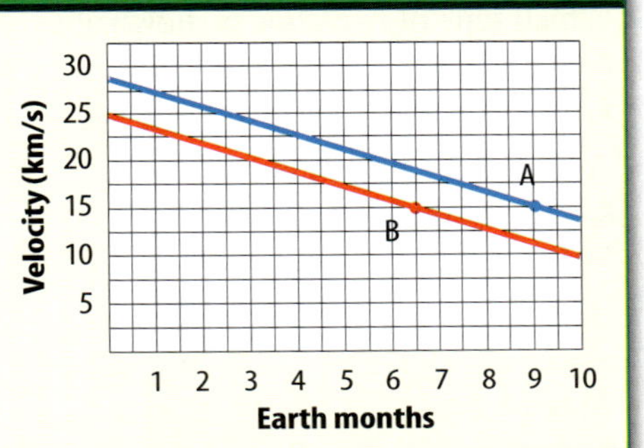

24. **Satellite Orbits** The graph above predicts the average velocities of satellites A and B in orbit around a hypothetical planet. Because of contact with the planet's atmosphere, their velocities are decreasing. At a velocity of 15 km/s their orbits will decay and they will spiral downwards to the surface. Using the graph, determine how long will it take for each satellite to reach this point?

25. **Calculate Fuel** A spacecraft carries 30 kg of hydrazine fuel and uses and average of 500 g/y. How many years could this fuel last?

26. **Space Distances** Find the distance in AUs to a star 68 light-years (LY) distant. (1 LY $= 6.3 \times 10^4$ AUs)

Science Online bookj.msscience.com/chapter_review

Assessment Resources

Reproducible Masters

Chapter *Fast File* Resources
Chapter Review, pp. 33–34
Chapter Tests, pp. 35–38
Assessment Transparency Activity, p. 45

Glencoe Science Web site
Chapter Review Test
Standardized Test Practice

Glencoe Technology
Assessment Transparency
ExamView® Pro Testmaker
MindJogger Videoquiz
Interactive Chalkboard

18. No; sound must travel through matter. Space is a vacuum, so sound will not travel across it.

19.

United States Space Probes

Probe	Launch Date	Planets or Objects Visited
Vikings 1 and *2*	1975	Mars
Galileo	1989	Venus, Europa, Jupiter
Lunar Prospector	1998	Earth's Moon
Pathfinder	1996	Mars

20. *Sputnik*, the shuttle, and *Hubble* all orbit Earth like satellites. *Voyager 2* and *Cassini* are space probes.

21. Both are sent into space to collect data and send it back to Earth. Space probes travel into space, whereas satellites orbit Earth or other bodies.

Performance Activities

22. Check to see that the elements are present and labeled correctly. Use **PASC**, p. 135.

Applying Math

National Math Standards

1, 4, 5, 9

23. The speed of light is 3×10^8 m/s or 3×10^5 km/s. By dividing this into 13 billion, gives 43,000 s. Then dividing again by 3,600 s/h, gives 11.9 h.

24. A = 9 months and B = 6.5 months.

25. If it has 30 kg and uses 500 g/y, dividing 30 by .5 kg gives 60 years.

26. The distance 6.3×10^4 AU/LY multiplied by 68 LY gives 428.4×10^4 or 4.28×10^6 AU's

FAST FILE

Answer Sheet A practice answer sheet can be found at bookj.msscience.com/answer_sheet.

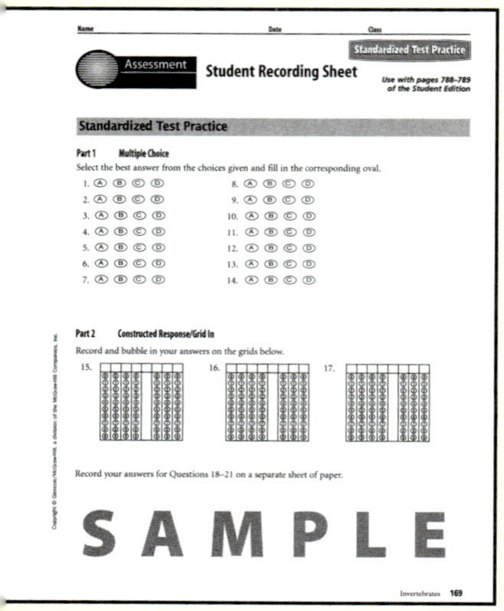
Name Date Class

Standardized Test Practice

Assessment Student Recording Sheet

Use with pages 788–789 of the Student Edition

Standardized Test Practice

Part 1 Multiple Choice

Select the best answer from the choices given and fill in the corresponding oval.

1. Ⓐ Ⓑ Ⓒ Ⓓ 8. Ⓐ Ⓑ Ⓒ Ⓓ
2. Ⓐ Ⓑ Ⓒ Ⓓ 9. Ⓐ Ⓑ Ⓒ Ⓓ
3. Ⓐ Ⓑ Ⓒ Ⓓ 10. Ⓐ Ⓑ Ⓒ Ⓓ
4. Ⓐ Ⓑ Ⓒ Ⓓ 11. Ⓐ Ⓑ Ⓒ Ⓓ
5. Ⓐ Ⓑ Ⓒ Ⓓ 12. Ⓐ Ⓑ Ⓒ Ⓓ
6. Ⓐ Ⓑ Ⓒ Ⓓ 13. Ⓐ Ⓑ Ⓒ Ⓓ
7. Ⓐ Ⓑ Ⓒ Ⓓ 14. Ⓐ Ⓑ Ⓒ Ⓓ

Part 2 Constructed Response/Grid In

Record and bubble in your answers on the grids below.

15. 16. 17.

Record your answers for Questions 18–21 on a separate sheet of paper.

SAMPLE

Invertebrates 169

Part 1 Multiple Choice

1. A
2. D
3. D
4. B
5. C
6. A
7. A
8. C
9. A
10. B
11. D

Part 2 Short Response

12. Space probes often travel far into the solar system.
13. It demonstrated that artificial satellites could be launched into Earth orbit and started the space race.
14. communication, positioning, weather, Earth monitoring, intelligence gathering

chapter 1 Standardized Test Practice

Part 1 Multiple Choice

Record your answers on the answer sheet provided by your teacher or on a sheet of paper.

Use the figure below to answer question 1.

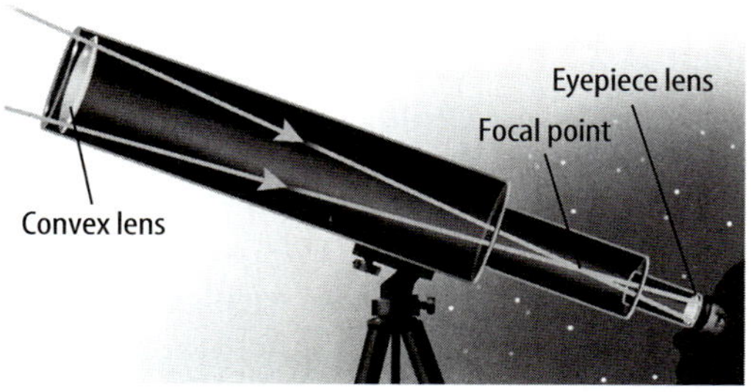

1. Which type of telescope is shown above?
   A. refracting **C.** reflecting
   B. radio **D.** space
2. Who was the first human in space?
   A. Edwin Aldrin **C.** Neil Armstrong
   B. John Glenn **D.** Yuri Gagarin
3. Which is an engine that can launch an object into space?
   A. space probe **C.** capsule
   B. shuttle **D.** rocket
4. Which is the speed of light in a vacuum?
   A. 300 km/s **C.** 3,000 km/s
   B. 300,000 km/s **D.** 30,00 km/s
5. Which of the following is an advantage of space telescopes?
   A. They are cheaper to build.
   B. They have fewer technical problems.
   C. They obtain higher quality images.
   D. They can be repaired easily.

Test-Taking Tip

Making Answers Do not mark the test booklet when taking the test. Be sure to mark ALL answers on your answer sheet and leave no blanks.

6. Which type of radiation has a shorter wavelength than visible light does?
   A. ultraviolet **C.** infrared
   B. microwaves **D.** radio waves
7. Which space probe visited Mars?
   A. *Viking 1* **C.** *Magellan*
   B. *Mariner 2* **D.** *Pioneer 10*
8. Which United States space program included several lunar landings?
   A. Gemini **C.** Apollo
   B. Mercury **D.** space shuttle

Examine the diagram below. Then answer questions 9–11.

9. What is the name of the curved path that the satellite follows?
   A. an orbit **C.** a revolution
   B. a rotation **D.** a track
10. Which force pulls the satellite toward Earth?
    A. the Moon's gravity
    B. Earth's gravity
    C. the Sun's gravity
    D. Earth's magnetic field
11. Imagine that the satellite in the diagram above started to orbit at a slower speed. Which of the following probably would happen to the satellite?
    A. It would fly off into space.
    B. It would crash into the Moon.
    C. It would crash into the Sun.
    D. It would crash into Earth.

Part 2 Short Response/Grid In

15. Radio telescopes use radio waves, whereas optical telescopes use visible light
16. 3.60
17. 3.23
18. 1.0 m above, −15.1° C − (−74.4°C) = 59.3°C; for 0.5 m above, −12.5°C − (−74.9°C) = 62.4°C; for 0.25 m above, −8.9°C − (−75.9°C) = 67.0°C
19. Atmospheric temperatures begin to increase after sunrise and continue to increase until after solar noon. They cool through the nighttime hours. Air nearest the surface has the greatest temperature fluctuations.

Part 2 Short Response/Grid In

Record your answers on the answer sheet provided by your teacher or on a sheet of paper.

12. Explain the difference between a space probe and a satellite that is orbiting Earth.

13. Why was the flight of *Sputnik 1* important?

14. List four ways that satellites are useful.

15. How are radio telescopes different from optical telescopes?

Use the table below to answer questions 16–19. The table includes data collected by *Mars Pathfinder* on the third Sol, or Martian day, of operation.

Sol 3 Temperature Data from Mars Pathfinder			
Proportion of Sol	Temperature (°C)		
	1.0 m above surface	0.5 m above surface	0.25 m above surface
3.07	−70.4	−70.7	−73.4
3.23	−74.4	−74.9	−75.9
3.33	−53.0	−51.9	−46.7
3.51	−22.3	−19.2	−15.7
3.60	−15.1	−12.5	−8.9
3.70	−26.1	−25.7	−24.0
3.92	−63.9	−64.5	−65.8

16. Which proportion of sol value corresponds to the warmest temperatures at all three heights?

17. Which proportion of sol value corresponds to the coldest temperatures at all three heights?

18. What is the range of the listed temperature values for each distance above the surface?

19. Explain the data in the table. Why do the temperatures vary in this way?

Part 3 Open Ended

Record your answers on a sheet of paper.

Use the diagram below to answer question 20.

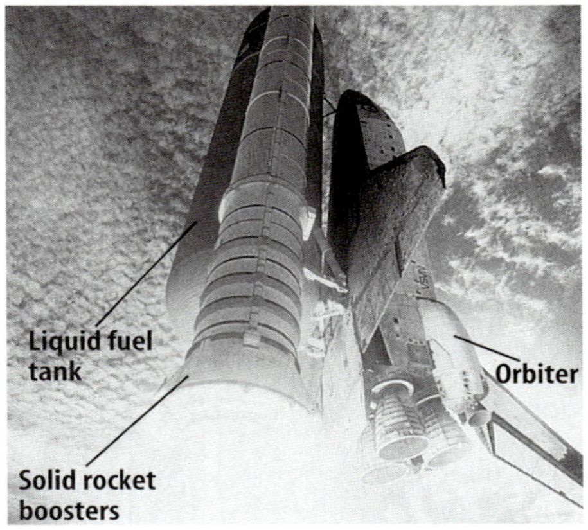

20. Explain the purpose of each of the labeled objects.

21. List four advancements in technology directly attributable to space exploration and how they have impacted everyday life on Earth.

22. What are the advantages of having reusable spacecraft? Are there any disadvantages? Explain.

23. What is the *International Space Station?* How is it used?

24. What are the advantages of international cooperation during space exploration? Are there disadvantages?

25. Explain how the voices of astronauts onboard the space shuttle can be heard on Earth.

26. List several benefits and costs of space exploration. Do you think that the benefits of space exploration outweigh the costs? Explain why you do or do not.

24. Advantages include sharing costs and improved relations between countries. Disadvantages might include less control over operations by any one country and the fact that many countries must fulfill their obligations on time.

25. The astronaut's voices are converted to radio waves, which are received on Earth and converted back to sound.

26. Benefits: new technology, scientific advances, possible future colonization and resource utilization; Cost: money spent, risk to human lives.

Rubrics

For more help evaluating open-ended assessment questions, see the rubric on p. 10T.

Part 3 Open Ended

20. The liquid fuel tank provides fuel for the three main engines. It is released as the orbiter nears orbit. The solid rocket boosters also burn during liftoff, but fall away sooner than the liquid tank. They are recovered and reused. The orbiter is the vehicle that orbits and returns to Earth.

21. Advances in medicine include better ways to diagnose heart and vision problems and better heart pacemakers. Cochlear implants help restore hearing. Image enhancement aids many fields including law enforcement and communication advances can control traffic signals as emergency vehicles approach intersections.

22. Advantages include lower costs and the ability to fly more frequently. Disadvantages include need for careful maintenance to maintain safety.

23. The *International Space Station* is an orbiting research platform. It is used for a variety of scientific endeavors.

chapter 2 Organizer

Section/Objectives	Standards		Labs/Features
	National	**State/Local**	
Chapter Opener	See pp. 9T–10T for a Key to Standards.		**Launch Lab:** Model Rotation and Revolution, p. 39 **Foldables,** p. 39
Section 1 Earth 2 sessions 1 block 1. **Examine** Earth's physical characteristics. 2. **Differentiate** between rotation and revolution. 3. **Discuss** what causes seasons to change.	National Content Standards: UCP.1–3, UCP.5, A.1–2, B.2, D.1		**Integrate Life Science,** p. 41 **MiniLAB:** Making Your Own Compass, p. 42 **Science Online,** p. 43 **Science Online,** p. 45
Section 2 The Moon—Earth's Satellite 2 sessions 1 block 4. **Identify** phases of the Moon and their cause. 5. **Explain** why solar and lunar eclipses occur. 6. **Infer** what the Moon's surface features may reveal about its history.	National Content Standards: UCP.1–3, UCP.5, A.1–2, B.2		**MiniLAB:** Comparing the Sun and the Moon, p. 47 **Science Online,** p. 49 **Integrate Career,** p. 51 **Visualizing the Moon's Surface,** p. 52 **Applying Science:** What will you use to survive on the Moon?, p. 53 **Lab:** Moon Phases and Eclipses, p. 55
Section 3 Exploring Earth's Moon 4 sessions 2 blocks 7. **Describe** recent discoveries about the Moon. 8. **Examine** facts about the Moon that might influence future space travel.	National Content Standards: UCP.1–3, UCP.5, A.1–2, D.1, D.3, G.3		**Science Online,** p. 57 **Lab:** Tilt and Temperature, p. 60 **Science and History:** The Mayan Calendar, p. 62

Glencoe Exclusive! TeacherWorks™ All-In-One Planner and Resource Center

Lab Materials	Reproducible Resources	Section Assessment	Technology
Launch Lab: basketball	**Chapter *FAST FILE* Resources** Foldables Worksheet, p. 15 Directed Reading Overview, p. 17 Note-taking Worksheets, pp. 31–34	Glencoe's Assessment Advantage	TeacherWorks includes: • Interactive Teacher Edition • Lesson Planner with calendar • Access to all program blacklines • Correlations to standards • Web links
MiniLAB: plastic foam cup, sewing needle, tape, plate, water, scissors, bar magnet	**Chapter *FAST FILE* Resources** Transparency Activity, p. 44 MiniLAB, p. 3 Enrichment, p. 28 Reinforcement, p. 25 Directed Reading, p. 18 Lab Activity, pp. 11–12, 13–14 Transparency Activity, pp. 47–48	**Performance** MiniLAB, p. 42 Applying Skills, p. 45 **Content** Section Review, p. 45	Section Focus Transparency Teaching Transparency Virtual Labs CD-ROM Guided Reading Audio Program Interactive Chalkboard CD-ROM
MiniLAB: chalk, pavement, string, meterstick **Lab:** unshaded light source, polystyrene ball, globe, pencil *Need materials?* Contact Science Kit at 1-800-828-7777 or www.sciencekit.com on the Internet.	**Chapter *FAST FILE* Resources** Transparency Activity, p. 45 MiniLAB, p. 4 Enrichment, p. 29 Reinforcement, p. 26 Directed Reading, p. 19 Lab Worksheets, pp. 7–8 **Cultural Diversity,** pp. 29, 71	**Portfolio** Science Journal, p. 47 Activity, p. 50 **Performance** MiniLAB, p. 47 Applying Science, p. 53 **Content** Section Review, p. 54	Section Focus Transparency Virtual Labs CD-ROM Guided Reading Audio Program Interactive Chalkboard CD-ROM Video Lab
Lab: tape, black construction paper, gooseneck lamp with 75-w bulb, Celsius alcohol thermometer, watch, protractor	**Chapter *FAST FILE* Resources** Transparency Activity, p. 46 Enrichment, p. 30 Reinforcement, p. 27 Directed Reading, pp. 19, 20 Lab Worksheets, pp. 9–10 **Lab Management and Safety,** p. 61	**Portfolio** Differentiated Instruction, p. 58 **Performance** Applying Skills, p. 59 **Content** Section Review, p. 59	Section Focus Transparency Virtual Labs CD-ROM Guided Reading Audio Program Interactive Chalkboard CD-ROM

End of Chapter Assessment

Blackline Masters	Technology	Professional Series
Chapter *FAST FILE* Resources Chapter Review, pp. 37–38 Chapter Tests, pp. 39–42 **Standardized Test Practice,** pp. 11–14	MindJogger Videoquiz Virtual Labs CD-ROM Exam*View*® Pro Testmaker TeacherWorks CD-ROM Interactive Chalkboard CD-ROM	**Performance Assessment in the Science Classroom (PASC)**

The Sun-Earth-Moon System

Transparencies

Section Focus

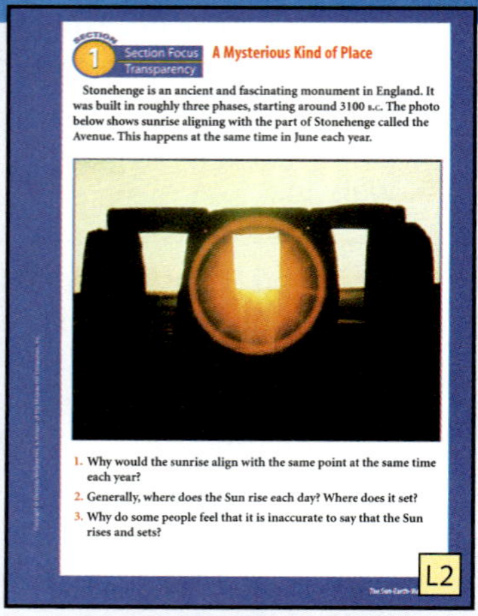
Section 1 Section Focus Transparency — A Mysterious Kind of Place

Stonehenge is an ancient and fascinating monument in England. It was built in roughly three phases, starting around 3100 B.C. The photo below shows sunrise aligning with the part of Stonehenge called the Avenue. This happens at the same time in June each year.

1. Why would the sunrise align with the same point at the same time each year?
2. Generally, where does the Sun rise each day? Where does it set?
3. Why do some people feel that it is inaccurate to say that the Sun rises and sets?

L2

Section 2 Section Focus Transparency — A Lovely Gibbous Earth

What would it be like to live on the Moon? We would need a lot of help and protection. There is no atmosphere on the Moon, and the temperatures are too extreme for life as we know it. But if we do build lunar living quarters in the future, we could enjoy seeing a lovely Earth in the sky.

1. If we lived on the Moon, could we observe phases of Earth similar to the phases of the Moon observed from Earth?
2. How could people living on the Moon protect themselves from the harsh conditions there?

L2

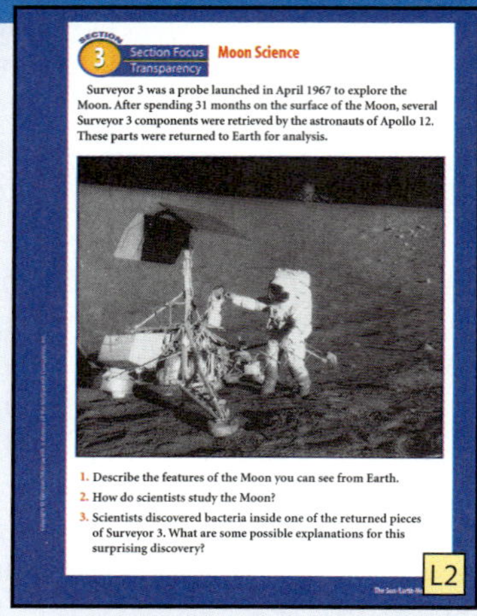
Section 3 Section Focus Transparency — Moon Science

Surveyor 3 was a probe launched in April 1967 to explore the Moon. After spending 31 months on the surface of the Moon, several Surveyor 3 components were retrieved by the astronauts of Apollo 12. These parts were returned to Earth for analysis.

1. Describe the features of the Moon you can see from Earth.
2. How do scientists study the Moon?
3. Scientists discovered bacteria inside one of the returned pieces of Surveyor 3. What are some possible explanations for this surprising discovery?

L2

This is a representation of key blackline masters available in the Teacher Classroom Resources. See Resource Manager boxes within the chapter for additional information.

Assessment

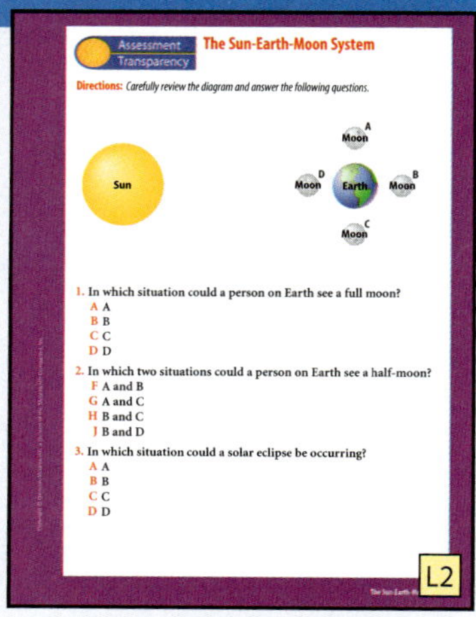
Assessment Transparency — The Sun-Earth-Moon System

Directions: Carefully review the diagram and answer the following questions.

1. In which situation could a person on Earth see a full moon?
   A A
   B B
   C C
   D D
2. In which two situations could a person on Earth see a half-moon?
   F A and B
   G A and C
   H B and C
   J B and D
3. In which situation could a solar eclipse be occurring?
   A A
   B B
   C C
   D D

L2

Teaching

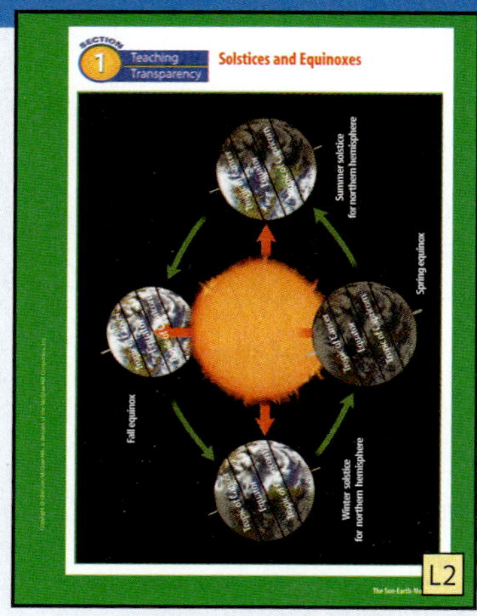
Section 1 Teaching Transparency — Solstices and Equinoxes

L2

Key to Teaching Strategies

The following designations will help you decide which activities are appropriate for your students.

- **L1** Level 1 activities should be appropriate for students with learning difficulties.
- **L2** Level 2 activities should be within the ability range of all students.
- **L3** Level 3 activities are designed for above-average students.
- **ELL** ELL activities should be within the ability range of English Language Learners.
- **COOP LEARN** Cooperative Learning activities are designed for small group work.
- **LS** Multiple Learning Styles logos, as described on page 6T, are used throughout to indicate strategies that address different learning styles.
- **P** These strategies represent student products that can be placed into a best-work portfolio.
- **PBL** Problem-Based Learning activities apply real-world situations to learning.

Hands-on Activities

Student Text Lab Worksheet

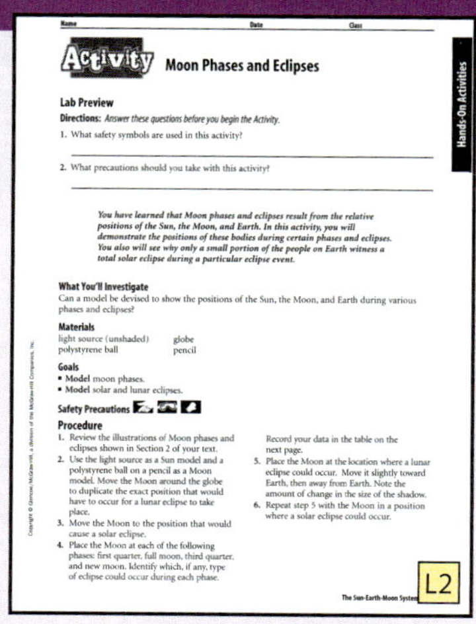
Activity — Moon Phases and Eclipses

Lab Preview

Directions: Answer these questions before you begin the Activity.

1. What safety symbols are used in this activity?
2. What precautions should you take with this activity?

L2

Laboratory Activities

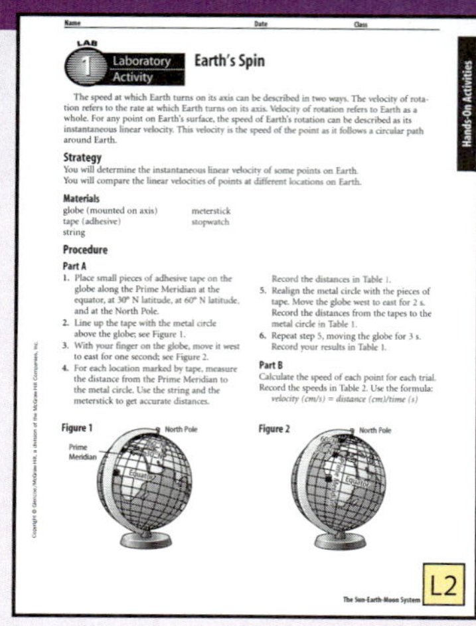
Lab 1 Laboratory Activity — Earth's Spin

L2

Resource Manager

Meeting Different Ability Levels

Content Outline

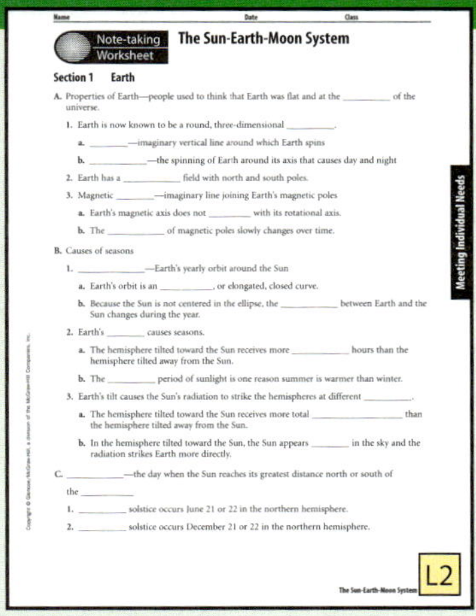

Reinforcement

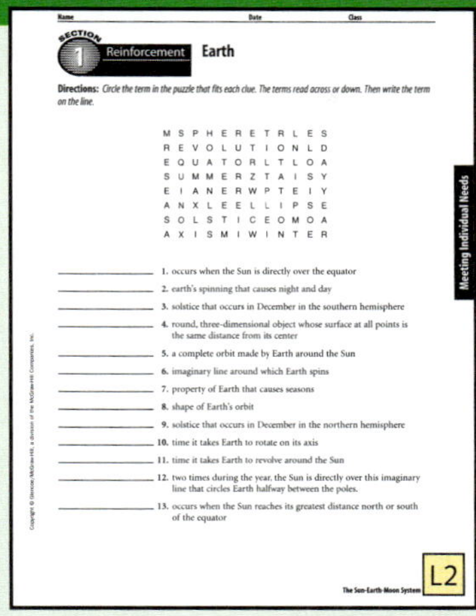

Enrichment

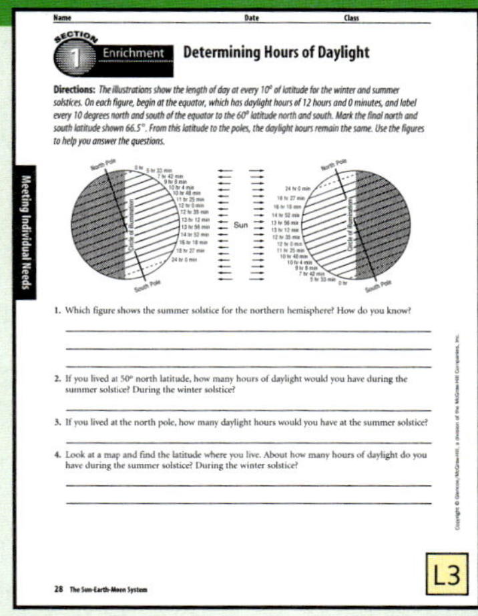

Directed Reading (English/Spanish)

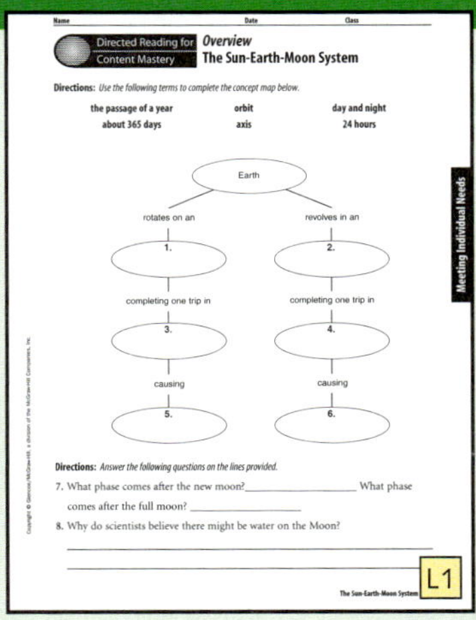

Study Guide

Reading Essentials

Assessment

Test Practice Workbook

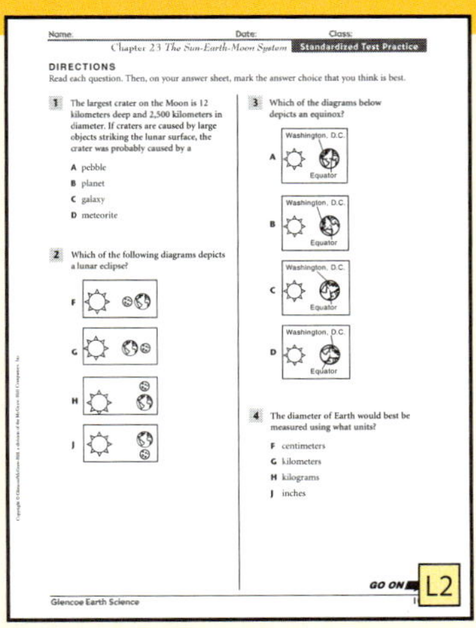

Chapter Review

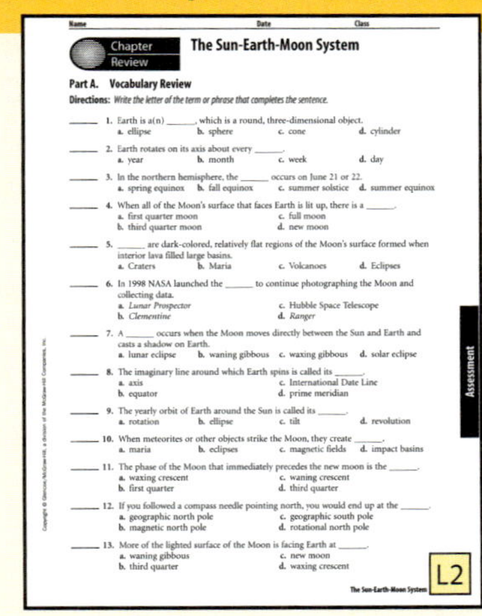

Chapter Tests

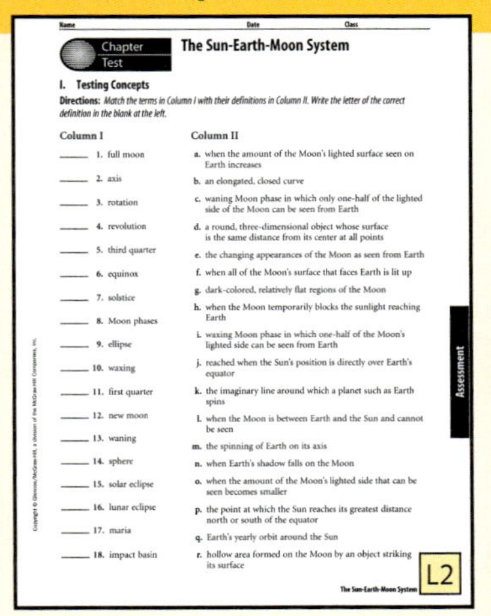

Science Content Background

section 1 Earth

Earth's Rotation

Scientists hypothesize that the gravitational attraction of the Moon is dragging on Earth, causing the length of a day to increase. Evidence from atomic clocks and ancient corals supports the hypothesis that Earth's rotation is slowing. Atomic clocks measure time by recording the frequency of electromagnetic waves given off by atoms. They gain or lose less than one second in over 1,000,000 years. Using these clocks, scientists have found that each successive day on Earth is getting longer. Corals deposit monthly growth lines on their shells in much the same way trees develop yearly growth rings. By studying growth lines on 375-million-year-old corals, scientists have determined that one year had 440 fewer days at the time these corals were growing.

Earth's Magnetic Field

The direction of Earth's magnetic field changes periodically over time. Minerals in rocks such as basalt align themselves according to the magnetic field orientation at the time they form. When Earth's magnetic field is reversed, new iron minerals being formed reflect that magnetic reversal. Charged particles from the Sun can set up electrical currents that flow toward Earth's poles. These currents can interact with gas in Earth's atmosphere. The gas then emits light to form the northern lights.

section 2 The Moon—Earth's Satellite

The Moon's Surface

About 4 billion years ago, large meteorites formed huge impact basins on the Moon's surface. By 3.3 billion years ago, lava flows had filled in the basins, forming the large, relatively flat maria seen today.

section 3 Exploring the Moon

Learning from Moon Explorations

Deviations in the lunar orbits of spacecraft led to the discovery of mascons, which are areas of high concentrations of mass. The presence of mascons indicates that the entire interior of the Moon cannot be liquid. Enough crust must be present to support mascons.

chapter content resources

Internet Resources

For additional content background, visit bookj.msscience.com to:

- access your book online
- find references to related articles in popular science magazines
- access Web links with related content background
- access current events with science journal topics

Print Resources

Celestial Delights, Francis Reddy and Greg Walz-Chojnocki, Celestial Arts, 2002

How to Identify the Night Sky, Storm Dunlop and Wil Tircon, Barnes & Noble Inc., 2003

A+ Projects in Astronomy, Janice Van Cleave, John Wiley & Sons, 2002

IDENTIFYING Misconceptions

Find Out What Students Think

Students may think that . . .

The Sun lights up different parts of the Moon as the Sun moves through the sky.

Students' experiences are with a large Sun and a small Moon, thus some students develop the idea that the Sun moves through the sky and shines on different parts of the Moon. In reality, the Moon shines because light is reflected from the surface facing the Sun. Because half of the Moon is always facing the Sun, half of its surface is always lit. As the Moon revolves around Earth, different parts of the side facing Earth are lit. The shape seen at any one time depends on the relative positions of Earth, Moon, and Sun.

Discussion

Organize the class into groups, and have them work together to draw the cycle of phases of the Moon from memory. Then have each student write a paragraph explaining how he or she thinks these changing shapes are produced. Review the paragraphs to help you identify misconceptions.

Promote Understanding

Activity

Have students observe and record the appearance of the Moon once each week for four weeks.

- Have students compare their drawings with those they completed as a group. Point out that as the cycle starts, the side of the Moon facing Earth is completely dark. Then a small portion of the side facing Earth is lighted. The portion of the side that is lighted grows larger until full Moon. After that, the lighted portion grows smaller until the Moon is completely dark again. In the northern hemisphere the cycle appears to move from right to left, but this is just the reverse in the southern hemisphere.
- Have students explain where the lighted side of the Moon "goes." If necessary, reinforce this using a light source and two balls—representing the Moon and the Earth. Underscore that half of the Moon is always lit; we just can't see it from Earth.

Telegraph Colour Library/FPG International

Assess

After completing the chapter, see *Identifying Misconceptions* in the Study Guide at the end of the chapter.

chapter 2

Chapter Vocabulary

sphere, p. 40
axis, p. 41
rotation, p. 41
revolution, p. 43
ellipse, p. 43
solstice, p. 44
equinox, p. 45
moon phase, p. 47
new moon, p. 47
waxing, p. 47
full moon, p. 47
waning, p. 47
solar eclipse, p. 49
lunar eclipse, p. 50
maria, p. 51
impact basin, p. 57

Science Journal Some students may not know that rotation brings morning and revolution brings summer. Check later to be sure they have learned this.

INTERACTIVE CHALKBOARD with Image Bank

PowerPoint® Presentations

This CD-ROM is an editable Microsoft® PowerPoint® presentation that includes:

- a pre-made presentation for every chapter
- interactive graphics
- animations
- audio clips
- image bank
- all new section and chapter questions
- Standardized Test Practice
- transparencies
- pre-lab questions for all labs
- Foldables directions
- links to bookj.msscience.com

chapter 2

The Sun-Earth-Moon System

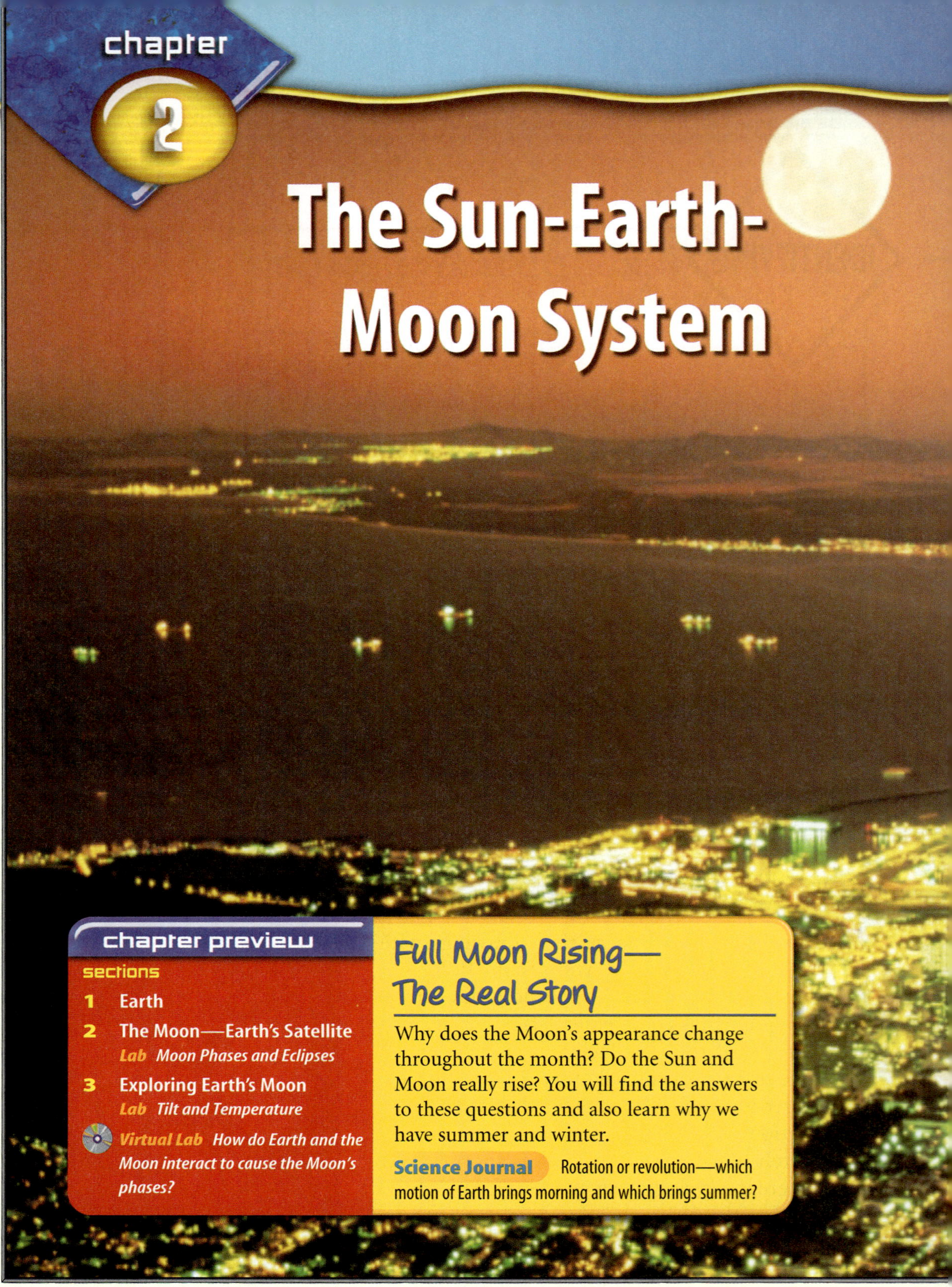

chapter preview

sections

1. Earth
2. The Moon—Earth's Satellite
   Lab Moon Phases and Eclipses
3. Exploring Earth's Moon
   Lab Tilt and Temperature

Virtual Lab How do Earth and the Moon interact to cause the Moon's phases?

Full Moon Rising—The Real Story

Why does the Moon's appearance change throughout the month? Do the Sun and Moon really rise? You will find the answers to these questions and also learn why we have summer and winter.

Science Journal Rotation or revolution—which motion of Earth brings morning and which brings summer?

Theme Connection

Systems and Interactions Earth's rotation and revolution in the Earth-Sun system are responsible for the day-night cycle and the seasons. Phases of the Moon and solar and lunar eclipses are caused by interactions among the Sun, the Moon, and Earth.

About the Photo

Nightlight In rural areas with no artificial light, a full moon can give almost enough light to read by. In brightly lit cities however, the Moon's light goes unnoticed. But even in cities, people can see the Moon on clear nights and notice how its shape changes.

Start-Up Activities

Launch LAB

Model Rotation and Revolution

The Sun rises in the morning; at least, it seems to. Instead, it is Earth that moves. The movements of Earth cause day and night, as well as the seasons. In this lab, you will explore Earth's movements.

1. Hold a basketball with one finger at the top and one at the bottom. Have a classmate gently spin the ball.
2. Explain how this models Earth's rotation.
3. Continue to hold the basketball and walk one complete circle around another student in your class.
4. How does this model Earth's revolution?
5. **Think Critically** Write a paragraph in your Science Journal describing how these movements of the basketball model Earth's rotation and revolution.

Science Online — Preview this chapter's content and activities at bookj.msscience.com

FOLDABLES™ Study Organizer

Earth and the Moon All on Earth can see and feel the movements of Earth and the Moon as they circle the Sun. Make the following Foldable to organize what you learn about these movements and their effects.

STEP 1 **Fold** a sheet of paper in half lengthwise.

STEP 2 **Fold** paper down 2.5 cm from the top. (Hint: From the tip of your index finger to your middle knuckle is about 2.5 cm.)

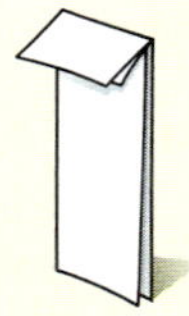

STEP 3 **Open and draw** lines along the 2.5-cm fold. **Label** as shown.

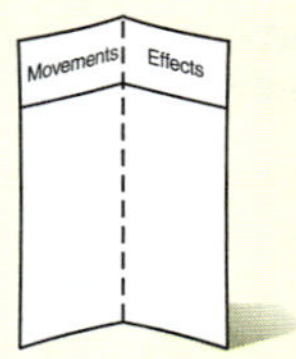

Summarize in a Table As you read the chapter, summarize the movements of Earth and the Moon in the left column and the effects of these movements in the right column.

J ◆ 39

Launch LAB

Purpose Use the Launch Lab to introduce students to the difference between rotation and revolution. L1 ELL COOP LEARN LS **Kinesthetic**

Preparation Obtain several basketballs from the physical education department.

Materials basketballs

Alternate Materials volleyballs, soccer balls

Teaching Strategy Have students work in pairs and take turns modeling rotation and revolution.

Think Critically

Both rotation and revolution are movements of Earth. Rotation is when an object spins around a fixed pole or axis. Revolution is when one object moves around another.

Assessment

Process Have students list at least two things that rotate and revolve. Rotate: CDs, tops, globes; revolve: planets, a person walking around a building.

FOLDABLES™ Study Organizer — Dinah Zike Study Fold

Student preparation materials for this Foldable are available in the Chapter *FAST FILE* Resources.

1 Motivate

Bellringer

INTERACTIVE CHALKBOARD PowerPoint® Presentations

Section Focus Transparencies also are available on the Interactive Chalkboard CD-ROM. L2 ELL

Section Focus Transparency 1 — A Mysterious Kind of Place

Stonehenge is an ancient and fascinating monument in England. It was built in roughly three phases, starting around 3100 B.C. The photo below shows sunrise aligning with the part of Stonehenge called the Avenue. This happens at the same time in June each year.

1. Why would the sunrise align with the same point at the same time each year?
2. Generally, where does the Sun rise each day? Where does it set?
3. Why do some people feel that it is inaccurate to say that the Sun rises and sets?

L2

The Sun-Earth-Moon System

Tie to Prior Knowledge

Earth's Properties Tell students that every planet has unique physical properties. Ask students to brainstorm physical properties of Earth. Start them out by writing "Earth is spherical" on the board. Accept other student responses. Review responses after finishing the section.

section 1 Earth

as you read

What You'll Learn

- **Examine** Earth's physical characteristics.
- **Differentiate** between rotation and revolution.
- **Discuss** what causes seasons to change.

Why It's Important

Your life follows the rhythm of Earth's movements.

Review Vocabulary
orbit: the path taken by an object revolving around another

New Vocabulary

- sphere
- axis
- rotation
- revolution
- ellipse
- solstice
- equinox

Properties of Earth

You awaken at daybreak to catch the Sun "rising" from the dark horizon. Then it begins its daily "journey" from east to west across the sky. Finally the Sun "sinks" out of view as night falls. Is the Sun moving—or are you?

It wasn't long ago that people thought Earth was the center of the universe. It was widely believed that the Sun revolved around Earth, which stood still. It is now common knowledge that the Sun only appears to be moving around Earth. Because Earth spins as it revolves around the Sun, it creates the illusion that the Sun is moving across the sky.

Another mistaken idea about Earth concerned its shape. Even as recently as the days of Christopher Columbus, many people believed Earth to be flat. Because of this, they were afraid that if they sailed far enough out to sea, they would fall off the edge of the world. How do you know this isn't true? How have scientists determined the true shape of Earth?

Spherical Shape A round, three-dimensional object is called a **sphere** (SFIHR). Its surface is the same distance from its center at all points. Some common examples of spheres are basketballs and tennis balls.

In the late twentieth century, artificial satellites and space probes sent back pictures showing that Earth is spherical. Much earlier, Aristotle, a Greek astronomer and philosopher who lived around 350 B.C., suspected that Earth was spherical. He observed that Earth cast a curved shadow on the Moon during an eclipse.

In addition to Aristotle, other individuals made observations that indicated Earth's spherical shape. Early sailors, for example, noticed that the tops of approaching ships appeared first on the horizon and the rest appeared gradually, as if they were coming over the crest of a hill, as shown in **Figure 1.**

Figure 1 For many years, sailors have observed that the tops of ships coming across the horizon appear first. This suggests that Earth is spherical, not flat, as was once widely believed.

40

Section 1 Resource Manager

Chapter FAST FILE Resources

Transparency Activity, pp. 44, 47–48
Directed Reading for Content Mastery, pp. 17, 18
Note-Taking Worksheets, pp. 3–31
MiniLab, p.3
Lab Activity, pp. 11–12, 13–14
Enrichment, p. 28
Reinforcement, p. 25

Additional Evidence Sailors also noticed changes in how the night sky looked. As they sailed north or south, the North Star moved higher or lower in the sky. The best explanation was a spherical Earth.

Today, most people know that Earth is spherical. They also know all objects are attracted by gravity to the center of a spherical Earth. Astronauts have clearly seen the spherical shape of Earth. However, it bulges slightly at the equator and is somewhat flattened at the poles, so it is not a perfect sphere.

Rotation Earth's **axis** is the imaginary vertical line around which Earth spins. This line cuts directly through the center of Earth, as shown in the illustration accompanying **Table 1.** The poles are located at the north and south ends of Earth's axis. The spinning of Earth on its axis, called **rotation,** causes day and night to occur. Here is how it works. As Earth rotates, you can see the Sun come into view at daybreak. Earth continues to spin, making it seem as if the Sun moves across the sky until it sets at night. During night, your area of Earth has rotated so that it is facing away from the Sun. Because of this, the Sun is no longer visible to you. Earth continues to rotate steadily, and eventually the Sun comes into view again the next morning. One complete rotation takes about 24 h, or one day. How many rotations does Earth complete during one year? As you can infer from **Table 1,** it completes about 365 rotations during its one-year journey around the Sun.

Reading Check *Why does the Sun seem to rise and set?*

INTEGRATE Life Science

Earth's Rotation
Suppose that Earth's rotation took twice as long as it does now. In your Science Journal, predict how conditions such as global temperatures, work schedules, plant growth, and other factors might change under these circumstances.

Table 1 Physical Properties of Earth

Diameter (pole to pole)	12,714 km
Diameter (equator)	12,756 km
Circumference (poles)	40,008 km
Circumference (equator)	40,075 km
Mass	5.98×10^{24} kg
Average density	5.52 g/cm^3
Average distance to the Sun	149,600,000 km
Period of rotation (1 day)	23 h, 56 min
Period of revolution (1 year)	365 days, 6 h, 9 min

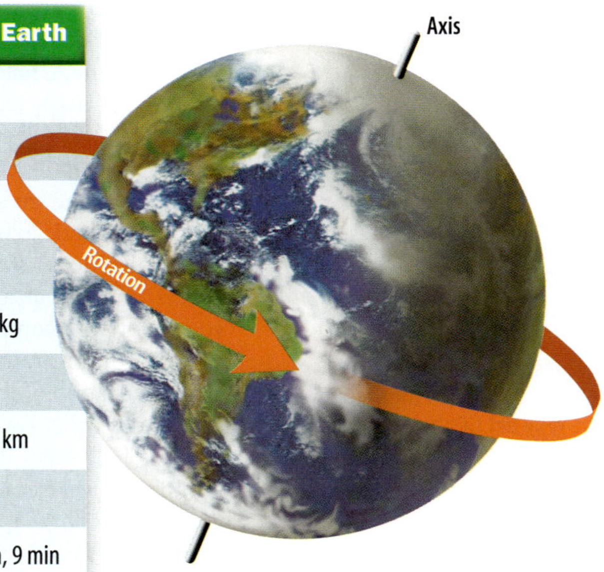

2 Teach

INTEGRATE Life Science

Earth's Rotation Students should infer that global daytime temperatures would be higher as a result of longer hours of sunlight, and nighttime temperatures would be cooler because of more continuous hours of darkness. Students might also suggest that people would have different daily habits, perhaps taking naps during the longer day, and having periods of activity in the longer night. In Alaska, the growing season is short, but the long hours of summer sunlight are good for the production of oats, barley, potatoes, hay, and cool-climate vegetables. Many people in Alaska try to maintain a "normal" 24-hour cycle, using additional lighting during the dark periods and black-out shades to provide artificial night during the light periods.

Reading Check

Answer because Earth rotates

Visual Learning

Table 1 Challenge students to use the information in **Table 1** to prove that Earth is not a perfect sphere. If Earth were a perfect sphere, the diameter would be the same at the equator and from pole to pole. L2

Quick Demo

Earth's Shape

Materials large round balloon, opaque marker

Estimated Time five minutes

Procedure Prepare balloon ahead of time, by inflating it and marking the poles and the equator. With a student holding the balloon on each side, compress it slightly using one hand at the top and one at the bottom. Point out the slight difference between the diameters of the balloon at the equator and at the poles.

Differentiated Instruction

Challenge Ask students to calculate the percent difference between the circumference of Earth at the equator and at the poles. Ask them to research what is thought to cause this flattening and present their findings to the class. L3

Cultural Diversity

Babylonian Astronomers Babylonian records of solar eclipses have been used to document the slowing of Earth's rotation. The eclipses could have occurred on those dates only if Earth were rotating slightly faster in the past than it is now. Have students research other areas of interest of these early astronomers. dates of appearance of the new moon (to begin months), lunar and solar calendars L3

Quick Demo

Magnetic Poles

Materials transparency of Earth, overhead projector, bar magnet, iron filings

Estimated Time 20 minutes

Procedure Label the geographic poles on the drawing of Earth. Place the transparency over a bar magnet on the overhead projector. Rotate the transparency so that Earth's poles form an angle of 11.5° with the magnet. Sprinkle iron filings over the transparency and tap it lightly. Ask students how this model is similar to Earth and its magnetic field. L2

Mini LAB

Purpose Students make simple compasses. L1 ELL LS **Kinesthetic**

Materials plastic foam cup, scissors, bar magnet, sewing needle, tape, plate, water

Teaching Strategy Have students use a compass to locate north before beginning.

Safety Precautions Warn students to use sharp objects carefully. Collect and count needles before the end of class.

Analysis

1. The needle lines up with Earth's magnetic field, with the needle oriented roughly north-south.
2. There are no landmarks on the open ocean. Simple compasses could have helped sailors navigate by indicating north.

Assessment

Process Ask students to write an operational definition of a compass. The needle of a compass helps you determine direction by pointing toward the north pole in the northern hemisphere.

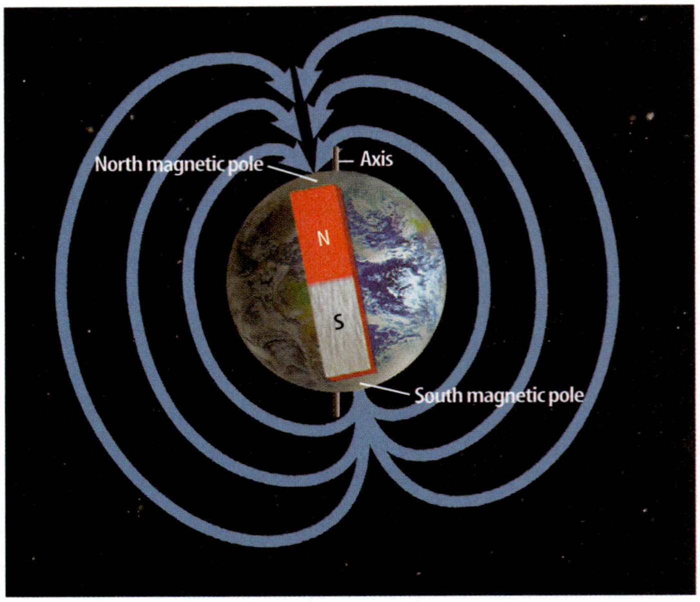

Figure 2 Earth's magnetic field is similar to that of a bar magnet, almost as if Earth contained a giant magnet. Earth's magnetic axis is angled 11.5 degrees from its rotational axis.

Mini LAB

Making Your Own Compass

Procedure

WARNING: *Use care when handling sharp objects.*

1. Cut off the bottom of a **plastic foam cup** to make a polystyrene disk.
2. Magnetize a **sewing needle** by continuously stroking the needle in the same direction with a **magnet** for 1 min.
3. **Tape** the needle to the center of the foam disk.
4. Fill a **plate** with **water** and float the disk, needle side up, in the water.

Analysis

1. What happened to the needle and disk when you placed them in the water? Why did this happen?
2. Infer how ancient sailors might have used magnets to help them navigate on the open seas.

Magnetic Field

INTEGRATE Physics Scientists hypothesize that the movement of material inside Earth's core, along with Earth's rotation, generates a magnetic field. This magnetic field is much like that of a bar magnet. Earth has a north and a south magnetic pole, just as a bar magnet has opposite magnetic poles at each of its ends. When you sprinkle iron shavings over a bar magnet, the shavings align with the magnetic field of the magnet. As you can see in **Figure 2,** Earth's magnetic field is similar—almost as if Earth contained a giant bar magnet. Earth's magnetic field protects you from harmful solar radiation by trapping many charged particles from the Sun.

Magnetic Axis When you observe a compass needle pointing north, you are seeing evidence of Earth's magnetic field. Earth's magnetic axis, the line joining its north and south magnetic poles, does not align with its rotational axis. The magnetic axis is inclined at an angle of 11.5° to the rotational axis. If you followed a compass needle, you would end up at the magnetic north pole rather than the rotational north pole.

The location of the magnetic poles has been shown to change slowly over time. The magnetic poles move around the rotational (geographic) poles in an irregular way. This movement can be significant over decades. Many maps include information about the position of the magnetic north pole at the time the map was made. Why would this information be important?

Text Question Answer

People using the map for navigation would need to adjust for the movement of the magnetic pole.

Curriculum Connection

Geography Challenge students to draw a map showing the location of both the geographic and magnetic north poles. The geographic north pole is in the Arctic Ocean, about 725 km (450 mi.) north of Greenland. When last measured, the magnetic north pole, to which magnetic compasses point, was in the Queen Elizabeth Islands of northern Canada, at about 78°27′N, 104°24′W. L3 LS **Logical-Mathematical**

What causes changing seasons?

Flowers bloom as the days get warmer. The Sun appears higher in the sky, and daylight lasts longer. Spring seems like a fresh, new beginning. What causes these wonderful changes?

Orbiting the Sun You learned earlier that Earth's rotation causes day and night. Another important motion is **revolution**, which is Earth's yearly orbit around the Sun. Just as the Moon is Earth's satellite, Earth is a satellite of the Sun. If Earth's orbit were a circle with the Sun at the center, Earth would maintain a constant distance from the Sun. However, this is not the case. Earth's orbit is an **ellipse** (ee LIHPS)—an elongated, closed curve. The Sun is not at the center of the ellipse but is a little toward one end. Because of this, the distance between Earth and the Sun changes during Earth's yearlong orbit. Earth gets closest to the Sun—about 147 million km away—around January 3. The farthest Earth gets from the Sun is about 152 million km away. This happens around July 4 each year.

Reading Check *What is an ellipse?*

Does this elliptical orbit cause seasonal temperatures on Earth? If it did, you would expect the warmest days to be in January. You know this isn't the case in the northern hemisphere, something else must cause the change.

Even though Earth is closest to the Sun in January, the change in distance is small. Earth is exposed to almost the same amount of Sun all year. But the amount of solar energy any one place on Earth receives varies greatly during the year. Next, you will learn why.

A Tilted Axis Earth's axis is tilted 23.5° from a line drawn perpendicular to the plane of its orbit. It is this tilt that causes seasons. The number of daylight hours is greater for the hemisphere, or half of Earth, that is tilted toward the Sun. Think of how early it gets dark in the winter compared to the summer. As shown in **Figure 3,** the hemisphere that is tilted toward the Sun receives more hours of sunlight each day than the hemisphere that is tilted away from the Sun. The longer period of sunlight is one reason summer is warmer than winter, but it is not the only reason.

Science Online

Topic: Ellipses
Visit bookj.msscience.com for Web links to information about orbits and ellipses.

Activity Scientists compare orbits by how close they come to being circular. To do this, they use a measurement called eccentricity. A circle has an eccentricity of zero. Ellipses have eccentricities that are greater than zero, but less than one. The closer the eccentricity is to one, the more elliptical the orbit.Compare the orbits of the four inner planets. List them in order of increasing eccentricity.

Figure 3 In summer, the northern hemisphere is tilted toward the Sun. Notice that the north pole is always lit during the summer.
Observe *Why is there a greater number of daylight hours in the summer than in the winter?*

Reading Check

Answer An elongated, closed curve

Visual Learning

Figure 3 How long do daylight hours last at the south pole when the northern hemisphere experiences summer? The south pole is always in darkness. L2

Caption Answer

Figure 3 because the hemisphere experiencing summer is tilted toward the Sun

Fun Fact

Earth's orientation in space changes with time. Its axis now points toward Polaris, the North Star. In about 12,000 years, it will point toward Vega. This change is the result of Earth's slow "wobble" on its axis.

Activity

Axis Angle Have students model the tilt of Earth's axis using plastic foam plates as the plane of Earth's orbit. Mark a small circle in the center to represent the Sun. Use a grape or small ball of clay to represent Earth and insert a toothpick to represent Earth's axis. Then stand a protractor on the plate and place the grape next to it so that the toothpick points to 66.5°. This will be 23.5° from the perpendicular. This is the tilt of Earth's orbit. Students can move the grape around the edge of the plate making certain to keep the axis at the same angle.

Science on Line Eccentricities in order of most circular to most elliptical are: Venus 0.00677323; Earth 0.01671022; Mars, 0.09341233; Mercury 0.20563069.

Teacher FYI

Midnight Sun At the Arctic Circle the Sun is visible 24 hours a day from about June 5 to July 5. The farther north you go, the length of the period of 24-hour daylight increases until at the north pole it lasts from about March 21 to September 21.

Differentiated Instruction

Learning Disabled Have students draw a picture or write a short paragraph describing their favorite season of the year. Encourage students to include science-connected observations, such as weather conditions, the condition of plant and animal life, and the look of the sky. Display their work in the classroom. L1 ELL IN **Visual-Spatial**

Quick Demo

Changing Seasons

Materials overhead projector

Estimated Time 10 minutes

Procedure Have students observe the light from an overhead projector as it strikes a wall straight on. Then slant the lens so that the light hits the wall at an angle. Explain how this is similar to the way sunlight strikes areas of Earth during different seasons. When light hits at a higher angle, it covers a smaller area and is more intense. When it hits at a slanting angle, it covers a larger area but is less intense. On Earth, sunlight hits at a higher angle in summer and at a slanting (lower) angle in winter.

Discussion

Seasons Remind students that seasons are opposite in the southern hemisphere. How would your life be different if midsummer came in January and midwinter came in July? Students might mention the change in how holidays are celebrated, when school is in session, or when vacations are taken. L2

LS **Interpersonal**

Radiation from the Sun Earth's tilt also causes the Sun's radiation to strike the hemispheres at different angles. Sunlight strikes the hemisphere tilted towards the Sun at a higher angle, that is, closer to 90 degrees, than the hemisphere tilted away. Thus it receives more total solar radiation than the hemisphere tilted away from the Sun, where sunlight strikes at a lower angle.

Summer occurs in the hemisphere tilted toward the Sun, when its radiation strikes Earth at a higher angle and for longer periods of time. The hemisphere receiving less radiation experiences winter.

Solstices

The **solstice** is the day when the Sun reaches its greatest distance north or south of the equator. In the northern hemisphere, the summer solstice occurs on June 21 or 22, and the winter solstice occurs on December 21 or 22. Both solstices are illustrated in **Figure 4.** In the southern hemisphere, the winter solstice is in June and the summer solstice is in December. Summer solstice is about the longest period of daylight of the year. After this, the number of daylight hours become less and less, until the winter solstice, about the shortest period of daylight of the year. Then the hours of daylight start to increase again.

Figure 4 During the summer solstice in the northern hemisphere, the Sun is directly over the tropic of Cancer, the latitude line at 23.5° N latitude. During the winter solstice, the Sun is directly over the tropic of Capricorn, the latitude line at 23.5° S latitude. At fall and spring equinoxes, the Sun is directly over the equator.

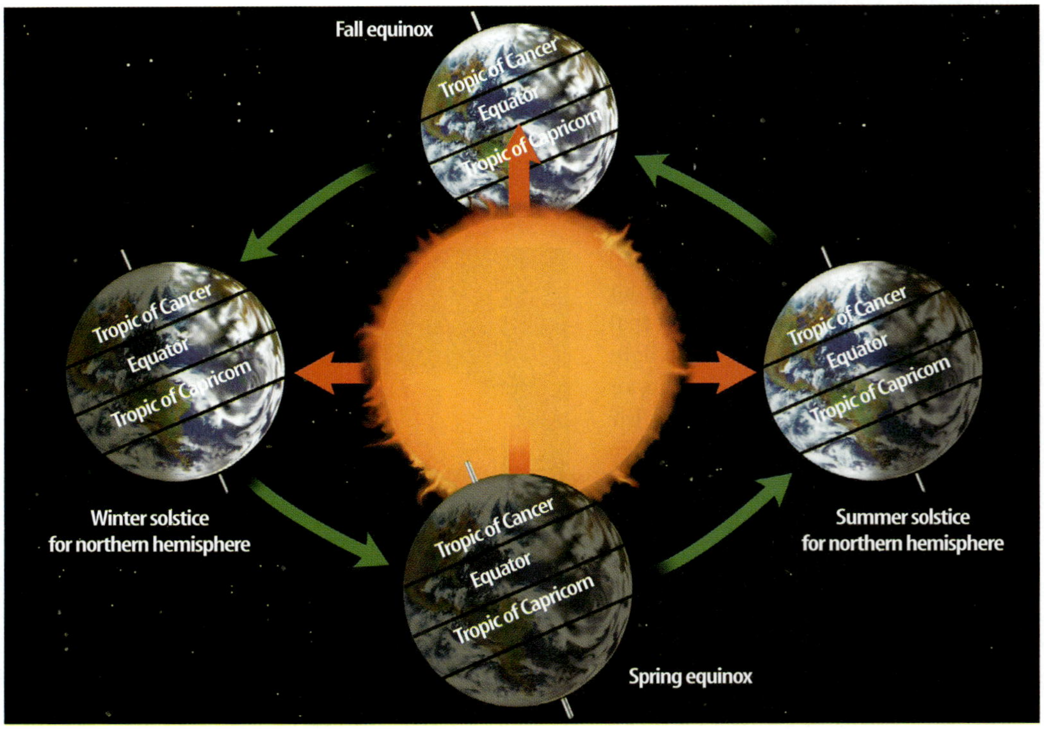

Curriculum Connection

History Discuss with students why we do certain things during certain seasons. For example, ask if they know why school in the U.S. usually lasts from fall through spring (in most areas). This custom dates back to when the U.S. was an agrarian society and children were needed to help with farm work in late spring and summer. Have students research the historical ties of other seasonal activities. L2

44 ◆ J CHAPTER 2 The Sun-Earth-Moon System

Equinoxes

An **equinox** (EE kwuh nahks) occurs when the Sun is directly above Earth's equator. Because of the tilt of Earth's axis, the Sun's position relative to the equator changes constantly. Most of the time, the Sun is either north or south of the equator, but two times each year it is directly over it, resulting in the spring and fall equinoxes. As you can see in **Figure 4,** at an equinox the Sun strikes the equator at the highest possible angle, 90°.

During an equinox, the number of daylight hours and nighttime hours is nearly equal all over the world. Also at this time, neither the northern hemisphere nor the southern hemisphere is tilted toward the Sun.

In the northern hemisphere, the Sun reaches the spring equinox on March 20 or 21, and the fall equinox occurs on September 22 or 23. In the southern hemisphere, the equinoxes are reversed. Spring occurs in September and fall occurs in March.

Earth Data Review As you have learned, Earth is a sphere that rotates on a tilted axis. This rotation causes day and night. Earth's tilted axis and its revolution around the Sun cause the seasons. One Earth revolution takes one year. In the next section, you will read how the Moon rotates on its axis and revolves around Earth.

Science Online

Topic: Seasons

Visit bookj.msscience.com for Web links to information about the seasons.

Activity Make a poster describing how the seasons differ in other parts of the world. Show how holidays might be celebrated differently and how farming might vary between hemispheres.

section 1 review

Summary

Properties of Earth

- Earth is a slightly flattened sphere that rotates around an imaginary line called an axis.
- Earth has a magnetic field, much like a bar magnet.
- The magnetic axis of Earth differs from its rotational axis.

Seasons

- Earth revolves around the Sun in an elliptical orbit.
- The tilt of Earth's axis and its revolution cause the seasons.
- Solstices are days when the Sun reaches its farthest points north or south of the equator.
- Equinoxes are the points when the Sun is directly over the equator.

Self Check

1. **Explain** why Aristotle thought Earth was spherical.
2. **Compare and contrast** rotation and revolution.
3. **Describe** how Earth's distance from the Sun changes throughout the year. When is Earth closest to the Sun?
4. **Explain** why it is summer in Earth's northern hemisphere at the same time it is winter in the southern hemisphere.
5. **Think Critically** **Table 1** lists Earth's distance from the Sun as an average. Why isn't an exact measurement available for this distance?

Applying Skills

6. **Classify** The terms *clockwise* and *counterclockwise* are used to indicate the direction of circular motion. How would you classify the motion of the Moon around Earth as you view it from above Earth's north pole? Now try to classify Earth's movement around the Sun.

Science Online bookj.msscience.com/self_check_quiz

3 Assess

DAILY INTERVENTION

Check for Understanding

Logical-Mathematical Have students use calculators to determine how much the shape of Earth differs from a perfect sphere. Have them find the percent difference between the polar and equatorial circumferences. L3

Reteach

Demonstrating Day Length Darken the room and place a globe next to a light source. Hold the globe tilted at an angle of 23.5° away from the light. Ask students to spin the globe slowly and notice how the light strikes each hemisphere. Repeat with the axis tilted toward the light. Remind students that periods of daylight are shorter in the northern hemisphere when it is tilted away from the Sun and longer when it is tilted toward the Sun.

Assessment

Oral Have each student write a question on the section content. Pair students and have each student ask his or her partner their question. Pair can exchange questions to extend the activity. Use **Performance Assessment in the Science Classroom,** p. 91. L2

section 1 review

1. Aristotle noted that Earth always cast a curved shadow on the Moon during an eclipse.
2. Both are movements. Rotation is a spinning motion of an object around a fixed axis, whereas revolution is the movement of one body around another.
3. Earth's orbit is elliptical. It is farthest from the Sun around July 4 and closest to the Sun around January 3.
4. The hemisphere tilted toward the Sun has summer; the hemisphere tilted away has winter.
5. Because Earth's orbit is elliptical, Earth's distance from the Sun changes as the planet revolves around it.
6. Both the movement of the Moon around Earth and Earth around the Sun are counterclockwise.

1 Motivate

INTERACTIVE CHALKBOARD
PowerPoint® Presentations

Bellringer

Section Focus Transparencies also are available on the Interactive Chalkboard CD-ROM. L2 ELL

Section 2 Section Focus Transparency **A Lovely Gibbous Earth**

What would it be like to live on the Moon? We would need a lot of help and protection. There is no atmosphere on the Moon, and the temperatures are too extreme for life as we know it. But if we do build lunar living quarters in the future, we could enjoy seeing a lovely Earth in the sky.

1. If we lived on the Moon, could we observe phases of Earth similar to the phases of the Moon observed from Earth?
2. How could people living on the Moon protect themselves from the harsh conditions there?

L2

Tie to Prior Knowledge

Moon Shapes Ask students to describe a full Moon. round disk in the night sky Have volunteers describe other Moon shapes they have seen and draw these shapes on the board. Explain that they will find out why the Moon seems to change shape in this section. L2

Text Question Answer

Because it is revolving around Earth at the same time as it is rotating

Caption Answer

Figure 5 Since rotation and revolution take about the same number of days we always see the same side of the Moon.

section 2

The Moon—Earth's Satellite

as you read

What You'll Learn

- **Identify** phases of the Moon and their cause.
- **Explain** why solar and lunar eclipses occur.
- **Infer** what the Moon's surface features may reveal about its history.

Why It's Important

Learning about the Moon can teach you about Earth.

Review Vocabulary
mantle: portion of the interior of a planet or moon lying between the crust and core

New Vocabulary
- moon phase
- new moon
- waxing
- full moon
- waning
- solar eclipse
- lunar eclipse
- maria

Motions of the Moon

Just as Earth rotates on its axis and revolves around the Sun, the Moon rotates on its axis and revolves around Earth. The Moon's revolution around Earth is responsible for the changes in its appearance. If the Moon rotates on its axis, why can't you see it spin around in space? The reason is that the Moon's rotation takes 27.3 days—the same amount of time it takes to revolve once around Earth. Because these two motions take the same amount of time, the same side of the Moon always faces Earth, as shown in **Figure 5.**

You can demonstrate this by having a friend hold a ball in front of you. Direct your friend to move the ball in a circle around you while keeping the same side of it facing you. Everyone else in the room will see all sides of the ball. You will see only one side. If the moon didn't rotate, we would see all of its surface during one month.

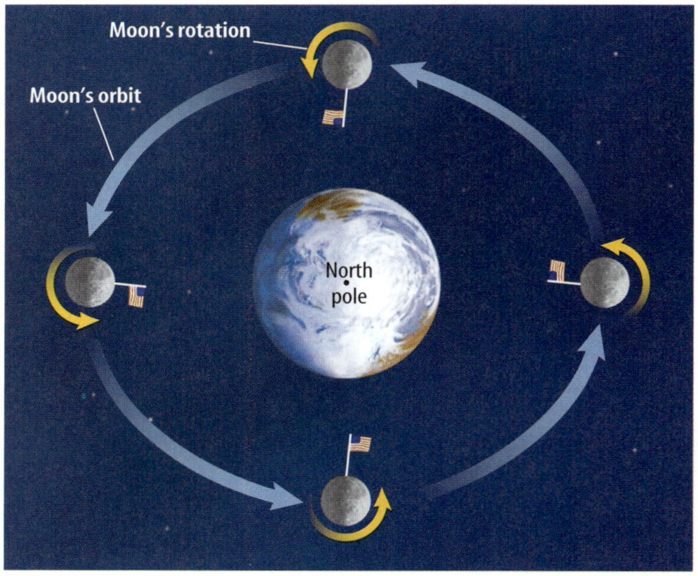

Figure 5 In about 27.3 days, the Moon orbits Earth. It also completes one rotation on its axis during the same period.
Think Critically *Explain how this affects which side of the Moon faces Earth.*

Section 2 Resource Manager

Chapter FAST FILE Resources
- Transparency Activity, p. 45
- Directed Reading for Content Mastery, p. 19
- MiniLAB, p. 4
- Enrichment, p. 29
- Reinforcement, p. 26
- Lab Worksheets, pp. 7–8

Reflection of the Sun The Moon seems to shine because its surface reflects sunlight. Just as half of Earth experiences day as the other half experiences night, half of the Moon is lighted while the other half is dark. As the Moon revolves around Earth, you see different portions of its lighted side, causing the Moon's appearance to change.

Phases of the Moon

Moon phases are the different forms that the Moon takes in its appearance from Earth. The phase depends on the relative positions of the Moon, Earth, and the Sun, as seen in **Figure 6** on the next page. A **new moon** occurs when the Moon is between Earth and the Sun. During a new moon, the lighted half of the Moon is facing the Sun and the dark side faces Earth. The Moon is in the sky, but it cannot be seen. The new moon rises and sets with the Sun.

Reading Check *Why can't you see a new moon?*

Waxing Phases After a new moon, the phases begin waxing. **Waxing** means that more of the illuminated half of the Moon can be seen each night. About 24 h after a new moon, you can see a thin slice of the Moon. This phase is called the waxing crescent. About a week after a new moon, you can see half of the lighted side of the Moon, or one quarter of the Moon's surface. This is the first quarter phase.

The phases continue to wax. When more than one quarter is visible, it is called waxing *gibbous* after the Latin word for "humpbacked." A **full moon** occurs when all of the Moon's surface facing Earth reflects light.

Waning Phases After a full moon, the phases are said to be waning. When the Moon's phases are **waning**, you see less of its illuminated half each night. Waning gibbous begins just after a full moon. When you can see only half of the lighted side, it is the third-quarter phase. The Moon continues to appear to shrink. Waning crescent occurs just before another new moon. Once again, you can see only a small slice of the Moon.

It takes about 29.5 days for the Moon to complete its cycle of phases. Recall that it takes about 27.3 days for the Moon to revolve around Earth. The discrepancy between these two numbers is due to Earth's revolution. The roughly two extra days are what it takes for the Sun, Earth, and Moon to return to their same relative positions.

Mini LAB

Comparing the Sun and the Moon

Procedure

1. Find an area where you can make a chalk mark on **pavement or similar surface.**
2. Tie a piece of **chalk** to one end of a 200-cm-long **string.**
3. Hold the other end of the string to the pavement.
4. Have a friend pull the string tight and walk around you, drawing a circle (the Sun) on the pavement.
5. Draw a 1-cm-diameter circle in the middle of the larger circle (the Moon).

Analysis

1. How big is the Sun compared to the Moon?
2. The diameter of the Sun is 1.39 million km. The diameter of Earth is 12,756 km. Draw two new circles modeling the sizes of the Sun and Earth. What scale did you use?

Try at Home

2 Teach

Mini LAB

Purpose Students compare the sizes of the Sun, the Moon, and Earth. L2 ELL LS **Logical-Mathematical**

Materials meterstick, a piece of chalk, string (200 cm)

Teaching Strategy Conduct this activity on a large, paved playground or parking lot.

Analysis

1. The Sun's diameter is about 400 times that of the Moon.
2. The Sun's diameter is about 109 times greater than that of Earth. Scales will vary. The diameters of new circles could be 109 cm (Sun) and 1 cm (Earth).

Assessment

Performance Tell students that the Moon's diameter is 3,476 km, and Earth's is 12,756 km. Have them draw new circles representing the relative sizes of Earth and the Moon. Use **Performance Assessment in the Science Classroom,** p. 127.

Reading Check

Answer because the side of the Moon facing Earth is not lighted

Fun Fact

The Moon shines by reflected sunlight. But it reflects only a portion of the light it receives. If it reflected all the light, it would appear 14 times brighter than it does.

Science Journal

Phases of the Moon Invite students to write an essay in their Science Journals about the way the Moon looks in the night sky. Essays can include science facts about the Moon as well as students' thoughts on the way the Moon makes them feel. Invite volunteers to share their essays in class. L2 LS **Intrapersonal** P

Visual Learning

Figure 6 What phase follows full moon? waning gibbous New moon? waxing crescent During which phase do you think a total solar eclipse could happen? new moon L2

Use Science Words

Word Usage Have students write sentences about the Moon's phases using the words wax and wane. Possible answer: Phases wax after a new moon. Once full moon occurs, the phases begin to wane. L2

IDENTIFYING Misconceptions

Moon Phases Students may think that the Sun lights up different parts of the Moon as the Sun moves through the sky. Refer to page F at the beginning of this chapter for teaching strategies that address this misconception.

Text Question Answer

The Moon passes directly between the Sun and Earth, blocking the light from the Sun.

Reading Check

Answer The shadow of the Moon falls on a small portion of Earth's surface; the Moon completely blocks the Sun.

Figure 6 The phases of the Moon change during a cycle that lasts about 29.5 days.

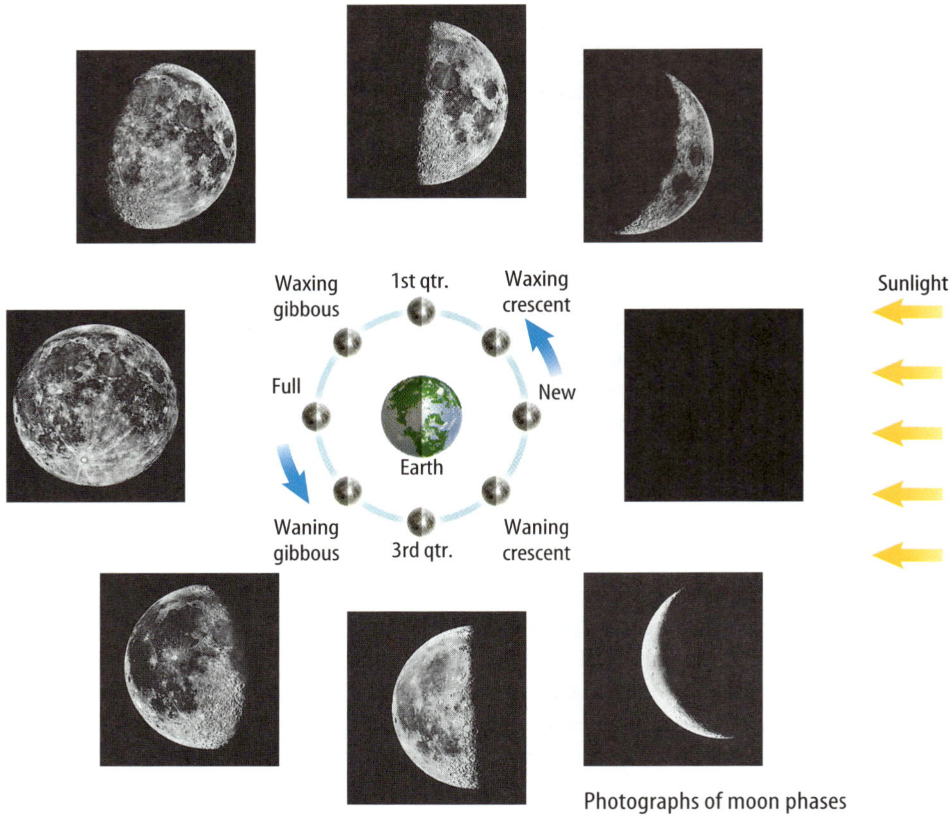

Photographs of moon phases are as seen from Earth.

Figure 7 The outer portion of the Sun's atmosphere is visible during a total solar eclipse. It looks like a halo around the Moon.

Eclipses

Imagine living 10,000 years ago. You are foraging for nuts and fruit when unexpectedly the Sun disappears from the sky. The darkness lasts only a short time, and the Sun soon returns to full brightness. You know something strange has happened, but you don't know why. It will be almost 8,000 years before anyone can explain what you just experienced.

The event just described was a total solar eclipse (ih KLIPS), shown in **Figure 7.** Today, most people know what causes such eclipses, but without this knowledge, they would have been terrifying events. During a solar eclipse, many animals act as if it is nighttime. Cows return to their barns and chickens go to sleep. What causes the day to become night and then change back into day?

Reading Check *What happens during a total solar eclipse?*

Active Reading

Write-Draw-Discuss This strategy encourages students to actively participate in reading and lectures, assimilating content creatively. Students write about an idea, clarify it, then make an illustration or drawing. Ask students to share responses with the class and display examples. Have students Write-Draw-Discuss about a concept in this section. L2

Teacher FYI

Solar Eclipse The Sun has a diameter about 400 times larger than that of the Moon. In order for the Moon to completely block out the Sun during a solar eclipse, the Sun must be about 400 times farther from Earth than the Moon.

48 ◆ J CHAPTER 2 The Sun-Earth-Moon System

What causes an eclipse? The revolution of the Moon causes eclipses. Eclipses occur when Earth or the Moon temporarily blocks the sunlight from reaching the other. Sometimes, during a new moon, the Moon's shadow falls on Earth and causes a solar eclipse. During a full moon, Earth's shadow can be cast on the Moon, resulting in a lunar eclipse.

An eclipse can occur only when the Sun, the Moon, and Earth are lined up perfectly. Because the Moon's orbit is not in the same plane as Earth's orbit around the Sun, lunar eclipses occur only a few times each year.

Eclipses of the Sun A **solar eclipse** occurs when the Moon moves directly between the Sun and Earth and casts its shadow over part of Earth, as seen in **Figure 8.** Depending on where you are on Earth, you may experience a total eclipse or a partial eclipse. The darkest portion of the Moon's shadow is called the umbra (UM bruh). A person standing within the umbra experiences a total solar eclipse. During a total solar eclipse, the only visible portion of the Sun is a pearly white glow around the edge of the eclipsing Moon.

Surrounding the umbra is a lighter shadow on Earth's surface called the penumbra (puh NUM bruh). Persons standing in the penumbra experience a partial solar eclipse. **WARNING:** *Regardless of which eclipse you view, never look directly at the Sun. The light can permanently damage your eyes.*

Science Online

Topic: Eclipses

Visit bookj.msscience.com for Web links to information about solar and lunar eclipses.

Activity Make a chart showing the dates when lunar and solar eclipses will be visible in your area. Include whether the eclipses will be total or partial.

Figure 8 Only a small area of Earth experiences a total solar eclipse during the eclipse event.

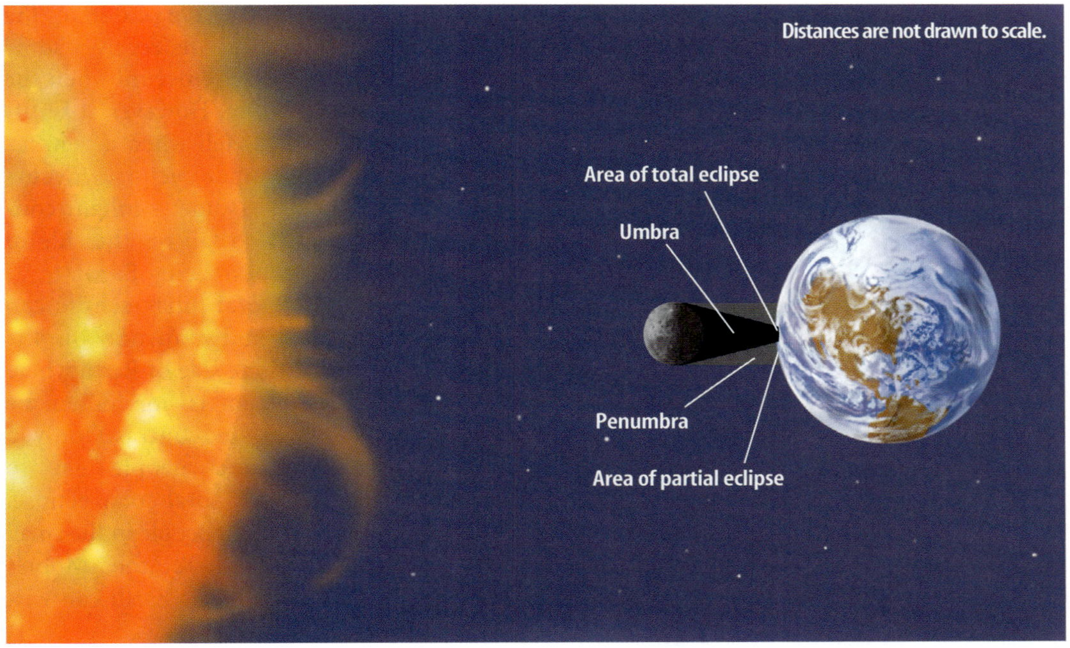

Quick Demo

Produce an Eclipse

Materials flashlight, two balls, one about 4 times larger than the other

Estimated Time 10 minutes

Procedure Dim the classroom lights. Have one student direct the beam of a flashlight on the larger ball. Then move the smaller ball through the beam until its shadow falls on the larger ball. Adjust the distance between the two balls until the shadow of the smaller ball completely darkens the larger. You can also demonstrate the difference between the Moon's and Earth's orbital planes, by moving the ball through the beam just above or below the small ball causing no shadow.

Use an Analogy

Tree Shadows To better understand how eclipses happen, have students consider the analogous situation of a lone tree casting a shadow on a bright, sunny day. The tree casts a shadow on a small area of the ground, just as the Moon casts a shadow on a small area of Earth's surface during a solar eclipse. For someone standing in the tree's shadow, the Sun is blocked out by the tree, just as the Sun is blocked by the Moon during a solar eclipse. L2

Discussion

Nocturnal Solar Eclipse Is it ever possible to see a solar eclipse at night? No, the Moon must cast a shadow on Earth during a solar eclipse. Thus, the place on Earth experiencing the solar eclipse must be on the day side of Earth, facing the Sun. L2

Fun Fact

No matter how large a full moon may appear when it is close to the horizon, it is really the same size as it is when overhead. To convince yourself that this is true, try bending over and looking at an apparently large full moon through your legs.

Curriculum Connection

Literature Have students work in small groups to research poems that mention moon phases or myths about solar or lunar eclipses. After gathering the information, encourage students to recite poems or to perform skits that illustrate some of the myths. L2 COOP LEARN LS **Linguistic and Auditory-Musical**

IDENTIFYING Misconceptions

Moonless Planets Some students may think that having one or more moons is a feature of *all* planets. Explain that not all planets have moons. In our solar system, neither Mercury nor Venus has a moon.

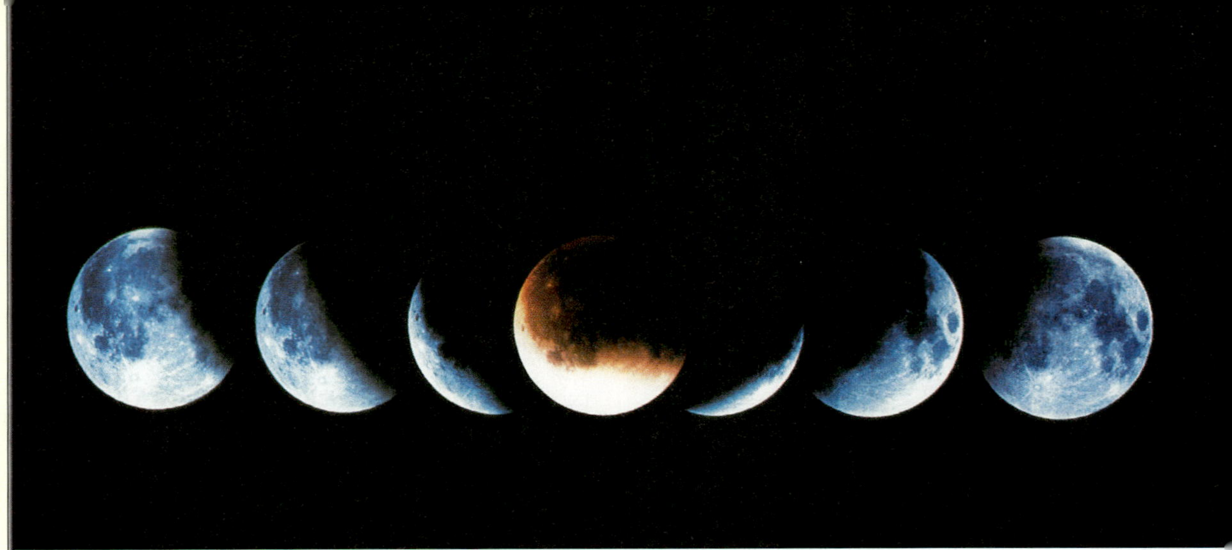

Discussion

Nocturnal Lunar Eclipse Why can a lunar eclipse only be seen on the night side of Earth? A lunar eclipse occurs when the Moon moves into Earth's shadow. Earth's shadow is cast on the night side of the planet, opposite the side facing the Sun. Thus, a lunar eclipse is only visible on the night side of Earth. (Also, lunar eclipses can only occur during a full moon, when the moon is on the opposite side of Earth from the Sun.) L2

Activity

Sketches Have students sketch and label the positions of the Sun, the Moon, and Earth during a solar and a lunar eclipse. L1 LS **Visual-Spatial** P

Figure 9 These photographs show the Moon moving from right to left into Earth's umbra, then out again.

Eclipses of the Moon When Earth's shadow falls on the Moon, a **lunar eclipse** occurs. A lunar eclipse begins when the Moon moves into Earth's penumbra. As the Moon continues to move, it enters Earth's umbra and you see a curved shadow on the Moon's surface, as in **Figure 9.** Upon moving completely into Earth's umbra, as shown in **Figure 10,** the Moon goes dark, signaling that a total lunar eclipse has occurred. Sometimes sunlight bent through Earth's atmosphere causes the eclipsed Moon to appear red.

A partial lunar eclipse occurs when only a portion of the Moon moves into Earth's umbra. The remainder of the Moon is in Earth's penumbra and, therefore, receives some direct sunlight. A penumbral lunar eclipse occurs when the Moon is totally within Earth's penumbra. However, it is difficult to tell when a penumbral lunar eclipse occurs because some sunlight continues to fall on the side of the Moon facing Earth.

A total lunar eclipse can be seen by anyone on the nighttime side of Earth where the Moon is not hidden by clouds. In contrast, only a lucky few people get to witness a total solar eclipse. Only those people in the small region where the Moon's umbra strikes Earth can witness one.

Figure 10 During a total lunar eclipse, Earth's shadow blocks light coming from the Sun.

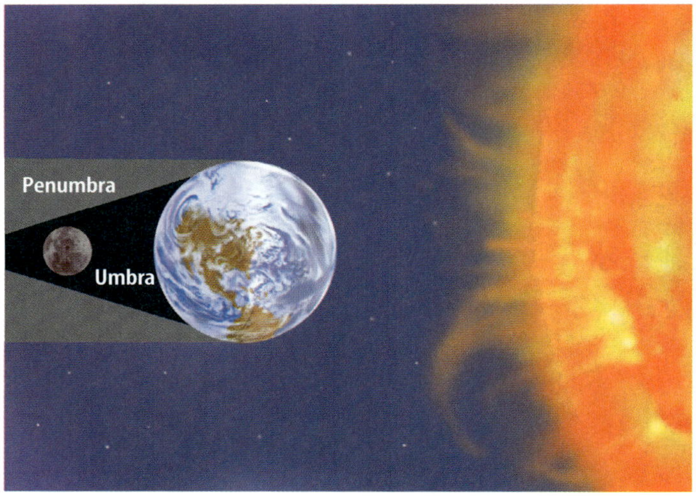

LAB DEMONSTRATION

Purpose to reinforce how solar eclipses occur

Materials marble, tennis ball

Preparation Obtain marbles and tennis balls.

Procedure Hold the tennis ball at arm's length, and hold the marble in front of the ball. Close one eye, and slowly move the marble toward your open eye. Observe what happens.

Expected Outcome As the marble nears the face, less and less of the ball is visible until it is no longer seen.

Assessment

What do the marble, the tennis ball, and your open eye represent? Marble–Moon; ball–Sun; open eye–observer on Earth Why did the Sun disappear? The moon is closer to Earth than the Sun. It is able to block the Sun when it passes directly between it and Earth. L2

The Moon's Surface

When you look at the Moon, as shown in **Figure 12** on the next page, you can see many depressions called craters. Meteorites, asteroids, and comets striking the Moon's surface created most of these craters, which formed early in the Moon's history. Upon impact, cracks may have formed in the Moon's crust, allowing lava to reach the surface and fill up the large craters. The resulting dark, flat regions are called **maria** (MAHR ee uh). The igneous rocks of the maria are 3 billion to 4 billion years old. So far, they are the youngest rocks to be found on the Moon. This indicates that craters formed after the Moon's surface originally cooled. The maria formed early enough in the Moon's history that molten material still remained in the Moon's interior. The Moon once must have been as geologically active as Earth is today. Before the Moon cooled to the current condition, the interior separated into distinct layers.

Inside the Moon

Earthquakes allow scientists to learn about Earth's interior. In a similar way, scientists use instruments such as the one in **Figure 11** to study moonquakes. The data they have received have led to the construction of several models of the Moon's interior. One such model, shown in **Figure 11,** suggests that the Moon's crust is about 60 km thick on the side facing Earth. On the far side, it is thought to be about 150 km thick. Under the crust, a solid mantle may extend to a depth of 1,000 km. A partly molten zone of the mantle may extend even farther down. Below this mantle may lie a solid, iron-rich core.

INTEGRATE Career

Seismology A seismologist is an Earth scientist who studies the propagation of seismic waves in geological materials. Usually this means studying earthquakes, but some seismologists apply their knowledge to studies of the Moon and planets. Seismologists usually study geology, physics, and applied mathematics in college and later specialize in seismology for an advanced degree.

Figure 11 Equipment, such as the seismograph left on the Moon by the *Apollo 12* mission, helps scientists study moonquakes.

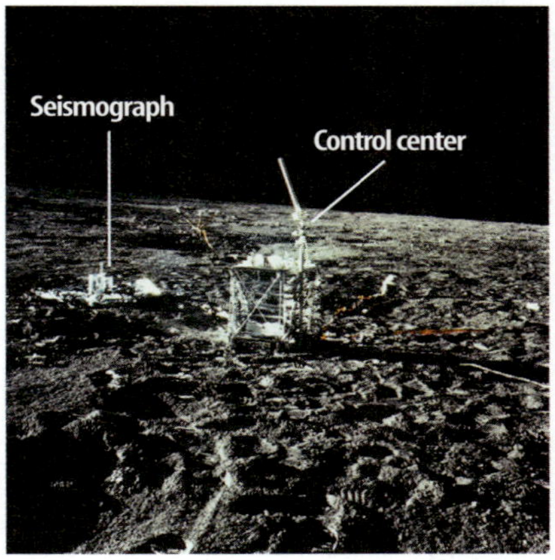

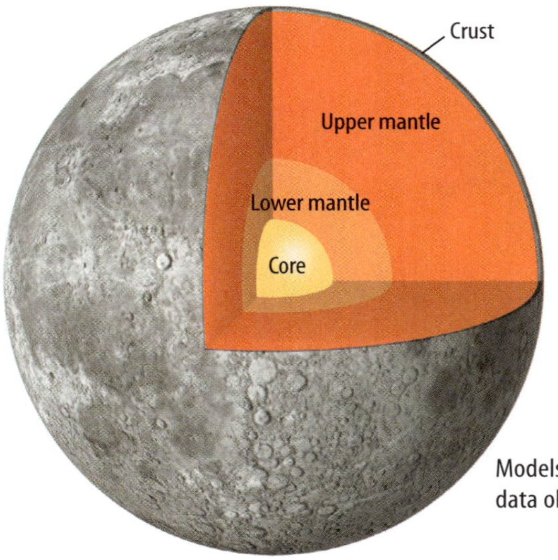

Models of the Moon's interior were created from data obtained by scientists studying moonquakes.

Activity

Moon Features Provide each group of four students a map of the Moon's surface. Have students identify and describe key surface features and explain how each might have formed. L2 COOP LEARN

Use Science Words

Word Origin Ask students to research the origin of the word *month* and write a short paragraph summarizing their findings. The word *month* comes from the same root as *Moon*. It is a period of time that can be measured by the movement of the Moon. L2

INTEGRATE Career

Seismology The primary role of a seismologist is the study of Earthquakes and related phenomena, including work with engineers to minimize damage caused by earthquakes. However most seismologists work in petroleum exploration. Often controlled explosions are detonated producing seismic waves that can indicate underlying oil-bearing structures. Seismologists also can detect and measure underground nuclear explosions and therefore aid in policing nuclear test ban treaties.

Discussion

Moonquakes What might cause quakes on the Moon? Possible answers: objects hitting the Moon's surface; geological processes, such as the movement of magma, inside the Moon L2

Cultural Diversity

Lunar Calendars The Nuer, a group of seminomadic people in Central Africa, base their calendar on full moons, each of which has a name that corresponds to an activity. For example, in *Kur*, make the year's first cattle camps; in *Dwat*, break camp and return home. Challenge students to discover other cultures that use lunar calendars. Chinese, Islamic, and Hebrew calendars are lunar calendars. L2

NATIONAL GEOGRAPHIC

Visualizing the Moon's Surface

Have students examine the pictures and read the captions. Then ask the following questions.

How does the landing site chosen for *Apollo 16* appear different from the other *Apollo* landing sites? *Apollo 16* is the only one with a landing site obviously outside of the maria. It is clearly placed in a mountainous region. However, *Apollo 14* also landed in highlands.

Which do you think formed first: the craters or the "seas"? Because there are few craters in the "sea" regions, the maria must have formed after the meteorite and asteroid bombardments that formed the craters.

What do the crater names seem to reflect? They are named for great scientists and philosophers.

Activity

Apollo Poems Have groups of students write poems about the *Apollo* missions. Each poem should convey information about the missions, such as where they landed and what they did. Have students read the poems aloud in class. L2 COOP LEARN LS **Linguistic**

NATIONAL GEOGRAPHIC **VISUALIZING THE MOON'S SURFACE**

Figure 12

By looking through binoculars, you can see many of the features on the surface of the Moon. These include craters that are hundreds of kilometers wide, light-colored mountains, and darker patches that early astronomers called maria (Latin for "seas"). However, as the NASA Apollo missions discovered, these so-called seas do not contain water. In fact, maria (singular, mare) are flat, dry areas formed by ancient lava flows. Some of the Moon's geographic features are shown below, along with the landing sites of Apollo missions sent to investigate Earth's closest neighbor in space.

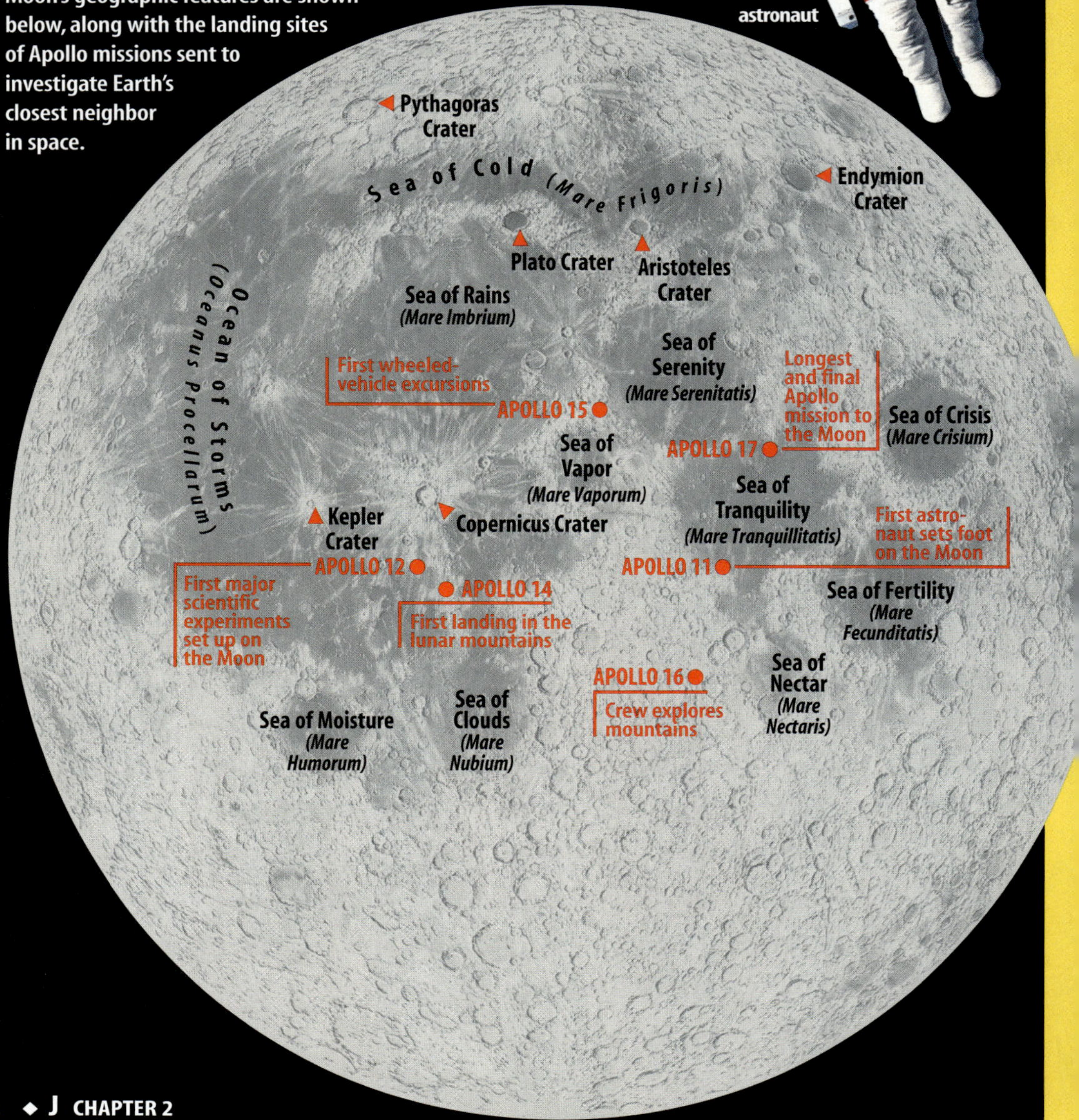

52 ◆ J CHAPTER 2

Differentiated Instruction

Challenge Ask students to choose five moon craters not shown in this image that are named for scientists. Have them research these people and what they contributed to modern science. Have them make a poster presenting the locations of the craters and what they found out about their namesakes. L2

Visually Impaired Have students model a portion of lunar surface using aluminum foil. Make outlines of craters by twisting folded strips of foil and taping them to circles drawn on cardboard. Then cover with a sheet of foil, smoothing foil in crater centers and allowing crimping and wrinkling on crater rims and between craters. L2

The Moon's Origin

Before the *Apollo* space missions in the 1960s and 1970s, there were three leading theories about the Moon's origin. According to one theory, the Moon was captured by Earth's gravity. Another held that the Moon and Earth condensed from the same cloud of dust and gas. An alternative theory proposed that Earth ejected molten material that became the Moon.

The Impact Theory The data gathered by the *Apollo* missions have led many scientists to support a new theory, known as the impact theory. It states that the Moon formed billions of years ago from condensing gas and debris thrown off when Earth collided with a Mars-sized object as shown in **Figure 13.**

Figure 13 According to the impact theory, a Mars-sized object collided with Earth around 4.6 billion years ago. Vaporized materials ejected by the collision began orbiting Earth and quickly consolidated into the Moon.

Applying Science

What will you use to survive on the Moon?

You have crash-landed on the Moon. It will take one day to reach a moon colony on foot. The side of the Moon that you are on will be facing away from the Sun during your entire trip. You manage to salvage the following items from your wrecked ship: food, rope, solar-powered heating unit, battery-operated heating unit, oxygen tanks, map of the constellations, compass, matches, water, solar-powered radio transmitter, three flashlights, signal mirror, and binoculars.

Identifying the Problem

The Moon lacks a magnetic field and has no atmosphere. How do the Moon's physical properties and the lack of sunlight affect your decisions?

Solving the Problem

1. Which items will be of no use to you? Which items will you take with you?
2. Describe why each of the salvaged items is useful or not useful.

Applying Science

Answers

1. Useless: solar-powered heating unit, compass, matches, solar-powered radio receiver and transmitter, signal mirror. Useful: food, rope, battery-operated heating unit, oxygen tanks, map of the constellations, water, flashlights, binoculars.
2. Useful: Need food, water, and oxygen to live; rope handy if someone falls in a crevice; use binoculars to see far away; use constellation map to navigate; flashlights provide light; batteries power flashlights and heating unit. Useless: signal mirror, solar-powered heating units and radios—no Sun; matches—no oxygen; compass—no magnetic field.

Fun Fact

Many people in our culture see the light and dark patterns on the Moon as the face of a man. Other cultures see different things. For example, the Maya saw a leaping rabbit.

Virtual Labs

Moon Phases *How do Earth and the Moon interact to cause the Moon's phases?*

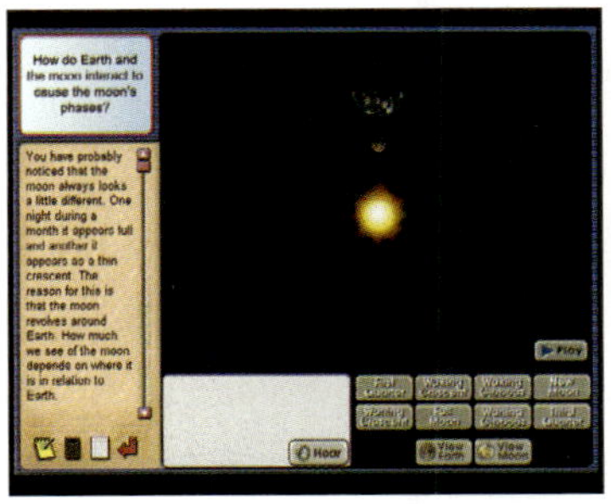

Curriculum Connection

Language Arts Have students research the origin and meaning of the phrase "once in a blue Moon." Students may find several explanations. Recently, the second full moon in a month has been called a blue moon. Because this happens infrequently the phrase "once in a blue moon" is used to indicate a rare event. L2 LS **Linguistic**

SECTION 2 The Moon—Earth's Satellite J ◆ 53

Reading Check

Answer These studies have helped scientists draw conclusions about Earth's shape, the movement of Earth and the Moon around the Sun, and the presence of surface features on bodies other than Earth.

3 Assess

DAILY INTERVENTION

Check for Understanding

Visual-Spatial Ask students to organize what they have learned about the Moon by drawing a spider concept map. Tell them to place the Moon in the center and arrange main topics radiating around it and subtopics branching from these. L2

Reteach

Eclipses Have groups of three students use a globe, a tennis ball, and a beach ball to model solar and lunar eclipses. L1

Figure 14 Moon rocks collected by astronauts provide scientists with information about the Moon and Earth.

The Moon in History Studying the Moon's phases and eclipses led to the conclusion that both Earth and the Moon were in motion around the Sun. The curved shadow Earth casts on the Moon indicated to early scientists that Earth was spherical. When Galileo first turned his telescope toward the Moon, he found a surface scarred by craters and maria. Before that time, many people believed that all planetary bodies were perfectly smooth and lacking surface features. Now, actual moon rocks are available for scientists to study, as seen in **Figure 14.** By doing so, they hope to learn more about Earth.

Reading Check *How has observing the Moon been important to science?*

section 2 review

Summary

Motions of the Moon

- The Moon rotates on its axis about once each month.
- The Moon also revolves around Earth about once every 27.3 days.
- The Moon shines because it reflects sunlight.

Phases of the Moon

- During the waxing phases, the illuminated portion of the Moon grows larger.
- During waning phases, the illuminated portion of the Moon grows smaller.
- Earth passing directly between the Sun and the Moon causes a lunar eclipse.
- The Moon passing between Earth and the Sun causes a solar eclipse.

Structure and Origin of the Moon

- The Moon's surface is covered with depressions called impact craters.
- Flat, dark regions within craters are called maria.
- The Moon may have formed as the result of a collision between Earth and a Mars-sized object.

Self Check

1. **Explain** how the Sun, Moon, and Earth are positioned relative to each other during a new moon and how this alignment changes to produce a full moon.
2. **Describe** what phase the Moon must be in to have a lunar eclipse. A solar eclipse?
3. **Define** the terms *umbra* and *penumbra* and explain how they relate to eclipses.
4. **Explain** why lunar eclipses are more common than solar eclipses and why so few people ever have a chance to view a total solar eclipse.
5. **Think Critically** What do the surface features and their distribution on the Moon's surface tell you about its history?

Applying Math

6. **Solve Simple Equations** The Moon travels in its orbit at about 3,400 km/h. Therefore, during a solar eclipse, its shadow sweeps at this speed from west to east. However, Earth rotates from west to east at about 1,670 km/h near the equator. At what speed does the shadow really move across this part of Earth's surface?

Science Online bookj.msscience.com/self_check_quiz

section 2 review

1. The Moon is between the Sun and Earth; Earth is between the Moon and the Sun.
2. The Moon must be in the full moon phase; new Moon phase.
3. The umbra is the dark central shadow; the penumbra is the light shadow around the umbra. Both are cast during an eclipse.
4. Lunar eclipses are more common because Earth's shadow is larger than the Moon's and can be seen anywhere on Earth. The Moon's smaller shadow sweeps only a small part of Earth during each eclipse.
5. Possible answers: There have been many impacts on its surface. The Moon may have been geologically active at one time.
6. The Moon's shadow travels at 3,400 km/h but Earth's at 1,670 km/h, so the shadow appears to travel only about 1,730 km/h.

LAB

Moon Phases and Eclipses

In this lab, you will demonstrate the positions of the Sun, the Moon, and Earth during certain phases and eclipses. You also will see why only a small portion of the people on Earth witness a total solar eclipse during a particular eclipse event.

Real-World Question

Can a model be devised to show the positions of the Sun, the Moon, and Earth during various phases and eclipses?

Goals

- **Model** moon phases.
- **Model** solar and lunar eclipses.

Materials

light source (unshaded)
globe
polystyrene ball
pencil

Safety Precautions

Procedure

1. Review the illustrations of moon phases and eclipses shown in Section 2.
2. Use the light source as a Sun model and a polystyrene ball on a pencil as a Moon model. Move the Moon around the globe to duplicate the exact position that would have to occur for a lunar eclipse to take place.
3. Move the Moon to the position that would cause a solar eclipse.
4. Place the Moon at each of the following phases: first quarter, full moon, third quarter, and new moon. Identify which, if any, type of eclipse could occur during each phase. Record your data.

Moon Phase Observations

Moon Phase	Observations
First quarter	no eclipse
Full moon	lunar eclipse
Third quarter	no eclipse
New moon	solar eclipse

5. Place the Moon at the location where a lunar eclipse could occur. Move it slightly toward Earth, then away from Earth. Note the amount of change in the size of the shadow.
6. Repeat step 5 with the Moon in a position where a solar eclipse could occur.

Conclude and Apply

1. **Identify** which phase(s) of the Moon make(s) it possible for an eclipse to occur.
2. **Describe** the effect of a small change in distance between Earth and the Moon on the size of the umbra and penumbra.
3. **Infer** why a lunar and a solar eclipse do not occur every month.
4. **Explain** why only a few people have experienced a total solar eclipse.
5. **Diagram** the positions of the Sun, Earth, and the Moon during a first-quarter moon.
6. **Infer** why it might be better to call a full moon a half moon.

Communicating Your Data

Communicate your answers to other students.

BENCH TESTED

Real-World Question

Purpose Students model the positions of Earth, the Moon, and the Sun during lunar phases and eclipses. L2 ELL COOP LEARN IS **Kinesthetic**

Process Skills sequence, communicate, observe and infer, compare and contrast, recognize and use spatial relationships, recognize cause and effect

Time Required 45 minutes

Safety Precautions Caution students to avoid touching the bulb.

Procedure

Teaching Strategy Guide students in adjusting the position of the ball representing the Moon closer or farther from the globe until it casts a noticeable shadow.

Conclude and Apply

1. lunar eclipse: full moon; solar eclipse: new moon
2. During solar eclipses, the closer the two bodies, the larger the Moon's umbra and penumbra.
3. The Sun, the Moon, and Earth must be lined up perfectly, with the Moon in Earth's orbital plane. Most months, the Moon is above or below Earth's orbital plane i.e., not directly between the Sun and Earth.
4. When the umbra falls on Earth, it covers only a small area.
5. The Sun, Earth, and the Moon form a right angle.
6. Only half of the Moon is ever visible at one time to any observer; the other half is shadowed on the side away from the observer.

Assessment

Performance Assign each student a phase of the Moon. Have them draw what the Moon looks like from Earth, as well as a diagram of the positions of the Sun, Earth, and the Moon during that phase. Use **Performance Assessment in the Science Classroom**, p. 127. L2

Communicating Your Data

Instruct students to write short lab reports that describe their goal during this experiment, the procedure they used, and what they concluded.

Section 3 Exploring Earth's Moon

1 Motivate

Bellringer

INTERACTIVE CHALKBOARD PowerPoint® Presentations

Section Focus Transparencies also are available on the Interactive Chalkboard CD-ROM. L2 ELL

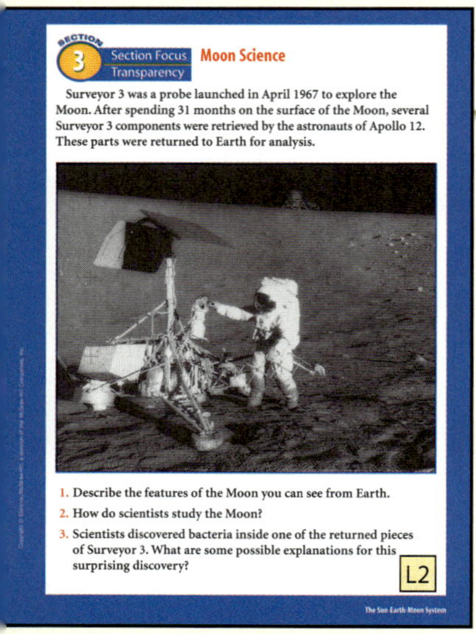
Section 3 Section Focus Transparency **Moon Science**

Surveyor 3 was a probe launched in April 1967 to explore the Moon. After spending 31 months on the surface of the Moon, several Surveyor 3 components were retrieved by the astronauts of Apollo 12. These parts were returned to Earth for analysis.

1. Describe the features of the Moon you can see from Earth.
2. How do scientists study the Moon?
3. Scientists discovered bacteria inside one of the returned pieces of Surveyor 3. What are some possible explanations for this surprising discovery?

L2

The Sun-Earth-Moon System

Tie to Prior Knowledge

Moon News Ask students to recall the latest "news" they have heard about the Moon. Possible answers: There may be water on the Moon; a space probe was sent to the Moon in recent years; astronauts might visit the Moon again. Tell students they will learn in this section about the first trip to the Moon and the possibility of future lunar missions. L2

as you read

***What* You'll Learn**

- **Describe** recent discoveries about the Moon.
- **Examine** facts about the Moon that might influence future space travel.

***Why* It's Important**

Continuing moon missions may result in discoveries about Earth's origin.

Review Vocabulary

comet: space object orbiting the Sun formed from dust and rock particles mixed with frozen water, methane, and ammonia

New Vocabulary

- impact basin

Missions to the Moon

The Moon has always fascinated humanity. People have made up stories about how it formed. Children's stories even suggested it was made of cheese. Of course, for centuries astronomers also have studied the Moon for clues to its makeup and origin. In 1959, the former Soviet Union launched the first *Luna* spacecraft, enabling up-close study of the Moon. Two years later, the United States began a similar program with the first *Ranger* spacecraft and a series of *Lunar Orbiters*. The spacecraft in these early missions took detailed photographs of the Moon.

The next step was the *Surveyor* spacecraft designed to take more detailed photographs and actually land on the Moon. Five of these spacecraft successfully touched down on the lunar surface and performed the first analysis of lunar soil. The goal of the *Surveyor* program was to prepare for landing astronauts on the Moon. This goal was achieved in 1969 by the astronauts of *Apollo 11*. By 1972, when the *Apollo* missions ended, 12 U.S. astronauts had walked on the Moon. A time line of these important moon missions can be seen in **Figure 15.**

Figure 15 This time line illustrates some of the most important events in the history of moon exploration.

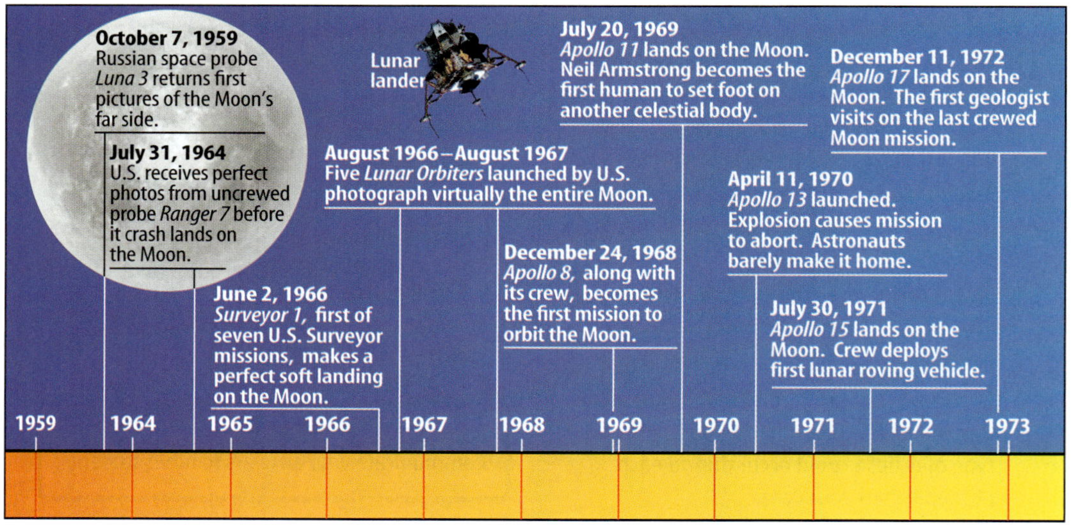

Section 3 Resource Manager

Chapter *FAST FILE* Resources

- Transparency Activity, p. 46
- Directed Reading for Content Mastery, pp. 19, 20
- Enrichment, p. 30
- Reinforcement, p. 27
- Lab Worksheets, pp. 9–10

Home and Community Involvement, p. 24

Reading and Writing Skill Activities, p. 11

Lab Management and Safety, p. 61

56 ◆ J CHAPTER 2 The Sun-Earth-Moon System

Surveying the Moon There is still much to learn about the Moon and, for this reason, the United States resumed its studies. In 1994, the *Clementine* was placed into lunar orbit. Its goal was to conduct a two-month survey of the Moon's surface. An important aspect of this study was collecting data on the mineral content of Moon rocks. In fact, this part of its mission was instrumental in naming the spacecraft. Clementine was the daughter of a miner in the ballad *My Darlin' Clementine.* While in orbit, *Clementine* also mapped features on the Moon's surface, including huge impact basins.

Reading Check *Why was* Clementine *placed in lunar orbit?*

Impact Basins When meteorites and other objects strike the Moon, they leave behind depressions in the Moon's surface. The depression left behind by an object striking the Moon is known as an **impact basin**, or impact crater. The South Pole-Aitken Basin is the oldest identifiable impact feature on the Moon's surface. At 12 km in depth and 2,500 km in diameter, it is also the largest and deepest impact basin in the solar system.

Impact basins at the poles were of special interest to scientists. Because the Sun's rays never strike directly, the crater bottoms remain always in shadow. Temperatures in shadowed areas, as shown in **Figure 16,** would be extremely low, probably never more than −173°C. Scientists hypothesize that any ice deposited by comets impacting the Moon throughout its history would remain in these shadowed areas. Indeed, early signals from *Clementine* indicated the presence of water. This was intriguing, because it could be a source of water for future moon colonies.

Science Online

Topic: The Far Side

Visit bookj.msscience.com for Web links to information about the far side of the Moon.

Activity Compare the image of the far side of the Moon with that of the near side shown in **Figure 12.** Make a list of all the differences you note and then compare them with lists made by other students.

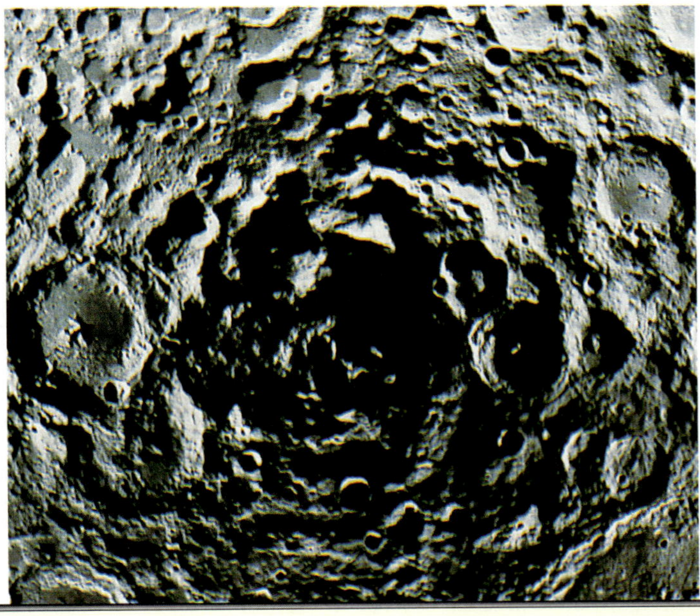

Figure 16 The South Pole-Aitken Basin is the largest of its kind found anywhere in the solar system. The deepest craters in the basin stay in shadow throughout the Moon's rotation. Ice deposits from impacting comets are thought to have collected at the bottom of these craters.

2 Teach

Inquiry Lab

Purpose To demonstrate that it is possible that sunlight does not reach the bottoms of some lunar craters at the Moon's south pole.

Possible Materials clay, paper maché, plastic foam balls, flashlight or other light source, glitter, aluminum foil

Estimated Time 1 hour (more time may be needed depending on what materials are used)

Teaching Strategies Advise students that they must try to match the approximate topography at the lunar pole. They can refer to dimensions given in the chapter and remember the model of the Aitken Basin in preparing their model and making their craters. L2

For additional inquiry activities, see *Science Inquiry Labs*.

Reading Check

Answer to conduct a two-month survey of the Moon's surface, including collecting data on the mineral content of lunar rocks

Make a Model

The Aitken Basin The view in **Figure 16** shows only the basin area. To put this in perspective, inflate a round balloon or punchball to a diameter of about 20 cm. Then draw a circle having a diameter of 14 cm centered on the tied end. Tell students that this is the approximate size of the Aitken basin at the Moon's south pole. Then press gently near the center of this circle with a pencil eraser, indenting the balloon no more than 1 mm. This represents the deepest part of the crater.

Differentiated Instruction

English-Language Learners Pair students with an English-speaking student and ask them to prepare a chart comparing the findings of *Clementine* and *Lunar Prospector.* Encourage them to use labeled diagrams and drawings to display their findings. For example, under *Clementine* they could show a diagram of the Moon that shows a thinner crust, especially under impact basins, on the side facing Earth.

Visual Learning

Figure 16 Have students contrast the appearance of the Moon's south pole, shown here, with that of Earth's north pole. Earth's north pole is covered with an ice sheet and snow. The Moon's south pole is bare rock. L2

Activity

Moon Observation Have students observe the Moon's surface during the next full moon. Have some students make their observations with the unaided eye, some use binoculars, and others use telescopes. Direct students to draw the Moon's surface as they see it, sketching any features they observe. Have students compare and contrast the drawings in class. Discuss which features are prominent, even without instruments, and which features can only be seen using binoculars or telescopes. L2 LS **Visual-Spatial**

Reading Check

Answer information on the Moon's surface that was used to make detailed lunar maps, the composition of the Moon's surface rock, and the thickness of lunar crust

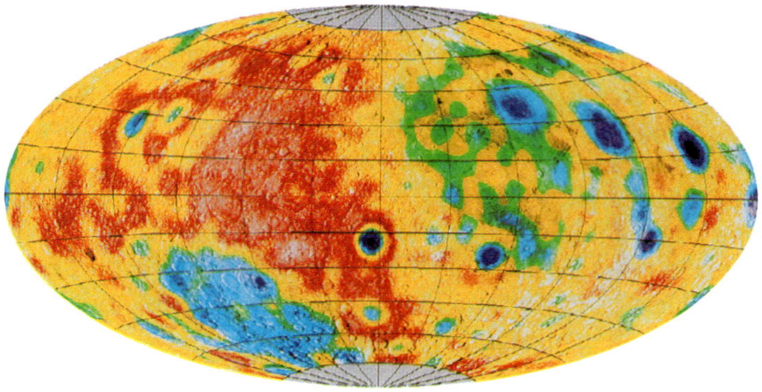

Figure 17 This computer-enhanced map based on *Clementine* data indicates the thickness of the Moon's crust. The crust of the side of the Moon facing Earth, shown mostly in red, is thinner than the crust on the far side of the Moon.

Mapping the Moon

A large part of *Clementine's* mission included taking high-resolution photographs so a detailed map of the Moon's surface could be compiled. *Clementine* carried cameras and other instruments to collect data at wavelengths ranging from infrared to ultraviolet. One camera could resolve features as small as 20 m across. One image resulting from *Clementine* data is shown in **Figure 17.** It shows that the crust on the side of the Moon that faces Earth is much thinner than the crust on the far side. Additional information shows that the Moon's crust is thinnest under impact basins. Based on analysis of the light data received from *Clementine,* a global map of the Moon also was created that shows its composition, as seen in **Figure 18.**

Reading Check *What information about the Moon did scientists learn from* Clementine?

The Lunar Prospector The success of Clementine opened the door for further moon missions. In 1998, NASA launched the desk-sized *Lunar Prospector,* shown in **Figure 18,** into lunar orbit. The spacecraft spent a year orbiting the Moon from pole to pole, once every two hours. The resulting maps confirmed the *Clementine* data. Also, data from *Lunar Prospector* confirmed that the Moon has a small, iron-rich core about 600 km in diameter. A small core supports the impact theory of how the Moon formed—only a small amount of iron could be blasted away from Earth.

Figure 18 *Lunar Prospector* performed high-resolution mapping of the lunar surface and had instruments that detected water ice at the lunar poles.

Differentiated Instruction

Challenge Have students research the mission of the *Lunar Prospector.* Then have them write a press release on the mission's successes and failures that have been released by NASA since the mission ended in July 1999. L3 LS **Linguistic** P

Teacher FYI

Owning the Moon Although both the United States and the former Soviet Union have sent spacecraft to the Moon, no nation owns the Moon. More than 90 nations signed a treaty in 1967 stating that neither the Moon nor any other body in space could be claimed by one nation or used for military purposes.

Icy Poles In addition to photographing the surface, *Lunar Prospector* carried instruments designed to map the Moon's gravity, magnetic field, and the abundances of 11 elements in the lunar crust. This provided scientists with data from the entire lunar surface rather than just the areas around the Moon's equator, which had been gathered earlier. Also, *Lunar Prospector* confirmed the findings of *Clementine* that water ice was present in deep craters at both lunar poles.

Later estimates concluded that as much as 3 billion metric tons of water ice was present at the poles, with a bit more at the north pole. Using data from *Lunar Prospector,* scientists prepared maps showing the location of water ice at each pole. **Figure 19** shows how water may be distributed at the Moon's north pole. At first it was thought that ice crystals were mixed with lunar soil, but most recent results suggest that the ice may be in the form of more compact deposits.

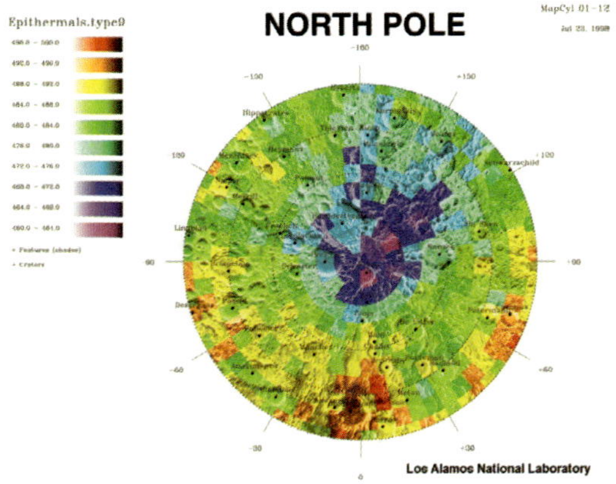

Figure 19 The *Lunar Prospector* data indicates that ice exists in crater shadows at the Moon's poles.

section 3 review

Summary

Missions to the Moon

- The first lunar surveys were done by *Luna,* launched by the former Soviet Union, and U.S.-launched *Ranger* and *Lunar Orbiters.*
- Five *Surveyor* probes landed on the Moon.
- U.S. Astronauts landed on and explored the Moon in the *Apollo* program.
- *Clementine,* a lunar orbiter, mapped the lunar surface and collected data on rocks.
- *Clementine* found that the lunar crust is thinner on the side facing Earth.
- Data from *Clementine* indicated that water ice could exist in shaded areas of impact basins.

Mapping the Moon

- *Lunar Prospector* orbited the Moon from pole to pole, collecting data that confirm *Clementine* results and that the Moon has a small iron-rich core.
- Data from *Lunar Prospector* indicate the presence of large quantities of water ice in craters at the lunar poles.

Self Check

1. **Name** the first U.S. spacecraft to successfully land on the Moon. What was the major purpose of this program?
2. **Explain** why scientists continue to study the Moon long after the *Apollo* program ended and list some of the types of data that have been collected.
3. **Explain** how water ice might be preserved in portions of deep impact craters.
4. **Describe** how the detection of a small iron-rich core supports the theory that the Moon was formed from a collision between Earth and a Mars-sized object.
5. **Think Critically** Why might the discovery of ice in impact basins at the Moon's poles be important to future space flights?

Applying Skills

6. **Infer** why it might be better to build a future moon base on a brightly lit plateau near a lunar pole in the vicinity of a deep crater. Why not build a base in the crater itself?

Science Online bookj.msscience.com/self_check_quiz

Discussion

Moon Exploration Should the United States continue to explore the Moon? Possible answers: Yes; exploration advances science; the Moon might be a base for further study of space or for staging space missions one day. No; studying the Moon has no relevance to everyday life; the money could be better used elsewhere.

3 Assess

DAILY INTERVENTION

Check for Understanding

Visual-Spatial Have students work in pairs to make concept maps describing the lunar missions mentioned in this section and include all their accomplishments. L2

Reteach

Ice in Craters Remind students that even on Earth, pockets of snow remain in shaded areas long after snow has melted everywhere else. This is true especially in ravines on the north side of mountains, where sunlight does not reach the ravine bottoms. Often skiers continue to ski in such ravines long after normal slopes have closed.

Assessment

Content Have students make a two-column chart categorizing the missions that landed on the Moon and those that orbited the Moon. Landed: *Surveyor, Apollo;* orbited: *Lunar Orbiter, Clementine, Lunar Prospector* L2

section 3 review

1. The first U.S. spacecraft to land on the Moon was *Surveyor* and the goal of this program was to prepare for astronauts landing on the Moon.
2. Because we still have much to learn about the Moon and understanding more about it will help us understand Earth and other planets; the Moon may even be a base from which we can explore space. Data collected have included surface mineralogy, geography, volcanic activity, geochemistry, and measurement of magnetic and gravitational fields.
3. Because deep craters at the polar regions never get direct sunlight and thus stay very cold; any ice deposited by comets would remain.
4. The fact that the core is small indicates that most of the material blasted from Earth came from the crust and mantle as would happen in an impact with a large object.
5. Future astronauts on the Moon would have a source of water.
6. You would want a base near the water in the crater, but need also light and power which you can get only in a well-lit area.

BENCH TESTED

Real-World Question

Purpose Students design and carry out an experiment to show how the angle at which sunlight strikes an area of Earth's surface determines the amount of heat received by that area. L2 ELL
LS **Logical-Mathematical**

Process Skills communicate, make and use tables, observe, infer, compare and contrast, recognize cause and effect, design an experiment, measure, hypothesize, separate and control variables, interpret data, formulate models

Time Required 45 minutes

Procedure

Alternate Materials A desk lamp held at the proper angle can be used instead of a gooseneck lamp.

Safety Precautions The light bulb and shade may be hot for some time after the lamp has been turned off.

Teaching Strategies Remind students to allow the thermometer to return to room temperature before each use.

LAB
TILT AND TEMPERATURE

If you walk on blacktop pavement at noon, you can feel the effect of solar energy. The Sun's rays hit at the highest angle at midday. Now consider the fact that Earth is tilted on its axis. How does this tilt affect the angle at which light rays strike an area on Earth? How is the angle of the light rays related to the amount of heat energy and the changing seasons?

Goals
- **Measure** the temperature change in a surface after light strikes it at different angles.
- **Describe** how the angle of light relates to seasons on Earth.

Materials
tape
black construction paper (one sheet)
gooseneck lamp with 75-watt bulb
Celsius thermometer
watch
protractor

Safety Precautions

WARNING: *Do not touch the lamp without safety gloves. The lightbulb and shade can be hot even when the lamp has been turned off. Handle the thermometer carefully. If it breaks, do not touch anything. Inform your teacher immediately.*

Real-World Question

How does the angle at which light strikes Earth affect the amount of heat energy received by any area on Earth?

Procedure

1. Choose three angles that you will use to aim the light at the paper.
2. **Determine** how long you will shine the light at each angle before you measure the temperature. You will measure the temperature at two times for each angle. Use the same time periods for each angle.
3. Copy the following data table into your Science Journal to record the temperature the paper reaches at each angle and time.

Temperature Data

Angle of Lamp	Initial Temperature	Temperature at ___ Minutes/Seconds	Temperature at ___ Minutes/Seconds
First angle			
Second angle	Answers will vary.		
Third angle			

4. Form a pocket out of a sheet of black construction paper and tape it to a desk or the floor.
5. Using the protractor, set the gooseneck lamp so that it will shine on the paper at one of the angles you chose.

Alternative Inquiry Lab

Studying Spherical Surfaces Students could develop an experiment to extend the investigation of the effect of the angle at which light strikes a three-dimensional surface has on temperature. This would more closely approximate Earth's curved surface. They might use spherical objects, such as melons or balls made out of various materials. They could mark the spheres to show the equator and the two tropics and indicate the poles with toothpicks or small dowels. They could measure temperature as in the original Lab, but they should devise a method to hold the sphere at the angles they choose to investigate.

6. Place the thermometer in the paper pocket. Turn on the lamp. Use the thermometer to measure the temperature of the paper at the end of the first time period. Continue shining the lamp on the paper until the second time period has passed. Measure the temperature again. Record your data in your data table.
7. Turn off the lamp until the paper cools to room temperature. Repeat steps 5 and 6 using your other two angles.

Conclude and Apply

1. **Describe** your experiment. Identify the variables in your experiment. Which were your independent and dependent variables?
2. **Graph** your data using a line graph. Describe what your graph tells you about the data.
3. **Describe** what happened to the temperature of the paper as you changed the angle of light.
4. **Predict** how your results might have been different if you used white paper. Explain why.
5. **Describe** how the results of this experiment apply to seasons on Earth.

Communicating Your Data

Compare your results with those of other students in your class. **Discuss** how the different angles and time periods affected the temperatures.

Conclude and Apply

Expected Outcome The surface area heats up more quickly when light hits it from a higher angle.

Answers to Questions

1. Students should mention that they varied the angle at which light struck the thermometer. The independent variable was the angle of light. The dependent variable was the increase in temperature of the paper.
2. Graphs should have time on the *y* axis and temperature on the *x* axis.
3. Temperature increased as the angle of the light approached 90°.
4. White paper would produce similar results, but temperature would not rise as high as with black paper.
5. During our summer, the Sun's rays strike the northern hemisphere at the highest angles, causing high summer temperatures. During our winter, the Sun's rays strike the northern hemisphere at lower angles, causing lower winter temperatures.

Error Analysis Possible causes of error include mismeasuring angles, changing the distance between the light and the paper, and not monitoring time correctly.

Assessment

Process Ask students to explain in writing why temperatures rise faster when light hits a surface area at a more direct angle.

Communicating Your Data

Compile data from all students in the class and display it on a class graph.

TIME

Content Background

The Mayan civilization flourished for centuries. As agriculturalists, wealth, and power depended entirely on what the Mayan people could produce from the land. Studying the passage of time and the cycle of the seasons was crucial to their culture. It was imperative, for example, that they be able to predict the onset of floods or periods of drought. This was the impetus behind the development of their calendars. The *Tzolkin* was based on cycles of 20 days, similar to our months. Important dates for planting and harvesting, along with religious celebrations, were recorded in this calendar.

The 365-day calendar, the *Haab*, was based on the Mayan's skillful observations of Earth's orbit, the phases of the Moon, and the movements of the visible planets and constellations. The good fortune or bad luck of each day was interpreted from the mythological stories based on these observations.

A third calendar measured the number of days that had passed since the year Zero. For the Mayans this was around August 2 in the year 3114 B.C.

Discussion

The Mayans considered their calendar so important that they built its structure into the Temple of Kukulkan. Do we do anything similar in our culture? Do we commemorate people or events in monuments? L2

Activity

Ask students to work in pairs to invent their own glyphs, one for each day of the week. Then have them explain their symbols and compare them with those devised by other students. L2

TIME SCIENCE AND HISTORY

SCIENCE CAN CHANGE THE COURSE OF HISTORY!

THE Mayan Calendar

Most people take for granted that a week is seven days, and that a year is 12 months. However there are other ways to divide time into useful units. Roughly 1,750 years ago, in what is now south Mexico and Central America, the Mayan people invented a calendar system based on careful observations of sun and moon cycles.

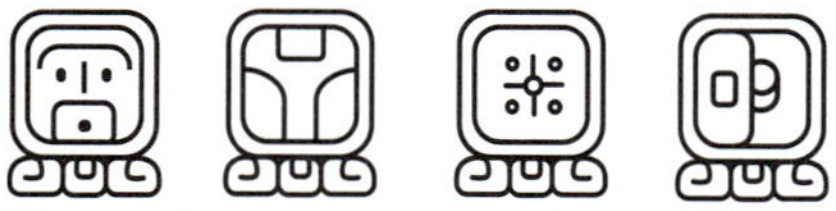

These glyphs represent four different days of the Tzolkin calendar.

In fact, the Maya had several calendars that they used at the same time.

Two calendars were most important—one was based on 260 days and the other on 365 days. The calendars were so accurate and useful that later civilizations, including the Aztecs, adopted them.

The 260-Day Calendar

This Mayan calendar, called the *Tzolkin* (tz uhl KIN), was used primarily to time planting, harvesting, drying, and storing of corn—their main crop. Each day of the *Tzolkin* had one of 20 names, as well as a number from 1 to 13 and a Mayan god associated with it.

The 365-Day Calendar

Another Mayan calendar, called the *Haab* (HAHB), was based on the orbit of Earth around the Sun. It was divided into 18 months with 20 days each, plus five extra days at the end of each year.

These calendars were used together making the Maya the most accurate reckoners of time before the modern period. In fact, they were only one day off every 6,000 years.

The Kukulkan, built around the year 1050 AD, in what is now Chichén Itzá, Mexico, was used by the Maya as a calendar. It had four stairways, each with 91 steps, a total of 365 including the platform on top.

Drawing Symbols The Maya created picture symbols for each day of their week. Historians call these symbols glyphs. Collaborate with another student to invent seven glyphs—one for each weekday. Compare them with other glyphs at msscience.com/time.

Science Online For more information, visit bookj.msscience.com/time

Resources for Teachers and Students

The Atlas of World Archaeology, edited by Paul G. Bahn, Checkmark Books, 2000

Ancient Mexico, The History and Culture of the Maya, Aztec and Other Pre-Columbian Peoples, by Longchena, Stewart, Tabori, and Chang, 1998

Atlas of Ancient America, by Coe, Snow, and Benson, Andromeda Oxford, Ltd., 1996

"Ancient Mesoamerica", Map Supplement, *National Geographic Magazine*, December 1997

Reviewing Main Ideas

Section 1 Earth

1. Earth is spherical and bulges slightly at its equator.
2. Earth rotates once per day and orbits the Sun in a little more than 365 days.
3. Earth has a magnetic field.
4. Seasons on Earth are caused by the tilt of Earth's axis as it orbits the Sun.

Section 2 The Moon—Earth's Satellite

1. Earth's Moon goes through phases that depend on the relative positions of the Sun, the Moon, and Earth.
2. Eclipses occur when Earth or the Moon temporarily blocks sunlight from reaching the other.
3. The Moon's maria are the result of ancient lava flows. Craters on the Moon's surface formed from impacts with meteorites, asteroids, and comets.

Section 3 Exploring Earth's Moon

1. The *Clementine* spacecraft took detailed photographs of the Moon's surface and collected data indicating the presence of water in deep craters.
2. NASA's *Lunar Prospector* spacecraft found additional evidence of ice.

Visualizing Main Ideas

Copy and complete the following concept map on the impact theory of the Moon's formation.

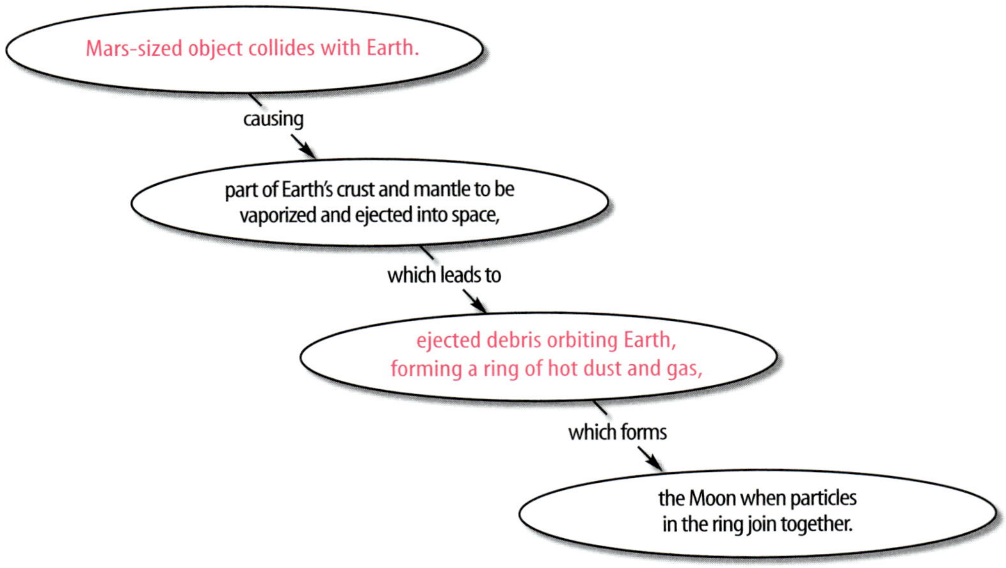

Identifying Misconceptions

Use the assessment as follow-up to page F at the beginning of this chapter.

Demonstration Have students demonstrate the phases of the Moon using a light source and two balls. L2

Expected Outcome Students will realize that the relative motions of the Sun-Earth-Moon model cause the phases of the Moon.

Reinforcement Have students locate information about lunar phases. Then have them mark the upcoming phases of the Moon on a calendar. L1

Assess

chapter 2 Study Guide

Reviewing Main Ideas

Summary statements can be used by students to review the major concepts of the chapter.

Visualizing Main Ideas

See student page.

Science Online

Visit bookj.msscience.com
/self_check_quiz
/interactive_tutor
/vocabulary_puzzlemaker
/chapter_review
/standardized_test

Assessment Transparency

For additional assessment questions, use the *Assessment Transparency* located in the transparency book.

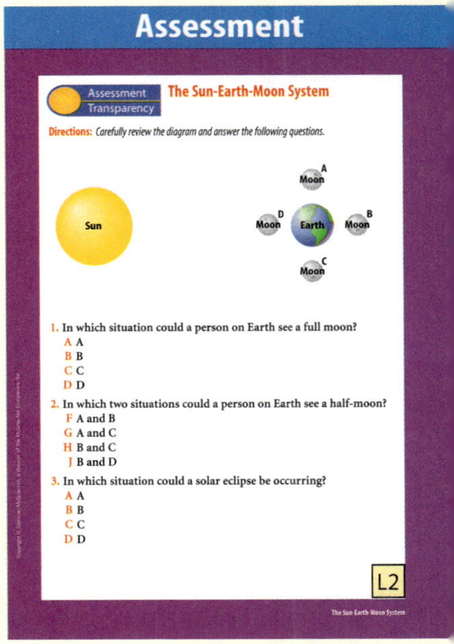

Assessment

Assessment Transparency — The Sun-Earth-Moon System

Directions: Carefully review the diagram and answer the following questions.

1. In which situation could a person on Earth see a full moon?
   A A
   B B
   C C
   D D
2. In which two situations could a person on Earth see a half-moon?
   F A and B
   G A and C
   H B and C
   J B and D
3. In which situation could a solar eclipse be occurring?
   A A
   B B
   C C
   D D

L2

Using Vocabulary

1. rotation
2. solstice
3. waning
4. impact basin
5. ellipse

Checking Concepts

6. D
7. A
8. C
9. C
10. A
11. D
12. B
13. A
14. B
15. D

Thinking Critically

16. If the Moon were between the observer and the Sun, phases would be observed. The specific phases would depend on the relative positions of the observer, the Moon, and the Sun.
17. The seasons would probably be more extreme, with warmer summers and colder winters.
18. If distance from the Sun caused seasons, then each season would occur at the same time in both hemispheres.
19. If the Moon were twice as far from Earth, there would never be a total solar eclipse. Partial eclipses would occur but probably less frequently.
20. Possible answers: Scientists will be able to choose a colony location that is relatively flat and near a water supply.

chapter 2 Review

Using Vocabulary

axis p. 41	new moon p. 47
ellipse p. 43	revolution p. 43
equinox p. 45	rotation p. 41
full moon p. 47	solar eclipse p. 49
impact basin p. 57	solstice p. 44
lunar eclipse p. 50	sphere p. 40
maria p. 51	waning p. 47
moon phase p. 47	waxing p. 47

Fill in the blanks with the correct vocabulary word or words.

1. The spinning of Earth around its axis is called ________.
2. The ________ is the point at which the Sun reaches its greatest distance north or south of the equator.
3. The Moon is said to be ________when less and less of the side facing Earth is lighted.
4. The depression left behind by an object striking the Moon is called a(n) ________.
5. Earth's orbit is a(n) ________.

Checking Concepts

Choose the word or phrase that best answers the question.

6. How long does it take for the Moon to rotate once?
   A) 24 hours **C)** 27.3 hours
   B) 365 days **D)** 27.3 days
7. Where is Earth's circumference greatest?
   A) equator **C)** poles
   B) mantle **D)** axis
8. Earth is closest to the Sun during which season in the northern hemisphere?
   A) spring **C)** winter
   B) summer **D)** fall
9. What causes the Sun to appear to rise and set?
   A) Earth's revolution
   B) the Sun's revolution
   C) Earth's rotation
   D) the Sun's rotation

Use the photo below to answer question 10.

10. What phase of the Moon is shown in the photo above?
   A) waning crescent **C)** third quarter
   B) waxing gibbous **D)** waning gibbous
11. How long does it take for the Moon to revolve once around Earth?
   A) 24 hours **C)** 27.3 hours
   B) 365 days **D)** 27.3 days
12. What is it called when the phases of the Moon appear to get larger?
   A) waning **C)** rotating
   B) waxing **D)** revolving
13. What kind of eclipse occurs when the Moon blocks sunlight from reaching Earth?
   A) solar **C)** full
   B) new **D)** lunar
14. What is the darkest part of the shadow during an eclipse?
   A) waxing gibbous **C)** waning gibbous
   B) umbra **D)** penumbra
15. What is the name for a depression on the Moon caused by an object striking its surface?
   A) eclipse **C)** phase
   B) moonquake **D)** impact basin

Science Online bookj.msscience.com/vocabulary_puzzlemaker

Use the ExamView® Pro Testmaker CD-ROM to:

- create multiple versions of tests
- create modified tests with one mouse click for inclusion students
- edit existing questions and add your own questions
- build tests aligned with state standards using built-in State Curriculum Tags
- change English tests to Spanish with one mouse click and vice versa

64 ◆ J CHAPTER REVIEW

Thinking Critically

16. **Predict** how the Moon would appear to an observer in space during its revolution. Would phases be observable? Explain.

17. **Predict** what the effect would be on Earth's seasons if the axis were tilted at 28.5° instead of 23.5°.

18. **Infer** Seasons in the two hemispheres are opposite. Explain how this supports the statement that seasons are NOT caused by Earth's changing distance from the Sun.

19. **Draw Conclusions** How would solar eclipses be different if the Moon were twice as far from Earth? Explain.

20. **Predict** how the information gathered by moon missions could be helpful in the future for people wanting to establish a colony on the Moon.

21. **Use Variables, Constants, and Controls** Describe a simple activity to show how the Moon's rotation and revolution work to keep the same side facing Earth at all times.

22. **Draw Conclusions** Gravity is weaker on the Moon than it is on Earth. Why might more craters be present on the far side of the Moon than on the side of the Moon facing Earth?

23. **Recognize Cause and Effect** During a new phase of the Moon, we cannot see it because no sunlight reaches the side we see. Yet sometimes when there is a thin crescent visible, we do see a faint image of the rest of the Moon. Explain what might cause this to happen.

24. **Describe** Earth's magnetic field. Include an explanation of how scientists believe it is generated and two ways in which it helps people on Earth.

Performance Activities

25. **Display** Draw a cross section of the Moon. Include the crust, outer and inner mantles, and possible core based on the information in this chapter. Indicate the presence of impact craters and show how the thickness of the crust varies from one side of the Moon to the other.

26. **Poem** Write a poem in which you describe the various surface features of the Moon. Be sure to include information on how these features formed.

Applying Math

27. **Orbital Tilt** The Moon's orbit is tilted at an angle of 5° to Earth's orbit around the sun. Using a protractor, draw the Moon's orbit around Earth. What fraction of a full circle (360°) is 5°?

Use the illustration below to answer question 28.

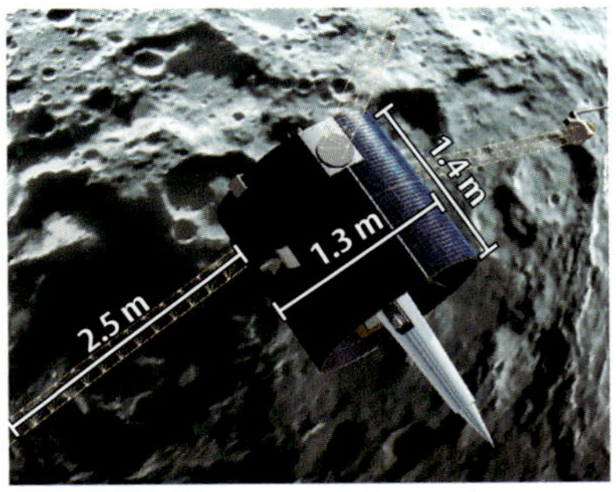

28. **Model to Scale** You are planning to make a scale model of the *Lunar Prospector* spacecraft, shown above. Assuming that the three instrument masts are of equal length, draw a labeled diagram of your model using a scale of 1 cm equals 30 cm.

29. **Spacecraft Velocity** The *Lunar Prospector* spacecraft shown above took 105 hours to reach the Moon. Assuming that the average distance from Earth to Moon is 384,000 km, calculate its average velocity on the trip.

21. Possible answer: Place an X on a basketball. As you walk around a classmate, turn the ball to keep the X facing him or her. As the ball revolves once, it rotates once.

22. Objects traveling between the Moon and Earth are more likely to be pulled toward Earth because of Earth's greater gravity.

23. The faint image of the Moon is illuminated by light reflected from Earth. Leonardo Da Vinci was the first to suggest this explanation.

24. Earth's magnetic field is produced by the movement of material within Earth's core as Earth rotates. It traps charged particles from the Sun that could endanger life and makes navigation possible using a compass.

Performance Activities

25. The model should be a hemisphere with dimensions roughly proportional to those shown in **Figure 11**. Model should show the crust thinner on one side.

26. Accept any reasonable effort. Make sure that the required information was included.

Applying Math

National Math Standards

1, 2, 3, 4, 9

27. It is 1/72 of 360 degrees.

28. Check diagram. Dimensions should be: body 4.3 cm × 4.6 cm; masts, 8.3 cm

29. Dividing 105 hours into 384,000 km gives a velocity of 3,657 km/hr.

Assessment Resources

Reproducible Masters

Chapter *Fast File* Resources
- Chapter Review, pp. 37–38
- Chapter Tests, pp. 39–42
- Assessment Transparency Activity, p. 49

Glencoe Science Web site
- Chapter Review Test
- Standardized Test Practice

Glencoe Technology
- Assessment Transparency
- Exam*View*® Pro Testmaker
- MindJogger Videoquiz
- Interactive Chalkboard

chapter 2 Standardized Test Practice

FAST FILE

Answer Sheet A practice answer sheet can be found at bookj.msscience.com/answer_sheet.

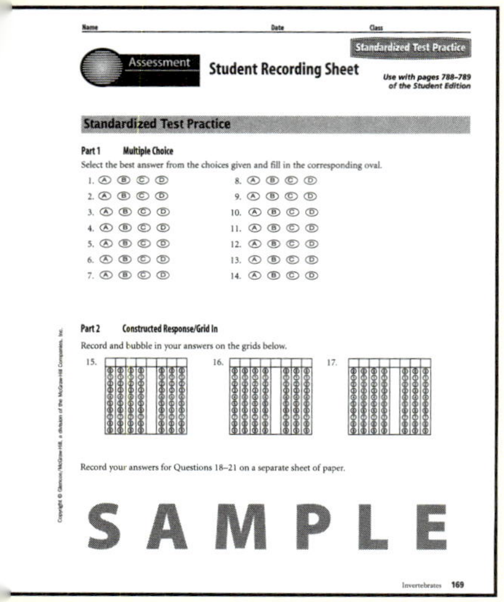
Assessment Student Recording Sheet

Standardized Test Practice

Use with pages 788–789 of the Student Edition

Standardized Test Practice

Part 1 Multiple Choice

Select the best answer from the choices given and fill in the corresponding oval.

Part 2 Constructed Response/Grid In

Record and bubble in your answers on the grids below.

Record your answers for Questions 18–21 on a separate sheet of paper.

SAMPLE

Part 1 Multiple Choice

1. C	**6.** C
2. D	**7.** D
3. A	**8.** D
4. D	**9.** B
5. C	

Part 2 Short Response

10. During summer in the northern hemisphere, the north pole is tilted toward the Sun. As Earth rotates, the north pole remains pointed towards the Sun, giving this area sunlight all 24 hours.

11. Earth's magnetic field traps charged particles that come from the Sun, helping to protect life on Earth from harmful solar radiation.

12. Solstices occur when the Sun reaches its greatest distance north or south of the equator, on June 21 or 22, and December 21 or 22. An equinox occurs when the Sun is directly above the equator. Equinoxes occur on March 20 or 21, and September 22 or 23.

chapter 2 Standardized Test Practice

Part 1 Multiple Choice

Record your answers on the answer sheet provided by your teacher or on a sheet of paper.

1. Which of the following terms would you use to describe the spinning of Earth on its axis?

A. revolution **C.** rotation
B. ellipse **D.** solstice

Use the illustration below to answer questions 2 and 3.

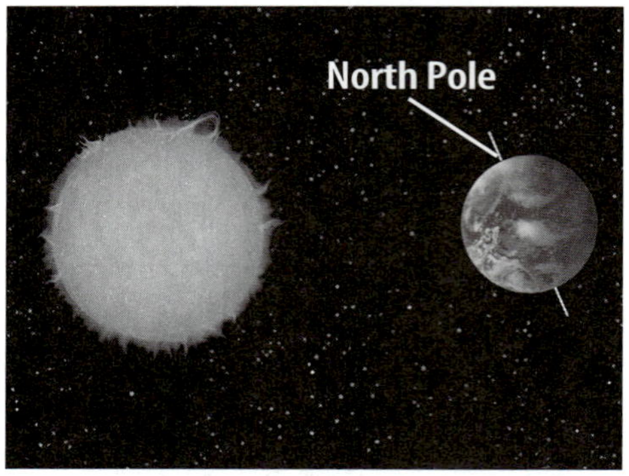

2. Which season is beginning for the southern hemisphere when Earth is in this position?

A. spring **C.** fall
B. summer **D.** winter

3. Which part of Earth receives the greatest total amount of solar radiation when Earth is in this position?

A. northern hemisphere
B. South Pole
C. southern hemisphere
D. equator

4. Which term describes the dark, flat areas on the Moon's surface which are made of cooled, hardened lava?

A. spheres **C.** highlands
B. moonquakes **D.** maria

Use the illustration below to answer questions 5 and 6.

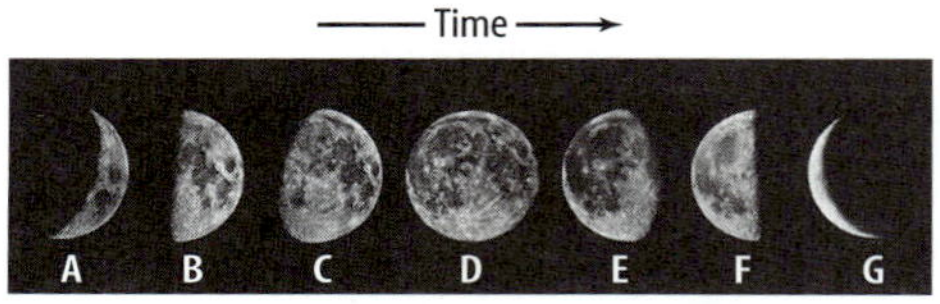

5. Which letter corresponds to the moon phase waning gibbous?

A. G **C.** E
B. C **D.** A

6. The Moon phase cycle lasts about 29.5 days. Given this information, about how long does it take the Moon to wax from new moon to full moon?

A. about 3 days **C.** about 2 weeks
B. about 1 week **D.** about 4 weeks

7. Where have large amounts of water been detected on the Moon?

A. highlands **C.** maria
B. lunar equator **D.** lunar poles

8. In what month is Earth closest to the Sun?

A. March **C.** July
B. September **D.** January

9. So far, where on the Moon have the youngest rocks been found?

A. lunar highlands **C.** lunar poles
B. maria **D.** lunar equator

Test-Taking Tip

Eliminate Choices If you don't know the answer to a multiple-choice question, eliminate as many incorrect choices as possible. Mark your best guess from the remaining answers before moving to the next question.

Question 5 Eliminate those phases that you know are not gibbous.

Part 2 Short Response/Grid In

13. Because the dark areas called maria formed after the Moon had cooled forming a crust, thus molten material was under this crust.

14. A solar eclipse.

15. The sky would darken gradually as the Moon covered the Sun.

16. The hemisphere tilted towards the Sun receives a greater amount of total radiation because there are longer hours of daylight. Also, the radiation is more effective because it strikes Earth at a higher angle.

17. Because we still needed to know much about the entire Moon, not just the equatorial regions.

18. An impact basin is a depression when an object struck the Moon. The largest is the Moon's Aitken Basin.

Part 2 Short Response/Grid In

Record your answers on the answer sheet provided by your teacher or on a sheet of paper.

10. Explain why the North Pole is always in sunlight during summer in the northern hemisphere.

11. Describe one positive effect of Earth's magnetic field.

12. Explain the difference between a solstice and an equinox. Give the dates of these events on Earth.

13. Explain how scientists know that the Moon was once geologically active.

Use the illustration below to answer questions 14 and 15.

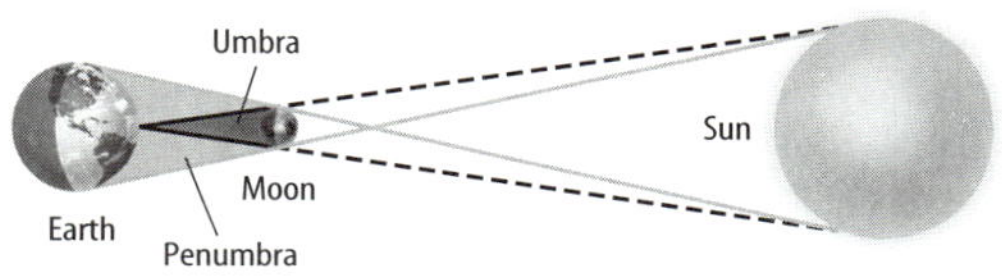

14. What type of eclipse is shown above?

15. Describe what a person standing in the Moon's umbra would see if he or she looked at the sky wearing protective eyewear.

16. The tilt of Earth on its axis causes seasons. Give two reasons why this tilt causes summer to be warmer than winter.

17. When the Apollo missions ended in 1972, 12 astronauts had visited the Moon and brought back samples of moon soil and rock. Explain why we continue to send orbiting spacecraft to study the Moon.

18. Define the term *impact basin*, and name the largest one known in the solar system.

Part 3 Open Ended

Record your answers on a sheet of paper.

Use the illustrations below to answer questions 19 and 20.

19. As a ship comes into view over the horizon, the top appears before the rest of the ship. How does this demonstrate that Earth is spherical?

20. If Earth were flat, how would an approaching ship appear differently?

21. Explain why eclipses of the Sun occur only occasionally despite the fact that the Moon's rotation causes it to pass between Earth and the Sun every month.

22. Recent data from the spacecraft *Lunar Prospector* indicate the presence of large quantities of water in shadowed areas of lunar impact basins. Describe the hypothesis that scientists have developed to explain how this water reached the moon and how it might be preserved.

23. Compare the impact theory of lunar formation with one of the older theories proposed before the *Apollo* mission.

24. Describe how scientists study the interior of the Moon and what they have learned so far.

25. Explain why Earth's magnetic north poles must be mapped and why these maps must be kept up-to-date.

Part 3 Open Ended

19. Seeing the top of an approaching ship first suggests that the bottom of the ship is just out of sight, beyond a curved surface over which the ship is traveling.

20. If Earth were flat, all parts of an approaching ship, from top to bottom, could be seen. The ship would first appear very small and gradually growing larger.

21. Eclipses do not occur every month because the alignment of the Sun, Moon, and Earth must be perfect.

22. Scientists hypothesize that comets deposited water early in the Moon's history and that it remains frozen because temperatures remain extremely low where sunlight cannot reach.

23. The modern hypothesis states that the Moon formed when a large object impacted Earth ejecting material that eventually condensed to form the Moon. Older theories said that Earth and Moon condensed from the same cloud of dust and gas, that the Moon was captured in Earth's gravity, and that a rapidly spinning Earth ejected material that became the Moon.

24. One way that scientists study the interior of the Moon is by using seismographs. They have learned that under the Moon's crust lies a solid mantle, a partly molten mantle, and possibly an iron-rich core.

25. Because compasses point to magnetic north and the magnetic poles move irregularly.

Rubrics

For more help evaluating open-ended assessment questions, see the rubric on p. 10T.

chapter 3 Organizer

Section/Objectives	Standards		Labs/Features
	National	State/Local	
Chapter Opener	See pp. 9T–10T for a Key to Standards.		**Launch Lab:** Model Crater Formation, p. 69 **Foldables,** p. 69
Section 1 The Solar System 2 sessions 1 block 1. **Compare** the Earth-centered and Sun-centered models of the solar system. 2. **Explain** that gravity holds the planets in their orbits around the Sun.	National Content Standards: UCP.1–3, UCP.5, A.1, A.2, B.2, G.3		**Science Online,** p. 71 **Integrate Physics,** p. 72 **Visualizing the Solar System's Formation,** p. 73 **Lab:** Planetary Orbits, p. 75
Section 2 The Inner Planets 2 sessions 1 block 3. **List** the inner planets in order from the Sun. 4. **Describe** each inner planet. 5. **Compare and contrast** Venus and Earth.	National Content Standards: UCP.1–3, UCP.5, A.1, A.2, D.3		**MiniLAB:** Inferring Effects of Gravity, p. 79 **Science Online,** p. 80 **Applying Math:** Diameter of Mars, p. 80
Section 3 The Outer Planets 2 sessions 1 block 6. **Describe** the characteristics of Jupiter, Saturn, Uranus, and Neptune. 7. **Explain** how Pluto differs from the other outer planets.	National Content Standards: UCP.1–3, UCP.5, A.1, A.2		**MiniLAB:** Modeling Planets, p. 84 **Integrate Language Arts,** p. 86
Section 4 Other Objects in the Solar System 3 sessions 1.5 blocks 8. **Describe** comets and how comets change when they approach the Sun. 9. **Distinguish** among comets, meteoroids, and asteroids. 10. **Explain** that objects from space sometimes impact Earth.	National Content Standards: UCP.1–3, UCP.5, A.1, A.2, G.3		**Lab:** Solar System Distance Model, p. 94 **Oops! Accidents in Science:** It Came from Outer Space, p. 96

Glencoe Exclusive! TeacherWorks™ All-In-One Planner and Resource Center

Lab Materials	Reproducible Resources	Section Assessment	Technology
Launch Lab: white flour, metal cake pan, colored powdered drink mix, marbles, bolts, nuts, metric ruler	**Chapter *FAST FILE* Resources** Foldables Worksheet, p. 17 Directed Reading Overview, p. 19 Note-taking Worksheets, pp. 35–38	GLENCOE'S ASSESSMENT ADVANTAGE	TeacherWorks includes: • Interactive Teacher Edition • Lesson Planner with calendar • Access to all program blacklines • Correlations to standards • Web links Video Lab
Lab: 2 thumbtacks or pins, cardboard, paper, metric ruler, string, pencil	**Chapter *FAST FILE* Resources** Transparency Activity, p. 48 Lab Worksheets, pp. 5–6 Enrichment, p. 31 Reinforcement, p. 27 Directed Reading, p. 20 **Cultural Diversity,** p. 29	**Portfolio** Science Journal, p. 72 **Performance** Applying Skills, p. 74 **Content** Section Review, p. 74	Section Focus Transparency Virtual Labs CD-ROM Guided Reading Audio Program Interactive Chalkboard CD-ROM
MiniLAB: calculator *Need materials?* Contact Science Kit at 1-800-828-7777 or www.sciencekit.com on the Internet.	**Chapter *FAST FILE* Resources** Transparency Activity, p. 49 MiniLAB, p. 3 Enrichment, p. 32 Reinforcement, p. 28 Directed Reading, p. 20 **Mathematics Skill Activities,** p. 1	**Portfolio** **Performance** MiniLAB, p. 79 Applying Math, p. 80 Applying Math, p. 81 **Content** Section Review, p. 81	Section Focus Transparency Virtual Labs CD-ROM Guided Reading Audio Program Interactive Chalkboard CD-ROM
MiniLAB: metric ruler, drawing compass, pencil, paper	**Chapter *FAST FILE* Resources** Transparency Activity, p. 50 MiniLAB, p. 4 Enrichment, p. 33 Reinforcement, p. 29 Directed Reading, p. 21	**Portfolio** Science Journal, p. 86 **Performance** MiniLAB, p. 84 Applying Skills, p. 87 **Content** Section Review, p. 87	Section Focus Transparency Virtual Labs CD-ROM Guided Reading Audio Program Interactive Chalkboard CD-ROM
Lab: meterstick, scissors, pencil, string, several sheets of notebook paper	**Chapter *FAST FILE* Resources** Transparency Activity, p. 51 Lab Worksheets, pp. 7–8 Transparency Activity, pp. 53–54 Enrichment, p. 34 Reinforcement, p. 30 Directed Reading, pp. 21, 22 **Lab Management and Safety,** p. 69	**Portfolio** Differentiated Instruction, p. 92 Assessment, p. 93 **Performance** Applying Math, p. 93 **Content** Section Review, p. 93	Section Focus Transparency Teaching Transparency Virtual Labs CD-ROM Guided Reading Audio Program Interactive Chalkboard CD-ROM

End of Chapter Assessment

GLENCOE'S ASSESSMENT ADVANTAGE

Blackline Masters	Technology	Professional Series
Chapter *FAST FILE* Resources Chapter Review, pp. 41–42 Chapter Tests, pp. 43–46 **Standardized Test Practice,** pp. 15–18	MindJogger Videoquiz Virtual Labs CD-ROM Exam*View*® Pro Testmaker TeacherWorks CD-ROM Interactive Chalkboard CD-ROM	**Performance Assessment in the Science Classroom (PASC)**

Transparencies

Section Focus

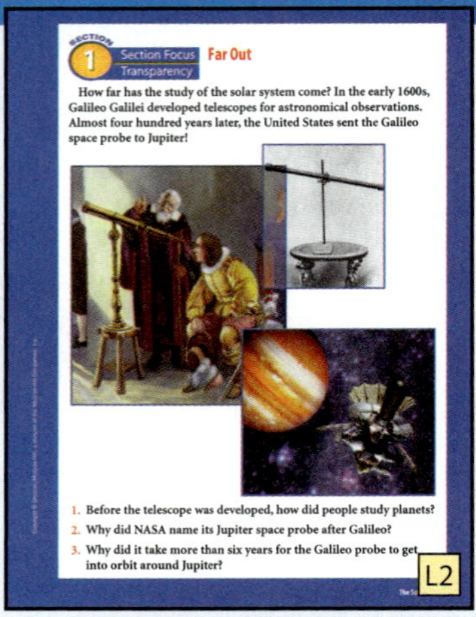

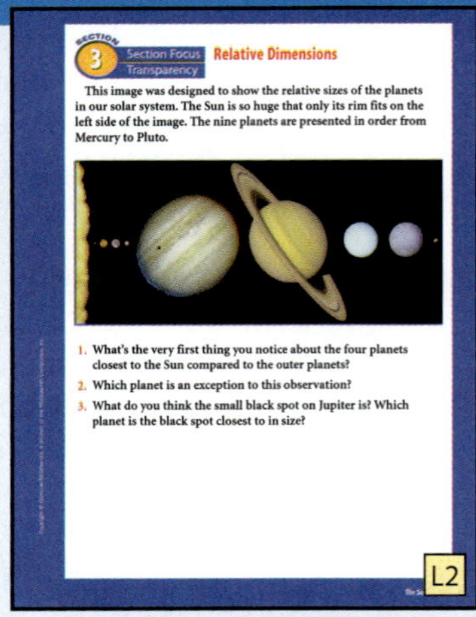

This is a representation of key blackline masters available in the Teacher Classroom Resources. See Resource Manager boxes within the chapter for additional information.

Assessment

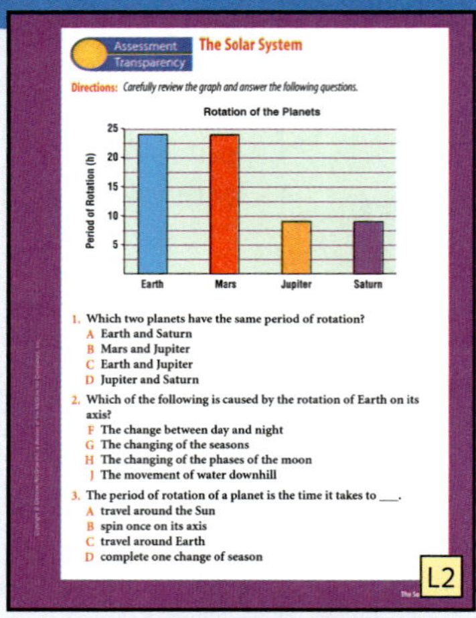

Teaching

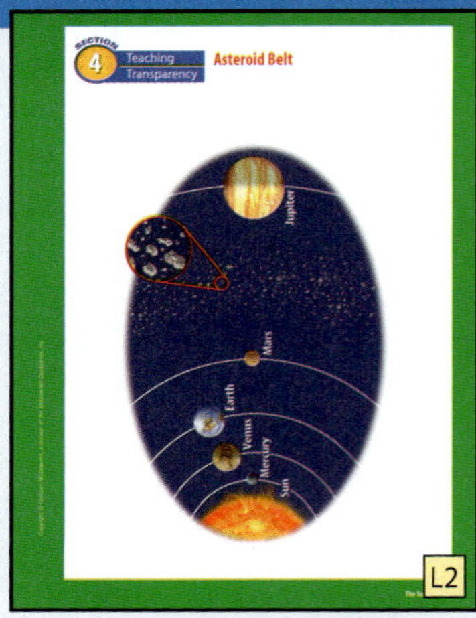

Key to Teaching Strategies

The following designations will help you decide which activities are appropriate for your students.

- **L1** Level 1 activities should be appropriate for students with learning difficulties.
- **L2** Level 2 activities should be within the ability range of all students.
- **L3** Level 3 activities are designed for above-average students.
- **ELL** ELL activities should be within the ability range of English Language Learners.
- **COOP LEARN** Cooperative Learning activities are designed for small group work.
- **LS** Multiple Learning Styles logos, as described on page 6T, are used throughout to indicate strategies that address different learning styles.
- **P** These strategies represent student products that can be placed into a best-work portfolio.
- **PBL** Problem-Based Learning activities apply real-world situations to learning.

Hands-on Activities

Student Text Lab Worksheet

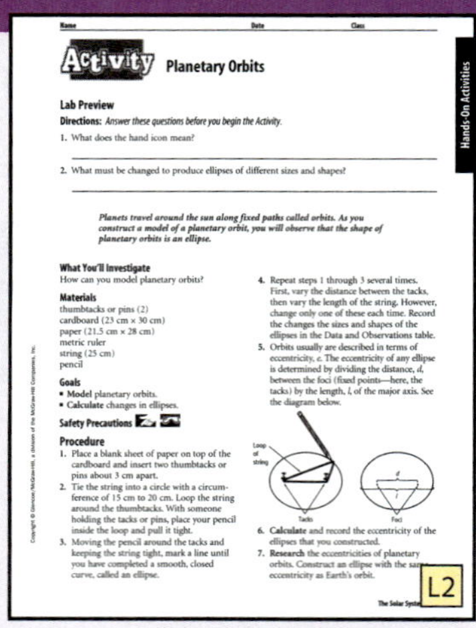

Laboratory Activities

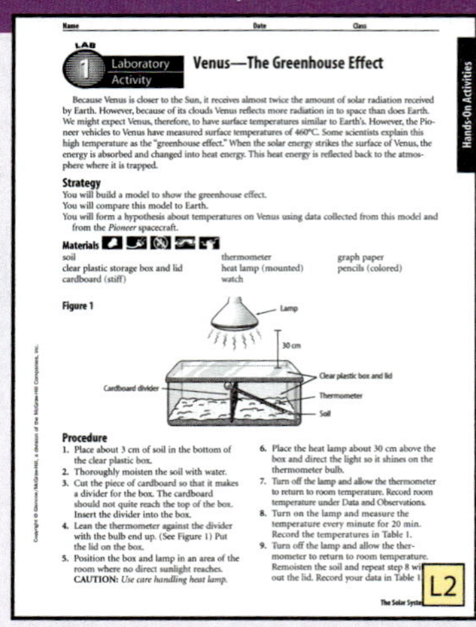

Meeting Different Ability Levels

Content Outline

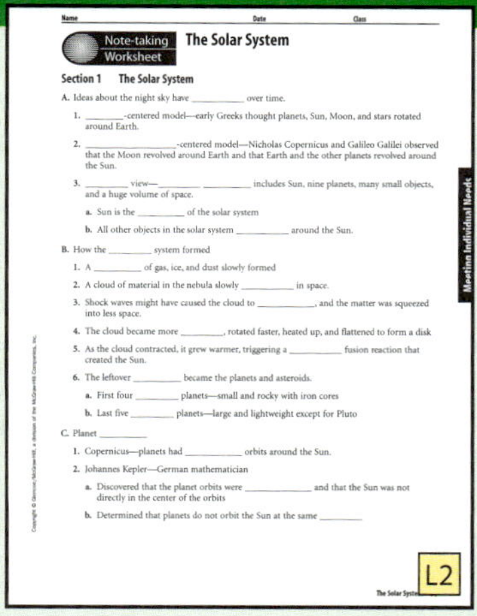

Reinforcement

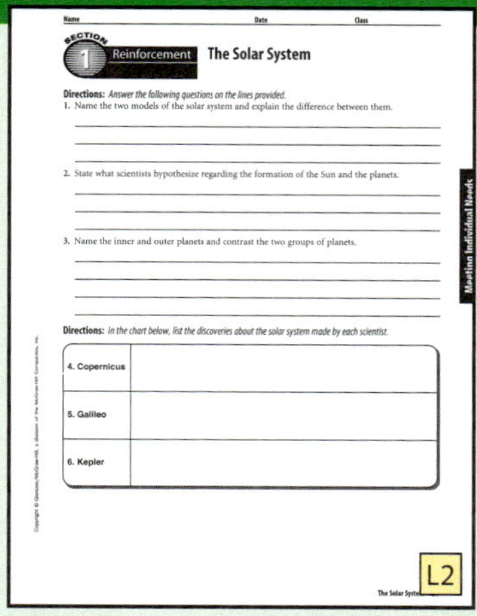

Enrichment

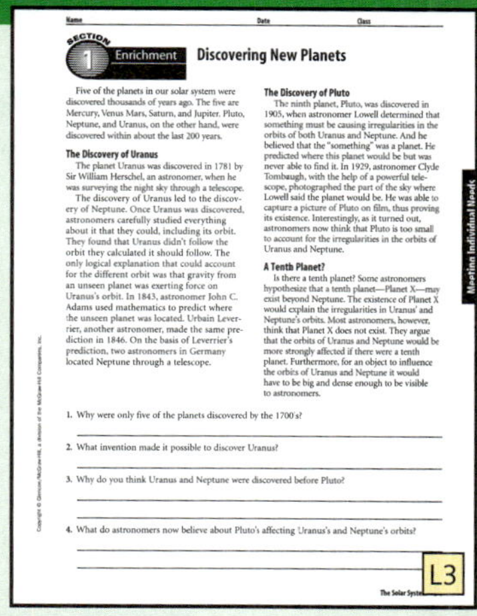

Directed Reading (English/Spanish)

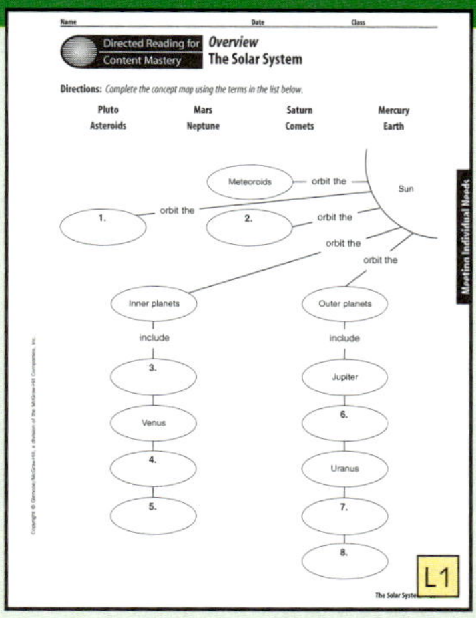

Study Guide

Reading Essentials

Assessment

Test Practice Workbook

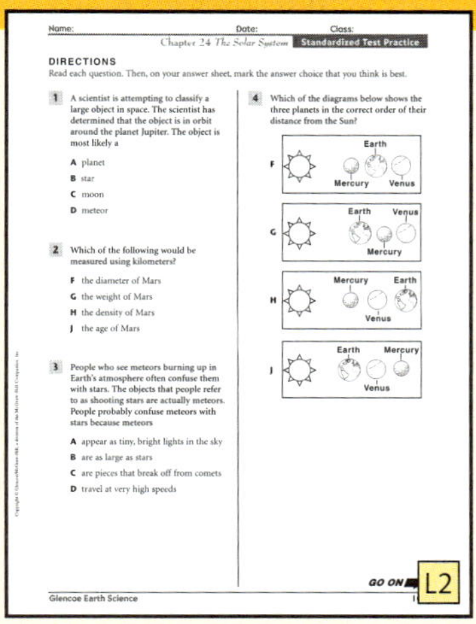

Chapter Review

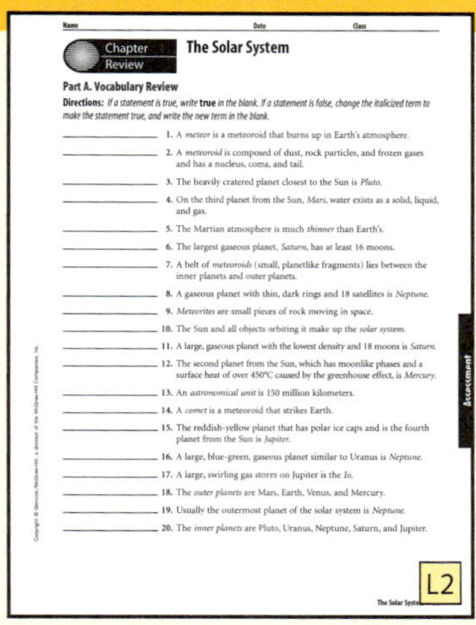

Chapter Tests

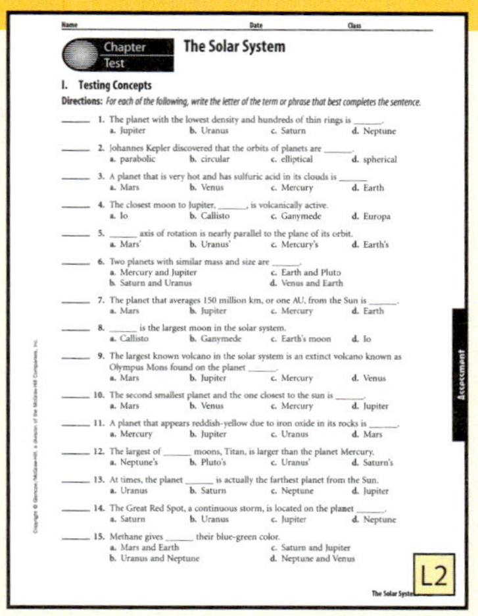

Science Content Background

section 1 The Solar System

Sun-Centered Model

Planets in the solar system usually appear to move across the background of stars toward the east. However, at certain times, planets appear to move toward the west. This is called retrograde motion. Copernicus cited the retrograde motion of Mars as one piece of evidence supporting his Sun-centered model of the solar system.

section 2 The Inner Planets

Inner Planets

Radar observations indicate that Mercury may have polar ice caps that exist inside craters where the Sun's light does not reach. The Caloris Basin, the result of an asteroid impact early in the formation of the solar system, is 1,300 km in diameter.

Venus is typically the brightest planet in the night sky, because its dense, cloud-covered atmosphere reflects most of the Sun's light back into space. Some of the Sun's energy, however, is re-radiated and then absorbed by the dense atmosphere. As a result, Venus has an average surface temperature of 472°C, hot enough to melt lead!

Mars today is a dry planet. However, dried-up riverbeds and deep channels indicate that Mars once had surface water. There is no confirmed evidence that the planet does, or ever has, supported life.

section 3 The Outer Planets

Outer Planets

The atmospheres of Jupiter and Saturn contain three cloud layers. The upper layer of clouds is composed of ammonia ice, the second layer is made of ammonium-hydrogen sulfide ice, and the lowest layer is mainly frozen water. Uranus does not produce as much heat as Jupiter and Saturn. However, some heat does exist which might be related to the cause of Uranus's magnetic field.

Hypotheses suggest that Pluto may not have formed in the orbit it now occupies or that Pluto and its moon, Charon, are large cometary members of the Kuiper Belt. Pluto and Neptune's moon Triton are the most similar objects in the solar system. The retrograde revolution of Triton may indicate that it was captured by the gravity of Neptune and once was a planet of the Sun, like Pluto.

section 4 Other Objects in the Solar System

Comets

Many comets orbit the Sun in the Oort Cloud some 50,000 AU from the Sun. Some comets take more than a million years to complete one orbit.

chapter content resources

Internet Resources

For additional content background, visit bookj.msscience.com to:

- access your book online
- find references to related articles in popular science magazines
- access Web links with related content background
- access current events with science journal topics

Print Resources

The Solar System by Michael A. Seeds, Brooks/Cole Publishing, 2003

Planetary Geoscience by A.D. Fortes and C. Vita-Finzi, Blackwell Publishing Co., 2004

Explorations: Stars, Galaxies and Planets, by Tom Arny, McGraw-Hill Higher Education, 2004

IDENTIFYING Misconceptions

Find Out What Students Think

Students may think that . . .

The Sun is not a star.

Students have learned about stars and about the Sun. But some students do not make the connection that the Sun is the closest star to Earth. The Sun is a main sequence star that is the primary source of energy for our solar system.

Demonstration

The Sun can be observed by using an index card and a thin, stiff piece of white cardboard. **WARNING:** *Warn students not to look directly at the Sun. Permanent eye damage can result.*

- Place the cardboard, which acts as a viewing screen, on the ground. Prop up one side of the screen so that it is perpendicular to incoming sunlight.
- Next, use a pin to poke a small hole in the index card. Hold the index card about 60 cm above the screen with the hole pointing toward the Sun. The Sun's image should appear on the screen.

After students have viewed the Sun's image, discuss what they have seen. Show pictures of the Sun, and emphasize that this body provides heat and light for our solar system.

Promote Understanding

Discussion

Have the class brainstorm a list of what they know about the Sun, including what they observed using the pin-hole viewer. Record students' ideas on the board. Next to this list, draw a diagram similar to that shown here. Then have students identify features of the star shown in the diagram and compare them to the features of the Sun. For example, they could state that stars have a core and the Sun has a core, stars have a photosphere and the Sun has a photosphere, and so on. After the comparison, have students draw conclusions as to whether the Sun is a star.

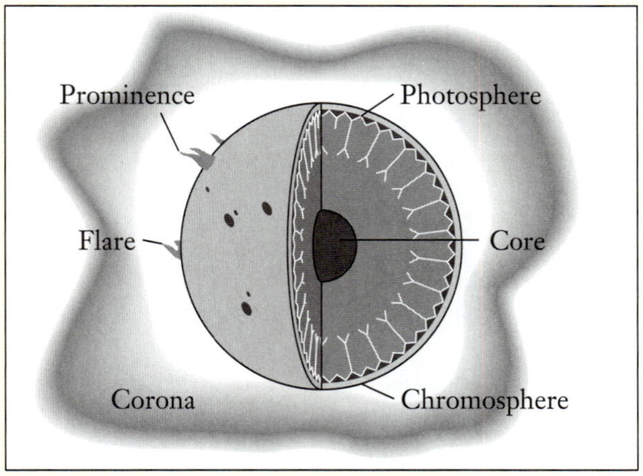

Assess

After completing the chapter, see *Identifying Misconceptions* in the Study Guide at the end of the chapter.

chapter 3

Chapter Vocabulary

solar system, p. 71
Mercury, p. 76
Venus, p. 77
Earth, p. 78
Mars, p. 78
Jupiter, p. 82
Great Red Spot, p. 82
Saturn, p. 84
Uranus, p. 85
Neptune, p. 86
Pluto, p. 87
comet, p. 90
meteor, p. 91
meteorite, p. 92
asteroid, p. 92

Science Journal Student descriptions will reveal prior knowledge and possible misconceptions about the solar system.

INTERACTIVE CHALKBOARD with Image Bank

PowerPoint® Presentations

This CD-ROM is an editable Microsoft® PowerPoint® presentation that includes:

- a pre-made presentation for every chapter
- interactive graphics
- animations
- audio clips
- image bank
- all new section and chapter questions
- Standardized Test Practice
- transparencies
- pre-lab questions for all labs
- Foldables directions
- links to bookj.msscience.com

chapter 3

The Solar System

chapter preview

sections

1 The Solar System
Lab Planetary Orbits
2 The Inner Planets
3 The Outer Planets
4 Other Objects in the Solar System
Lab Solar System Distance Model
Virtual Lab What are the dimensions of the solar system?

How is space explored?

You've seen the Sun and the Moon. You also might have observed some of the planets. But to get a really good look at the solar system from Earth, telescopes are needed. The optical telescope at the Keck Observatory in Hawaii allows scientists a close-up view.

Science Journal If you could command the Keck telescope, what would you view? Describe what you would see.

Theme Connection

Scale and Structure The solar system includes many objects of different sizes and structures. This theme is emphasized in the contrast between inner planets, which are small and rocky, and outer planets, most of which are large and gaseous.

About the Photo

Keck Observatory The photo shows the Keck Observatory on top of Mauna Kea in Hawaii. The observatory houses both optical and infrared telescopes.

Start-Up Activities

Launch LAB

Model Crater Formation

Some objects in the solar system have many craters. The Moon is covered with them. The planet Mercury also has a cratered landscape. Even Earth has some craters. All of these craters formed when rocks from space hit the surface of the planet or moon. In this lab, you'll explore crater formation.

1. Place white flour into a metal cake pan to a depth of 3 cm.
2. Cover the flour with 1 cm of colored powdered drink mix or different colors of gelatin powder.
3. From different heights, ranging from 10 cm to 25 cm, drop various-sized marbles into the pan.
4. **Think Critically** Make drawings in your Science Journal that show what happened to the surface of the powder when marbles were dropped from different heights.

Science Online — Preview this chapter's content and activities at bookj.msscience.com

FOLDABLES Study Organizer

The Solar System Make the following Foldable to help you identify what you already know, what you want to know, and what you learned about the solar system.

STEP 1 **Fold** a vertical sheet of paper from side to side. Make the front edge about 1.25 cm shorter than the back edge.

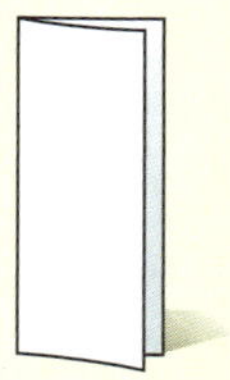

STEP 2 **Turn** lengthwise and **fold** into thirds.

STEP 3 **Unfold and cut** only the top layer along both folds to make three tabs.

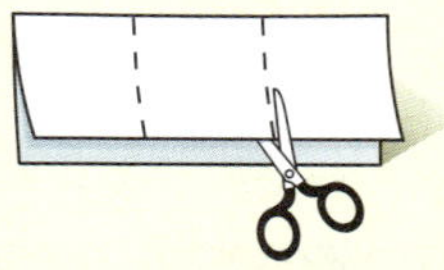

STEP 4 **Label** each tab.

Identify Questions Before you read the chapter, write what you already know about the solar system under the left tab of your Foldable. Write questions about what you'd like to know under the center tab. After you read the chapter, list what you learned under the right tab.

Launch LAB

Purpose Use the Launch Lab to introduce students to the effects of impacts on planetary surfaces. L2 ELL COOP LEARN LS **Visual-Spatial**

Materials metal cake pan, white flour, colored powdered drink mix or gelatin powder, marbles, nuts, bolts, metric ruler

Teaching Strategies Have students work in pairs or, if materials are in short supply, in groups of four.

Safety Precautions Students should wear safety goggles while dropping marbles. Students with wheat allergies should not participate in this activity.

Troubleshooting Have students place the pans on large sheets of newspaper for easier clean-up.

Think Critically

A crater forms as the object impacts the powder. White flour splashes outward over the dark powder surface.

Assessment

Performance Ask students to determine which craters were caused by the biggest marbles. For craters that overlap, have them determine which ones were formed first. Use **Performance Assessment in the Science Classroom**, p. 89.

FOLDABLES Study Organizer — Dinah Zike Study Fold

Student preparation materials for this Foldable are available in the Chapter *Fast File* Resources.

1 Motivate

Bellringer

Section Focus Transparencies also are available on the Interactive Chalkboard CD-ROM. L2 ELL

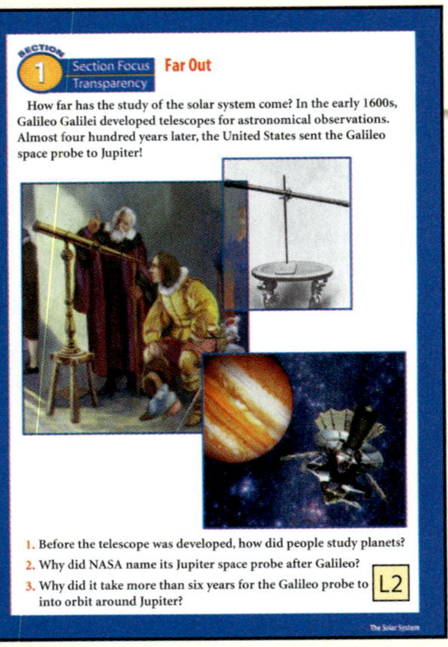

Section 1 Section Focus Transparency **Far Out**

How far has the study of the solar system come? In the early 1600s, Galileo Galilei developed telescopes for astronomical observations. Almost four hundred years later, the United States sent the Galileo space probe to Jupiter!

1. Before the telescope was developed, how did people study planets?
2. Why did NASA name its Jupiter space probe after Galileo?
3. Why did it take more than six years for the Galileo probe to into orbit around Jupiter? L2

The Solar System

Tie to Prior Knowledge

Apparent Movement Have students think about the apparent movement of the Sun in the sky each day. They should remember that it appears to rise in the east, move across the sky, and set in the west. Discuss with students why this apparent movement of the Sun once led people to think that Earth was the center of the universe. It seemed as though everything moved around Earth.

section 1 The Solar System

as you read

What You'll Learn

- **Compare** models of the solar system.
- **Explain** that gravity holds planets in orbits around the Sun.

Why It's Important

New technology has come from exploring the solar system.

Review Vocabulary

system: a portion of the universe and all of its components, processes, and interactions

New Vocabulary

- solar system

Ideas About the Solar System

People have been looking at the night sky for thousands of years. Early observers noted the changing positions of the planets and developed ideas about the solar system based on their observations and beliefs. Today, people know that objects in the solar system orbit the Sun. People also know that the Sun's gravity holds the solar system together, just as Earth's gravity holds the Moon in its orbit around Earth. This wasn't always accepted as fact.

Earth-Centered Model Many early Greek scientists thought the planets, the Sun, and the Moon were fixed in separate spheres that rotated around Earth. The stars were thought to be in another sphere that also rotated around Earth. This is called the Earth-centered model of the solar system. It included Earth, the Moon, the Sun, five planets—Mercury, Venus, Mars, Jupiter, and Saturn—and the sphere of stars.

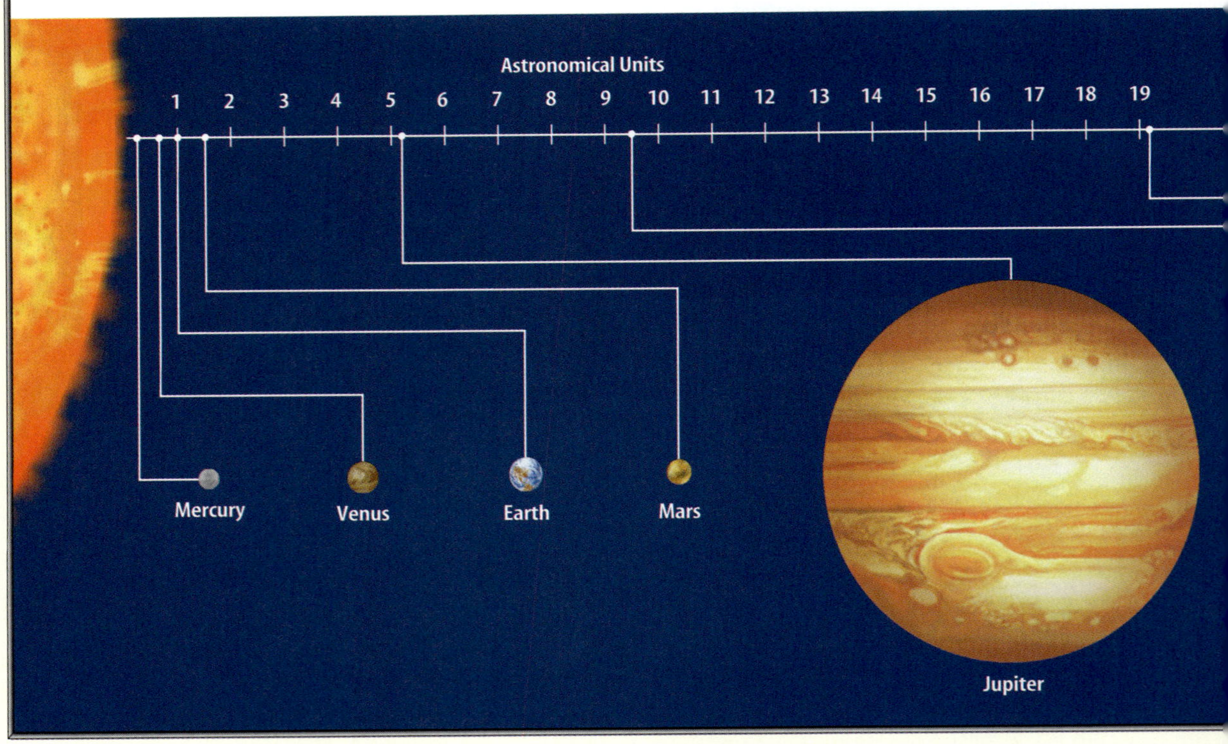

Section 1 Resource Manager

Chapter FAST FILE Resources

- Transparency Activity, p. 48
- Directed Reading for Content Mastery, pp. 19, 20
- Note-taking Worksheets, pp. 35–38
- Enrichment, p. 31
- Reinforcement, p. 27
- Lab Worksheets, pp. 5–6

Home and Community Involvement, p. 48

Reading and Writing Skill Activities, p. 47

Cultural Diversity, p. 29

Sun-Centered Model People believed the idea of an Earth-centered solar system for centuries. Then in 1543, Nicholas Copernicus published a different view. Copernicus stated that the Moon revolved around Earth and that Earth and the other planets revolved around the Sun. He also stated that the daily movement of the planets and the stars was caused by Earth's rotation. This is the Sun-centered model of the solar system.

Using his telescope, Galileo Galilei observed that Venus went through a full cycle of phases like the Moon's. He also observed that the apparent diameter of Venus was smallest when the phase was near full. This only could be explained if Venus were orbiting the Sun. Galileo concluded that the Sun is the center of the solar system.

Modern View of the Solar System We now know that the **solar system** is made up of nine planets, including Earth, and many smaller objects that orbit the Sun. The nine planets and the Sun are shown in **Figure 1.** Notice how small Earth is compared with some of the other planets and the Sun.

The solar system includes a huge volume of space that stretches in all directions from the Sun. Because the Sun contains 99.86 percent of the mass of the solar system, its gravity is immense. The Sun's gravity holds the planets and other objects in the solar system in their orbits.

Science Online

Topic: Solar System
Visit bookj.msscience.com for Web links to information about the solar system.

Activity Make a list of objects in the solar system. Write a one-sentence description of each object on your list.

Figure 1 Each of the nine planets in the solar system is unique. The distances between the planets and the Sun are shown on the scale. One astronomical unit (AU) is the average distance between Earth and the Sun.

2 Teach

Make a Model

Solar Systems Have students use common objects to make three-dimensional models of Sun-centered solar systems. Display the models in class. L2 COOP LEARN

LS **Visual-Spatial**

Discussion

Number of Planets Remind students that we now know the solar system has nine planets. Is it possible that there could be other planets beyond Pluto? Accept reasonable answers. Some may say no, that telescopes would have spotted other planets by now. Others may say that it is hard to detect objects at the outer reaches of the solar system, making the existence of undiscovered bodies possible. Explain that astronomers have studied the movements of Pluto to determine whether the gravity of an object farther out is affecting the tiny planet.

Visual Learning

Figure 1 Why couldn't the same scale be used to show all the planets and relative distances between them? If the same scale were used, the sizes of the planets would be too small to recognize or the overall size of the art would extend well beyond the available space.

Teacher FYI

Hybrid Model The Earth-centered model is known as the geocentric model, and the Sun-centered model is known as the heliocentric model. Tycho Brahe (1546–1601) proposed a combination of the two called the hybrid model. In the hybrid model, the Sun and Moon went around Earth and everything else went around the Sun.

INTEGRATE Physics

Rotational Motion The principle described on the student page is the conservation of angular momentum. This principle states that the total angular momentum of a system is constant as long as no external torques are applied. As the nebula contracts, the rate of rotation must increase to maintain constant angular momentum.

Fun Fact

Because of the conservation of angular momentum, a figure skater can spin as fast as 300 rpm (rotations per minute).

Reading Check

Answer the process during which atoms with lower mass fuse to form a higher mass element and release energy

Use Science Words

Word Origin Have students find the origin of the word *solar* (from the Latin *sol* meaning "Sun"). Have them list and define terms other than "solar system" that use solar. Possible answers: solar cell (device that converts sunlight into electricity), solar battery (system connecting many solar cells), solar flare (eruption of gases from the Sun's surface)

INTEGRATE Physics

Rotational Motion You might have noticed that when a twirling ice skater pulls in her arms, she spins faster. The same thing occurs when a cloud of gas, ice, and dust in a nebula contracts. As mass moves toward the center of the cloud, the cloud rotates faster.

How the Solar System Formed

Scientists hypothesize that the solar system formed from part of a nebula of gas, ice, and dust, like the one shown in **Figure 2.** Follow the steps shown in **Figures 3A** through **3D** to learn how this might have happened. A nearby star might have exploded or nearby O- or B-type stars formed, and the shock waves produced by these events could have caused the cloud to start contracting. As it contracted, the nebula likely fragmented into smaller and smaller pieces. The density in the cloud fragments became greater, and the attraction of gravity pulled more gas and dust toward several centers of contraction. This in turn caused them to flatten into disks with dense centers. As the cloud fragments continued to contract, they began to rotate faster and faster.

As each cloud fragment contracted, its temperature increased. Eventually, the temperature in the core of one of these cloud fragments reached about 10 million degrees Celsius and nuclear fusion began. A star was born—the beginning of the Sun.

It is unlikely that the Sun formed alone. A cluster of stars like the Sun, or smaller, likely formed from fragments of the original cloud. The Sun probably escaped from this cluster and has since revolved around the galaxy about 20 times.

Reading Check *What is nuclear fusion?*

Planet Formation Not all of the nearby gas, ice, and dust was drawn into the core of the cloud fragment. The matter that did not get pulled into the center collided and stuck together to form the planets and asteroids. Close to the Sun, the temperature was hot, and the easily vaporized elements could not condense into solids. This is why lighter elements are scarcer in the planets near the Sun than in planets farther out in the solar system.

Figure 2 Systems of planets, such as the solar system, form in areas of space like this, called a nebula.

The inner planets of the solar system—Mercury, Venus, Earth, and Mars—are small, rocky planets with iron cores. The outer planets are Jupiter, Saturn, Uranus, Neptune, and Pluto. Pluto, a small planet, is the only outer planet made mostly of rock and ice. The other outer planets are much larger and are made mostly of lighter substances such as hydrogen, helium, methane, and ammonia.

Science Journal

Solar System Models Have students write one-page reports in their Science Journals that describe what evidence led Copernicus to propose the Sun-centered model of the solar system. Some planets experience retrograde motion. Ask students to explain how this evidence refuted the Earth-centered model. Retrograde motion is difficult to explain with the Earth-centered model. L3 P

Differentiated Instruction

Learning Disabled Have students work in small groups to make mobiles of the solar system to display in class. Place learning disabled students in groups where they can contribute their ideas and knowledge. On the back side of each planet model, students should include facts such as temperature extremes, atmospheric conditions, average distance of orbit from the sun and number of moons. L1 COOP LEARN LS Visual-Spatial

NATIONAL GEOGRAPHIC

VISUALIZING THE SOLAR SYSTEM'S FORMATION

Figure 3

Through careful observations, astronomers have found clues that help explain how the solar system may have formed. A More than 4.6 billion years ago, the solar system was a cloud fragment of gas, ice, and dust. B Gradually, this cloud fragment contracted into a large, tightly packed, spinning disk. The disk's center was so hot and dense that nuclear fusion reactions began to occur, and the Sun was born. C Eventually, the rest of the material in the disk cooled enough to clump into scattered solids. D Finally, these clumps collided and combined to become the nine planets that make up the solar system today.

NATIONAL GEOGRAPHIC

Visualizing the Solar System's Formation

Have students examine the pictures and read the captions. Then ask the following questions.

How may gravity have affected the formation of the solar system? Gravity pulled the cloud of gas, ice, and dust together.

How would the solar system have been different if the disk's center had not been as hot and as dense as it was? Nuclear fusion reactions might not have occurred to form the Sun. Without light and heat energy, life would not have been possible anywhere in the solar system.

IDENTIFYING Misconceptions

The Closest Star Students may think the Sun is not a star. Refer to page F at the beginning of this chapter for teaching strategies that address this misconception.

Differentiated Instruction

Challenge Have students research other solar systems that have been discovered in recent years and how they differ from our solar system. Ask students to present oral reports on their findings. L3 LS **Linguistic**

SECTION 1 The Solar System J ◆ 73

3 Assess

DAILY INTERVENTION

Check for Understanding

Interpersonal Divide students into two groups for a class debate. Ask students to imagine that they live in the 16th century and know only what people at that time knew. Have one group support the Sun-centered model of the solar system and the other group support the Earth-centered model of the solar system. L2

Reteach

Solar System Cards Have students refer to **Figure 3** and make a series of cards, each of which lists one step in the formation of the solar system. Pass the set of cards around and have students take turns putting the steps in order. L2

Assessment

Process Ask students to use **Figure 1** to determine how Earth compares in size to the other planets. Four planets are larger and four are smaller than Earth. Use **Performance Assessment in the Science Classroom,** p. 99.

Table 1 Average Orbital Speed

Planet	Average Orbital Speed (km/s)
Mercury	48
Venus	35
Earth	30
Mars	24
Jupiter	13
Saturn	9.7
Uranus	6.8
Neptune	5.4
Pluto	4.7

Johannes Kepler

Motions of the Planets

INTEGRATE Physics

When Nicholas Copernicus developed his Sun-centered model of the solar system, he thought that the planets orbited the Sun in circles. In the early 1600s, German mathematician Johannes Kepler began studying the orbits of the planets. He discovered that the shapes of the orbits are not circular. They are oval shaped, or elliptical. His calculations further showed that the Sun is not at the center of the orbits but is slightly offset.

Kepler also discovered that the planets travel at different speeds in their orbits around the Sun, as shown in **Table 1.** You can see that the planets closer to the Sun travel faster than planets farther away from the Sun. Because of their slower speeds and the longer distances they must travel, the outer planets take much longer to orbit the Sun than the inner planets do.

Copernicus's ideas, considered radical at the time, led to the birth of modern astronomy. Early scientists didn't have technology such as space probes to learn about the planets. Nevertheless, they developed theories about the solar system that still are used today.

section 1 review

Summary

Ideas About the Solar System

- The planets in the solar system revolve around the Sun.
- The Sun's immense gravity holds the planets in their orbits.

How the Solar System Formed

- The solar system formed from a piece of a nebula of gas, ice, and dust.
- As the piece of nebula contracted, nuclear fusion began at its center and the Sun was born.

Motion of the Planets

- The planets' orbits are elliptical.
- Planets that are closer to the Sun revolve faster than those that are farther away from the Sun.

Self Check

1. **Describe** the Sun-centered model of the solar system. What holds the solar system together?
2. **Explain** how the planets in the solar system formed.
3. **Infer** why life is unlikely on the outer planets in spite of the presence of water, methane, and ammonia—materials needed for life to develop.
4. **List** two reasons why the outer planets take longer to orbit the Sun than the inner planets do.
5. **Think Critically** Would a year on the planet Neptune be longer or shorter than an earth year? Explain.

Applying Skills

6. **Concept Map** Make a concept map that compares and contrasts the Earth-centered model with the Sun-centered model of the solar system.

Science Online bookj.msscience.com/self_check_quiz

section 1 review

1. All objects in the solar system orbit the Sun; gravity
2. About 4.6 billion years ago, a large cloud fragment of gas, ice, and dust began to condense. The center of the fragment formed the Sun. The planets and other objects formed from the outer portions of the cloud fragment.
3. The outer planets are cold.
4. Outer planets travel at a slower speed and have larger orbits.
5. Longer; Neptune's orbit is much larger, and it travels more slowly.
6. Concept maps should include information about the central body of each model, and how the model explained the movement of objects in the night sky.

LAB

Planetary Orbits

Planets travel around the Sun along paths called orbits. As you construct a model of a planetary orbit, you will observe that the shape of planetary orbits is an ellipse.

Real-World Question

How can you model planetary orbits?

Goals

- **Model** planetary orbits.
- **Calculate** the eccentricity of ellipses.

Materials

thumbtacks or pins (2)	metric ruler
cardboard (23 cm × 30 cm)	string (25 cm)
paper (21.5 cm × 28 cm)	pencil

Safety Precautions

Procedure

1. Place a blank sheet of paper on top of the cardboard and insert two thumbtacks or pins about 3 cm apart.
2. Tie the string into a circle with a circumference of 15 cm to 20 cm. Loop the string around the thumbtacks. With someone holding the tacks or pins, place your pencil inside the loop and pull it tight.
3. Moving the pencil around the tacks and keeping the string tight, mark a line until you have completed a smooth, closed curve.
4. Repeat steps 1 through 3 several times. First, vary the distance between the tacks, then vary the length of the string. However, change only one of these each time. Make a data table to record the changes in the sizes and shapes of the ellipses.
5. Orbits usually are described in terms of eccentricity, *e*, which is determined by dividing the distance, *d*, between the foci (fixed points—here, the tacks) by the length, *l*, of the major axis. See the diagram below.

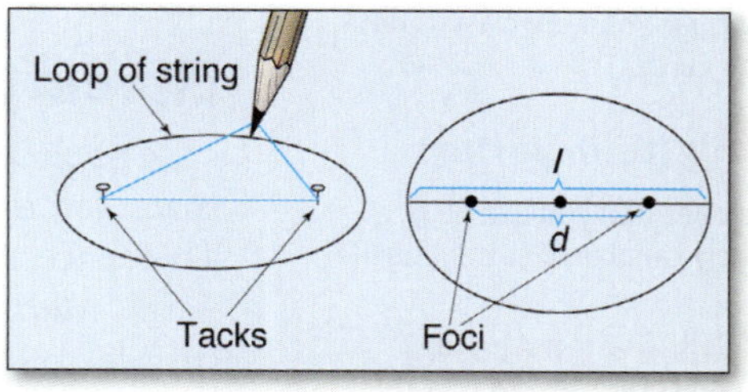

6. **Calculate** and record the eccentricity of the ellipses that you constructed.
7. **Research** the eccentricities of planetary orbits. Construct an ellipse with the same eccentricity as Earth's orbit.

Conclude and Apply

1. **Analyze** the effect that a change in the length of the string or the distance between the tacks has on the shape of the ellipse.
2. **Hypothesize** what must be done to the string or placement of tacks to decrease the eccentricity of a constructed ellipse.
3. **Describe** the shape of Earth's orbit. Where is the Sun located within the orbit?

Communicating Your Data

Compare your results with those of other students. **For more help, refer to the** Science Skill Handbook.

BENCH TESTED

Real-World Question

Purpose Students model planetary orbits and calculate the eccentricity of ellipses. L2 ELL
LS **Logical-Mathematical**

Process Skills observe, use numbers, compare and contrast, form a hypothesis, make and use tables, communicate

Time Required 40 minutes

Procedure

Teaching Strategy Provide students with several values for *d* and *l* and allow them to practice solving for *e* using the equation $e = d/l$.

Conclude and Apply

1. Increasing the length of the string or decreasing the distance between the tacks makes the shape more circular. Decreasing the string's length or increasing the distance between the foci makes the shape more elliptical.
2. Lengthen the string or move the tacks closer to each other.
3. Earth's orbit is an ellipse that has very low eccentricity. The Sun is located at one of the foci.

Assessment

Process Have students compare and contrast drawing ellipses and circles. A circle is drawn in the same way as an ellipse, but both foci are at the same point.

Sample data:

Ellipse	*d* (cm)	*l* (cm)	*e* (d/L)
# 1	3	15.6	0.19
# 2	5	13.5	0.37
# 3	5	8.7	0.57
Earth	0.48	28	0.017

Communicating Your Data

Students' results should be similar. Have them identify which ellipse produced the most circular orbit. The one with foci closest together.

1 Motivate

Bellringer

Section Focus Transparencies also are available on the Interactive Chalkboard CD-ROM. L2 ELL

Section 2 Section Focus Transparency *Really Far Out*

When you see images of Earth from space, it's fun to try to pick out the United States and the area where you live. But where does Earth live? This NASA image is of our galaxy, which is called the Milky Way. Our entire solar system is just a tiny part of the Milky Way.

1. How might other planets in the solar system be studied?
2. What characteristics of our environment nurture life on Earth?
3. Why might it be difficult to study Mercury, the planet closest to the Sun?

L2

The Solar System

Tie to Prior Knowledge

Planet Table Make four columns on the board, with the name of one of the four inner planets at the top of each column. Have students brainstorm things they know about each planet. Add each item to the list. Save the list and review it with the class after reading this section.

section 2

The Inner Planets

as you read

What You'll Learn

- **List** the inner planets in order from the Sun.
- **Describe** each inner planet.
- **Compare and contrast** Venus and Earth.

Why It's Important

The planet that you live on is uniquely capable of sustaining life.

Review Vocabulary
space probe: an instrument that is sent to space to gather information and send it back to Earth

New Vocabulary
- Mercury
- Venus
- Earth
- Mars

Inner Planets

Today, people know more about the solar system than ever before. Better telescopes allow astronomers to observe the planets from Earth and space. In addition, space probes have explored much of the solar system. Prepare to take a tour of the solar system through the eyes of some space probes.

Mercury

The closest planet to the Sun is **Mercury.** The first American spacecraft mission to Mercury was in 1974–1975 by *Mariner 10.* The spacecraft flew by the planet and sent pictures back to Earth. *Mariner 10* photographed only 45 percent of Mercury's surface, so scientists don't know what the other 55 percent looks like. What they do know is that the surface of Mercury has many craters and looks much like Earth's Moon. It also has cliffs as high as 3 km on its surface. These cliffs might have formed at a time when Mercury shrank in diameter, as seen in **Figure 4.**

Why would Mercury have shrunk? *Mariner 10* detected a weak magnetic field around Mercury. This indicates that the planet has an iron core. Some scientists hypothesize that Mercury's crust solidified while the iron core was still hot and molten. As the core started to solidify, it contracted. The cliffs resulted from breaks in the crust caused by this contraction.

Figure 4 Large cliffs on Mercury might have formed when the crust of the planet broke as the planet contracted.

Mercury has many craters.

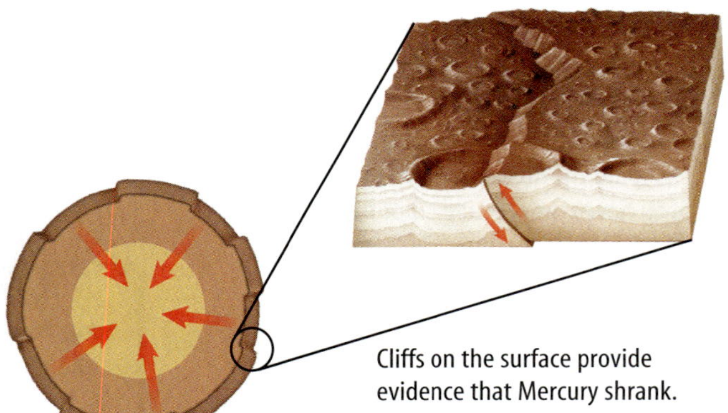
Cliffs on the surface provide evidence that Mercury shrank.

Section 2 Resource Manager

Chapter *FAST FILE* Resources
Transparency Activity, p. 49
Directed Reading for Content Mastery, p. 20
Enrichment, p. 32
MiniLAB, p. 3
Reinforcement, p. 28

Mathematics Skill Activities, p. 1
Reading and Writing Skill Activities, p. 11
Earth Science Critical Thinking/Problem Solving, p. 8
Performance Assessment in the Science Classroom, p. 46

76 ◆ J CHAPTER 3 The Solar System

Does Mercury have an atmosphere? Because of Mercury's low gravitational pull and high daytime temperatures, most gases that could form an atmosphere escape into space. *Mariner 10* found traces of hydrogen and helium gas that were first thought to be an atmosphere. However, these gases are now known to be temporarily taken from the solar wind.

Earth-based observations have found traces of sodium and potassium around Mercury. These atoms probably come from rocks in the planet's crust. Therefore, Mercury has no true atmosphere. This lack of atmosphere and its nearness to the Sun cause Mercury to have great extremes in temperature. Mercury's temperature can reach 425°C during the day, and it can drop to −170°C at night.

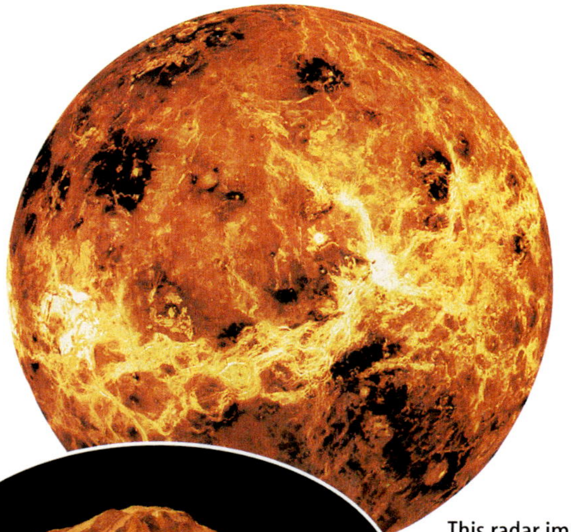

This radar image of Venus's surface was made from data acquired by *Magellan*.

Maat Mons is the highest volcano on Venus. Lava flows extend for hundreds of kilometers across the plains.

Figure 5 Venus is the second planet from the Sun.

Venus

The second planet from the Sun is **Venus,** shown in **Figure 5.** Venus is sometimes called Earth's twin because its size and mass are similar to Earth's. In 1962, *Mariner 2* flew past Venus and sent back information about Venus's atmosphere and rotation. The former Soviet Union landed the first probe on the surface of Venus in 1970. *Venera 7*, however, stopped working in less than an hour because of the high temperature and pressure. Additional *Venera* probes photographed and mapped the surface of Venus. Between 1990 and 1994, the U.S. *Magellan* probe used its radar to make the most detailed maps yet of Venus's surface. It collected radar images of 98 percent of Venus's surface. Notice the huge volcano in **Figure 5.**

Clouds on Venus are so dense that only a small percentage of the sunlight that strikes the top of the clouds reaches the planet's surface. The sunlight that does get through warms Venus's surface, which then gives off heat to the atmosphere. Much of this heat is absorbed by carbon dioxide gas in Venus's atmosphere. This causes a greenhouse effect similar to, but more intense than, Earth's greenhouse effect. Due to this intense greenhouse effect, the temperature on the surface of Venus is between 450°C and 475°C.

2 Teach

Use Science Words

Word Origin Venus is named after a mythic Italian goddess. Ask students to find out what qualities were attributed to this goddess. Venus is the ancient goddess of love and beauty.

Fun Fact

Because of the effect of tidal forces, Mercury rotates 1.5 times for each revolution around the Sun. This causes Mercury's day to be twice as long as its year.

Quick Demo

A Day on Mercury

Materials globe, shaded light source, stick-on paper note

Estimated Time five minutes

Procedure Model the relationship between Mercury's day and year for students. Place a stick-on note at the intersection of the globe's equator and prime meridian. This marks the starting position for a day. Revolve the globe around the shaded light source. For every 1/4 revolution, rotate the globe 3/8 of a turn (1.5 rotations in a whole revolution is 3/8 rotation in 1/4 revolution). Continue this procedure until you have revolved around the light source twice. Students will see that the stick-on note has returned to its original illuminated position.

Teacher FYI

Color of Planets The composition of the atmosphere or the surface of a planet produces its apparent color. Venus's yellow color is due to its clouds of sulfuric acid. Earth's oceans of liquid water help give it a distinctive blue color. Mars, the red planet, gets its color from iron oxide in weathered rocks on its surface.

Differentiated Instruction

Challenge Direct interested students to research the *Mariner 10* mission to Mercury. Students can access bookj.msscience.com for more information. Have these students work together to organize their findings and present them to the class. L2 COOP LEARN LS **Linguistic**

Discussion

Mission to Mars The two closest planets to Earth are Venus and Mars. But when astronauts from Earth visit another planet, it will be Mars rather than Venus. Why is Mars a more likely candidate for a visit than Venus? Venus has very high temperature and atmospheric pressure, which have caused space probes to malfunction. Students should infer that surface conditions on Mars are less hostile for humans.

Caption Answer

Figure 6 Earth's surface has abundant liquid water, and it supports life.

Quick Demo

Size Comparisons

Materials poster board, scissors, metric ruler

Estimated Time 10 minutes

Procedure Demonstrate the general difference in size between inner and outer planets by cutting circles out of poster board that represent Jupiter and Earth. The radius of the circle representing Jupiter should be 11.2 times the radius of the circle representing Earth. Pass the two circles around the class.

Visual Learning

Figure 7 Ask students to identify features on Earth that are similar to those shown in the figure. Earth has polar ice caps; although smaller, Mauna Loa in Hawaii is similar to Olympus Mons; the East African Rift Valley is similar to Valles Marineris.

Figure 6 More than 70 percent of Earth's surface is covered by liquid water.
Explain *how Earth is unique.*

Earth

Figure 6 shows **Earth,** the third planet from the Sun. The average distance from Earth to the Sun is 150 million km, or one astronomical unit (AU). Unlike other planets, Earth has abundant liquid water and supports life. Earth's atmosphere causes most meteors to burn up before they reach the surface, and it protects life-forms from the effects of the Sun's intense radiation.

Mars

Look at **Figure 7.** Can you guess why **Mars,** the fourth planet from the Sun, is called the red planet? Iron oxide in soil on its surface gives it a reddish color. Other features visible from Earth are Mars's polar ice caps and changes in the coloring of the planet's surface. The ice caps are made of frozen water covered by a layer of frozen carbon dioxide.

Most of the information scientists have about Mars came from *Mariner 9,* the *Viking* probes, *Mars Pathfinder, Mars Global Surveyor,* and *Mars Odyssey. Mariner 9* orbited Mars in 1971 and 1972. It revealed long channels on the planet that might have been carved by flowing water. *Mariner 9* also discovered the largest volcano in the solar system, Olympus Mons, shown in **Figure 7.** Olympus Mons is probably extinct. Large rift valleys that formed in the Martian crust also were discovered. One such valley, Valles Marineris, is shown in **Figure 7.**

Figure 7 Many features on Mars are similar to those on Earth.

Mars is often called the "red planet."

Olympus Mons is the largest volcano in the solar system.

Valles Marineris is more than 4,000 km long, up to 200 km wide, and more than 7 km deep.

Curriculum Connection

Literature Invite students to read a classic work of fiction that deals with Martians, such as *War of the Worlds* by H.G. Wells, originally published in 1898. Encourage students to discuss in class what they've read, concentrating on the differences between what the authors thought Mars was like and what we know of it today. L2 LS **Linguistic**

Active Reading

Pair of Pairs This strategy encourages students as partners to respond to a question. They brainstorm together, writing down their ideas. After a few minutes, they combine with another pair and share responses. Finally, the teacher asks the groups of four to share any responses. Have students use the Pair of Pairs strategy on a Think Critically question in a Section Assessment in this chapter.

78 ◆ J CHAPTER 3 The Solar System

The *Viking* Probes The *Viking 1* and *2* probes arrived at Mars in 1976. Each spacecraft consisted of an orbiter and a lander. The *Viking 1* and *2* orbiters photographed the entire surface of Mars from their orbits, while the *Viking 1* and *2* landers touched down on the planet's surface. The landers carried equipment to detect possible life on Mars. Some of this equipment was designed to analyze gases given off by Martian soil. These experiments found no conclusive evidence of life on Mars.

Pathfinder, Global Surveyor,* and *Odyssey The *Mars Pathfinder* carried a robot rover named Sojourner with equipment that allowed it to analyze samples of Martian rock and soil. Data from these tests indicated that iron in Mars's crust might have been leached out by groundwater. Cameras onboard *Global Surveyor* showed features that look like gullies formed by flowing water and deposits of sediment carried by the water flows. The features, shown in **Figure 8,** are young enough that scientists are considering the idea that liquid groundwater might exist on Mars and that it sometimes reaches the surface. It is also possible that volcanic activity might melt frost beneath the Martian surface. The features compare to those formed by flash floods on Earth, such as on Mount St. Helens.

More recently, instruments on another probe called *Mars Odyssey* detected frozen water on Mars. The water occurs as frost beneath a thin layer of soil. The frost is common in the far northern and far southern parts of Mars.

Reading Check *What evidence indicates that Mars has water?*

Mini LAB

Inferring Effects of Gravity

Procedure

1. Suppose you are a crane operator who is sent to Mars to help build a Mars colony.
2. You know that your crane can lift 44,500 N on Earth and Mars, but the gravity on Mars is only 40 percent of Earth's gravity.
3. Determine how much mass your crane could lift on Earth and Mars.

Analysis

1. How can what you have discovered be an advantage over construction on Earth?
2. How might construction advantages change the overall design of the Mars colony?

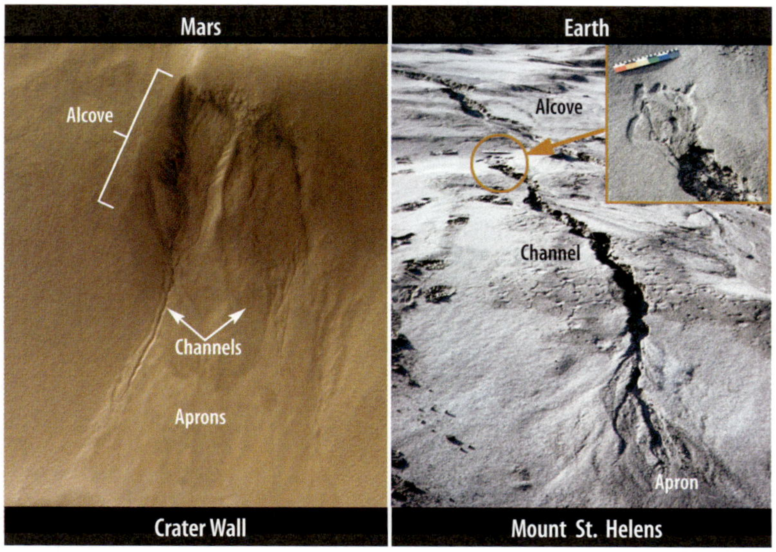

Figure 8 Compare the features found on Mars with those found on an area of Mount St. Helens in Washington state that experienced a flash flood.

Mini LAB

Purpose Students compare the mass that a crane could lift on Earth and Mars. L2 LS **Logical-Mathematical**

Teaching Strategy Remind students that the acceleration due to gravity on Earth is $9.8 m/s^2$.

- $F = ma$; $m = F/a$ $m_e = 44{,}500 N/9.8 m/s^2 =$ 4,540.8 kg on Earth
- acceleration due to gravity on Mars $= 0.4 \times 9.8 m/s^2 = 3.92 m/s^2$
- $m_m = 44{,}500 N/3.92 m/s^2 =$ 11,352 kg on Mars

Analysis

1. Because Mars's gravity is weaker, the same crane could lift more massive objects on Mars than on Earth.
2. The structure could be made of larger sections. More massive objects could be lifted higher.

Assessment

Process Have students infer how construction would be affected on a planet with greater gravity than Earth's. Only items less massive than those used on Earth could be lifted. Use **Performance Assessment in the Science Classroom,** p. 89.

Reading Check

Answer Mars has surface features that could have been formed by flowing water. Rock and soil analysis shows that iron in Mars's crust might have been leached by groundwater.

Differentiated Instruction

Behavioral Disorder Divide the students into groups to work on the MiniLAB. Provide calculators to students with behavior problems. Have them perform the calculations for the group. L2

SECTION 2 The Inner Planets J ◆ 79

Inquiry Lab

Gas-Exchange Experiment

Purpose to research and model the gas-exchange experiment that was performed by the *Viking Landers*

Possible Materials nail, plastic water jugs, plastic tubing, duct tape, soil, sugar, molasses, package of yeast, warm water, limewater, beakers

Estimated Time two days for research and planning; three days to observe the experiment

Teaching Strategies

- Allow students to develop their own designs under your supervision.
- One possible procedure is to insert tubing into the caps of two water jugs and seal with tape. Fill each jug with a shallow layer of soil. Add 50 g of sugar and 20 mL of molasses to each jug. Empty a package of yeast into one jug, and add warm water to cover the soil in both jugs. Snap on the lids, and drop the other ends of the two pieces of plastic tubing into separate beakers of limewater. The yeast in one jug will produce carbon dioxide and cause the limewater to become cloudy. This is a positive test for life processes. L2

For additional inquiry activities, see *Science Inquiry Labs*.

Applying Math

National Math Standards

Correlation to Mathematics Objectives

1, 9

Answers to Practice Problems

1. (12,756 km) × (0.949) = 12,105 km
2. (12,756 km) × (0.382) = 4,873 km

Science Online

Topic: Mars Exploration

Visit bookj.msscience.com for Web links to information about future missions to Mars.

Activity Make a timeline that shows when each probe is scheduled to reach Mars. Include the mission objectives for each probe on your timeline.

Mars's Atmosphere The *Viking* and *Global Surveyor* probes analyzed gases in the Martian atmosphere and determined atmospheric pressure and temperature. They found that Mars's atmosphere is much thinner than Earth's. It is composed mostly of carbon dioxide, with some nitrogen and argon. Surface temperatures range from −125°C to 35°C. The temperature difference between day and night results in strong winds on the planet, which can cause global dust storms during certain seasons. This information will help in planning possible human exploration of Mars in the future.

Martian Seasons Mars's axis of rotation is tilted 25°, which is close to Earth's tilt of 23.5°. Because of this, Mars goes through seasons as it orbits the Sun, just like Earth does. The polar ice caps on Mars change with the season. During winter, carbon dioxide ice accumulates and makes the ice cap larger. During summer, carbon dioxide ice changes to carbon dioxide gas and the ice cap shrinks. As one ice cap gets larger, the other ice cap gets smaller. The color of the ice caps and other areas on Mars also changes with the season. The movement of dust and sand during dust storms causes the changing colors.

Applying Math — Use Percentages

DIAMETER OF MARS The diameter of Earth is 12,756 km. The diameter of Mars is 53.3 percent of the diameter of Earth. Calculate the diameter of Mars.

Solution

1. *This is what you know:*
   - diameter of Earth: 12,756 km
   - percent of Earth's diameter: 53.3%
   - decimal equivalent: 0.533 (53.3% ÷ 100)
2. *This is what you need to find:* diameter of Mars
3. *This is the procedure you need to use:* Multiply the diameter of Earth by the decimal equivalent.
   (12,756 km) × (0.533) = 6,799 km

Practice Problems

1. Use the same procedure to calculate the diameter of Venus. Its diameter is 94.9 percent of the diameter of Earth.
2. Calculate the diameter of Mercury. Its diameter is 38.2 percent of the diameter of Earth.

Science Online — For more practice, visit bookj.msscience.com/math_practice

Differentiated Instruction

Challenge Have students research famous astronomers from different eras. Students can work in groups of five or six to write plays in which these astronomers are brought together to discuss their ideas. Encourage students to perform their plays for the class. L3 COOP LEARN

Martian Moons Mars has two small, irregularly shaped moons that are heavily cratered. Phobos, shown in **Figure 9,** is about 25 km in length, and Deimos is about 13 km in length. Deimos orbits Mars once every 31 h, while Phobos speeds around Mars once every 7 h.

Phobos has many interesting surface features. Grooves and chains of smaller craters seem to radiate out from the large Stickney Crater. Some of the grooves are 700 m across and 90 m deep. These features probably are the result of the large impact that formed the Stickney Crater.

Deimos is the outer of Mars's two moons. It is among the smallest known moons in the solar system. Its surface is smoother in appearance than that of Phobos because some of its craters have partially filled with soil and rock.

As you toured the inner planets through the eyes of the space probes, you saw how each planet is unique. Refer to **Table 3** following Section 3 for a summary of the planets. Mercury, Venus, Earth, and Mars are different from the outer planets, which you'll explore in the next section.

Figure 9 Phobos orbits Mars once every 7 h.
Infer *why Phobos has so many craters.*

section 2 review

Summary

Mercury
- Mercury is extremely hot during the day and extremely cold at night.
- Its surface has many craters.

Venus
- Venus's size and mass are similar to Earth's.
- Temperatures on Venus are between 450°C and 475°C.

Earth
- Earth is the only planet known to support life.

Mars
- Mars has polar ice caps, channels that might have been carved by water, and the largest volcano in the solar system, Olympus Mons.

Self Check

1. **Explain** why Mercury's surface temperature varies so much from day to night.
2. **List** important characteristics for each inner planet.
3. **Infer** why life is unlikely on Venus.
4. **Identify** the inner planet that is farthest from the Sun. Identify the one that is closest to the Sun.
5. **Think Critically** Aside from Earth, which inner planet could humans visit most easily? Explain.

Applying Math

6. **Use Statistics** The inner planets have the following average densities: Mercury, 5.43 g/cm^3; Venus, 5.24 g/cm^3; Earth, 5.52 g/cm^3; and Mars, 3.94 g/cm^3. Which planet has the highest density? Which has the lowest? Calculate the range of these data.

Science Online bookj.msscience.com/self_check_quiz

Caption Answer

Figure 9 Possible Answers: It is old-a lot of meteorites could have impacted it. It has no significant processes of erosion to obliterate craters. It has no atmosphere to help break up small meteorites before they hit its surface.

3 Assess

DAILY INTERVENTION

Check for Understanding

Auditory-Musical Work as a class to create an impromptu song or poem about the inner planets. Include important facts and concepts in your composition. L2

Reteach

Inner Planets Review the information about the inner planets that students gave at the beginning of this section. Ask if they would like to change or add to this information. L1

Assessment

Performance Ask students to write a list of the equipment they would need to live on the surface of Mars. As students call out items, have them explain why each was included. Use **Performance Assessment in the Science Classroom,** p. 173.

section 2 review

1. Mercury does not have an atmosphere to hold heat, and it is close to the Sun.
2. Mercury, no atmosphere, many craters, high cliffs; Venus, high temperatures, carbon dioxide atmosphere, sulfuric acid clouds; Earth, abundant surface water, life exists, nitrogen-oxygen atmosphere; Mars, polar caps, reddish surface, thin atmosphere
3. Temperatures are too high.
4. Mars, Mercury.
5. Humans could most easily visit Mars. The conditions on Mars are not as extreme as those on the other planets.
6. Earth; Mars; range = 5.52 g/cm^3 − 3.94 g/cm^3 = 1.58 g/cm^3

SECTION 2 The Inner Planets J ◆ 81

The Outer Planets

1 Motivate

INTERACTIVE CHALKBOARD
PowerPoint® Presentations

Bellringer

Section Focus Transparencies also are available on the Interactive Chalkboard CD-ROM. L2 ELL

SECTION 3 Section Focus Transparency **Relative Dimensions**

This image was designed to show the relative sizes of the planets in our solar system. The Sun is so huge that only its rim fits on the left side of the image. The nine planets are presented in order from Mercury to Pluto.

1. What's the very first thing you notice about the four planets closest to the Sun compared to the outer planets?
2. Which planet is an exception to this observation?
3. What do you think the small black spot on Jupiter is? Which planet is the black spot closest to in size?

L2

The Solar System

Tie to Prior Knowledge

Outer Planets Ask students to recall an important characteristic of one of the outer planets. Students might recall Saturn's rings or Jupiter's Great Red Spot. Explain that some of these characteristics, such as Saturn's rings, are visible from Earth.

as you read

What You'll Learn

- **Describe** the characteristics of Jupiter, Saturn, Uranus, and Neptune.
- **Explain** how Pluto differs from the other outer planets.

Why It's Important

Studying the outer planets will help scientists understand Earth.

Review Vocabulary

moon: a natural satellite of a planet that is held in its orbit around the planet by the planet's gravitational pull

New Vocabulary

- Jupiter
- Great Red Spot
- Saturn
- Uranus
- Neptune
- Pluto

Outer Planets

You might have heard about *Voyager, Galileo,* and *Cassini.* They were not the first probes to the outer planets, but they gathered a lot of new information about them. Follow the spacecrafts as you read about their journeys to the outer planets.

Jupiter

In 1979, *Voyager 1* and *Voyager 2* flew past **Jupiter,** the fifth planet from the Sun. *Galileo* reached Jupiter in 1995, and *Cassini* flew past Jupiter on its way to Saturn in 2000. The spacecrafts gathered new information about Jupiter. The *Voyager* probes revealed that Jupiter has faint dust rings around it and that one of its moons has active volcanoes on it.

Jupiter's Atmosphere Jupiter is composed mostly of hydrogen and helium, with some ammonia, methane, and water vapor. Scientists hypothesize that the atmosphere of hydrogen and helium changes to an ocean of liquid hydrogen and helium toward the middle of the planet. Below this liquid layer might be a rocky core. The extreme pressure and temperature, however, would make the core different from any rock on Earth.

You've probably seen pictures from the probes of Jupiter's colorful clouds. In **Figure 10,** you can see bands of white, red, tan, and brown clouds in its atmosphere. Continuous storms of swirling, high-pressure gas have been observed on Jupiter. The **Great Red Spot** is the most spectacular of these storms.

Figure 10 Jupiter is the largest planet in the solar system.

Notice the colorful bands of clouds in Jupiter's atmosphere.

The Great Red Spot is a giant storm about 25,000 km in size from east to west.

Section 3 Resource Manager

Chapter *Fast File* Resources
- Transparency Activity, p. 50
- Direct Reading for Content Mastery, p. 21
- MiniLAB, p. 4
- Enrichment, p. 33
- Reinforcement, p. 29

Physical Science Critical Thinking/Problem Solving, p. 4

82 ◆ J CHAPTER 3 The Solar System

Table 2 Large Moons of Jupiter

Io The most volcanically active object in the solar system; sulfurous compounds give it its distinctive reddish and orange colors; has a thin oxygen, sulfur, and sulfur dioxide atmosphere.	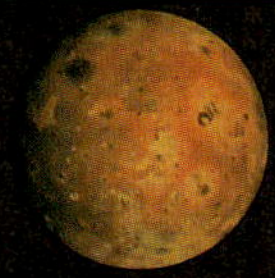
Europa Rocky interior is covered by a 100-km-thick crust of ice, which has a network of cracks; an ocean might exist under the ice crust; has a thin oxygen atmosphere.	
Ganymede Has a crust of ice about 500 km thick, covered with grooves; crust might surround an ocean of water or slushy ice; has a rocky core and a thin oxygen atmosphere.	
Callisto Has a heavily cratered crust of ice and rock several hundred kilometers thick; crust might surround a salty ocean around a rock core; has a thin atmosphere of carbon dioxide.	

Moons of Jupiter At least 61 moons orbit Jupiter. In 1610, the astronomer Galileo Galilei was the first person to see Jupiter's four largest moons, shown in **Table 2.** Io (I oh) is the closest large moon to Jupiter. Jupiter's tremendous gravitational force and the gravity of Europa, Jupiter's next large moon, pull on Io. This force heats up Io, causing it to be the most volcanically active object in the solar system. You can see a volcano erupting on Io in **Figure 11.** Europa is composed mostly of rock with a thick, smooth crust of ice. Under the ice might be an ocean as deep as 200 km. If this ocean of water exists, it will be the only place in the solar system, other than Earth and possibly Ganymede and Callisto, where liquid water exists in large quantities. Next is Ganymede, the largest moon in the solar system—larger even than the planet Mercury. Callisto, the last of Jupiter's large moons, is composed mostly of ice and rock. Studying these moons adds to knowledge about the origin of Earth and the rest of the solar system.

Figure 11 *Voyager 2* photographed the eruption of this volcano on Io in July 1979.

2 Teach

Activity

Planet Flash Cards Have pairs of students make flash cards of the planets in the solar system. They can draw the planet on one side of the card and write anything they know about it on the other side. Students should keep their flash cards handy so they can correct or update their information as they continue to read this chapter. L2 COOP LEARN LS **Interpersonal**

Visual Learning

Table 2 Have students compare and contrast the large moons of Jupiter.

Activity

Planet Quiz Show Form teams of four students to play "Planet Quiz Show." Have a student read statements from this section and the previous section and then call on a team to form the correct question. Award points for correct questions. L2 COOP LEARN LS **Linguistic**

Teacher FYI

Jupiter Impact In 1994, astronomers witnessed comet fragments crash into Jupiter's atmosphere. The impact left huge holes in Jupiter's clouds that lasted for weeks. No one knows what the fragments encountered under the clouds because no one has ever seen beneath Jupiter's cloud tops.

Teacher FYI

Saturn's Rings In 1619, Galileo became the first person to see Saturn's rings. But with his weak telescope, the rings looked like handles on the sides of the planet. The "handles" weren't recognized as rings until stronger telescopes were available decades later.

Mini LAB

Purpose Students design and draw a scale model that compares the planets. L1 ELL LS **Logical-Mathematical**

Materials metric ruler, drawing compass, pencil, paper

Teaching Strategy Have students use Earth's diameter as a basis for the scale. If Earth's diameter equals 1 unit, dividing the other planets' diameters by Earth's diameter yields the number of units for each planet.

Safety Precautions Caution students to use the compass carefully.

Analysis

1. Jupiter is largest. Pluto is smallest.
2. Answers will vary. If students select a scale diameter for Earth that is too large, they will not be able to draw the larger planets on their paper. One centimeter is a good scale diameter for Earth. This will allow students to draw Pluto with a diameter of 1.8 mm and Jupiter with a diameter of about 11.2 cm.

Assessment

Performance To further assess students' understanding of scale models, have them make planet models out of balls of clay. Use **Performance Assessment in the Science Classroom,** p. 123.

Try at Home

Figure 12 Saturn's rings are composed of pieces of rock and ice.

Mini LAB

Modeling Planets

Procedure

1. Research the planets to determine how the sizes of the planets in the solar system compare with each other.
2. Select a scale for the diameter of Earth.
3. Make a model by drawing a circle with this diameter on **paper.**
4. Using Earth's diameter as 1.0 unit, draw each of the other planets to scale.

Analysis

1. Which planet is largest? Which is smallest?
2. Which scale diameter did you select for Earth? Was this a good choice? Why or why not?

Try at Home

Saturn

The *Voyager* probes next surveyed Saturn in 1980 and 1981. *Cassini* is scheduled to reach Saturn on July 1, 2004. **Saturn** is the sixth planet from the Sun. It is the second-largest planet in the solar system, but it has the lowest density.

Saturn's Atmosphere Similar to Jupiter, Saturn is a large, gaseous planet. It has a thick outer atmosphere composed mostly of hydrogen and helium. Saturn's atmosphere also contains ammonia, methane, and water vapor. As you go deeper into Saturn's atmosphere, the gases gradually change to liquid hydrogen and helium. Below its atmosphere and liquid layer, Saturn might have a small, rocky core.

Rings and Moons The *Voyager* probes gathered new information about Saturn's ring system. The probes showed that Saturn has several broad rings. Each large ring is composed of thousands of thin ringlets. **Figure 12** shows that Saturn's rings are composed of countless ice and rock particles. These particles range in size from a speck of dust to tens of meters across. Saturn's ring system is the most complex one in the solar system.

At least 31 moons orbit Saturn. Saturn's gravity holds these moons in their orbits around Saturn, just like the Sun's gravity holds the planets in their orbits around the Sun. The largest of Saturn's moons, Titan, is larger than the planet Mercury. It has an atmosphere of nitrogen, argon, and methane. Thick clouds make it difficult for scientists to study the surface of Titan.

Cultural Diversity

Sumerian Calendar The ancient Sumerians, whose civilization in Mesopotamia reached its height about 3500 B.C., were among the earliest astronomers. They observed movement of the Sun and Moon and made a 28-day lunar calendar. This calendar allowed the Sumerians to predict the cycle of seasons. It also alerted farmers when it was time to plant and harvest crops. Invite students to research astronomy in places such as ancient China, Egypt, and Central America. Have them design a classroom bulletin board summarizing the accomplishments of ancient astronomers.

84 ◆ J CHAPTER 3 The Solar System

Uranus

Beyond Saturn, *Voyager 2* flew by Uranus in 1986. **Uranus** (YOOR uh nus) is the seventh planet from the Sun and was discovered in 1781. It is a large, gaseous planet with at least 21 moons and a system of thin, dark rings. Uranus's largest moon, Titania, has many craters and deep valleys. The valleys on this moon indicate that some process reshaped its surface after it formed. Uranus's 11 rings surround the planet's equator.

Figure 13 The atmosphere of Uranus gives the planet its distinct bluish-green color.

Uranus's Characteristics The atmosphere of Uranus is composed of hydrogen, helium, and some methane. Methane gives the planet the bluish-green color that you see in **Figure 13.** Methane absorbs the red and yellow light, and the clouds reflect the green and blue. Few cloud bands and storm systems can be seen on Uranus. Evidence suggests that under its atmosphere, Uranus has a mantle of liquid water, methane, and ammonia surrounding a rocky core.

Figure 14 shows one of the most unusual features of Uranus. Its axis of rotation is tilted on its side compared with the other planets. The axes of rotation of the other planets, except Pluto, are nearly perpendicular to the planes of their orbits. However, Uranus's axis of rotation is nearly parallel to the plane of its orbit. Some scientists believe a collision with another object tipped Uranus on its side.

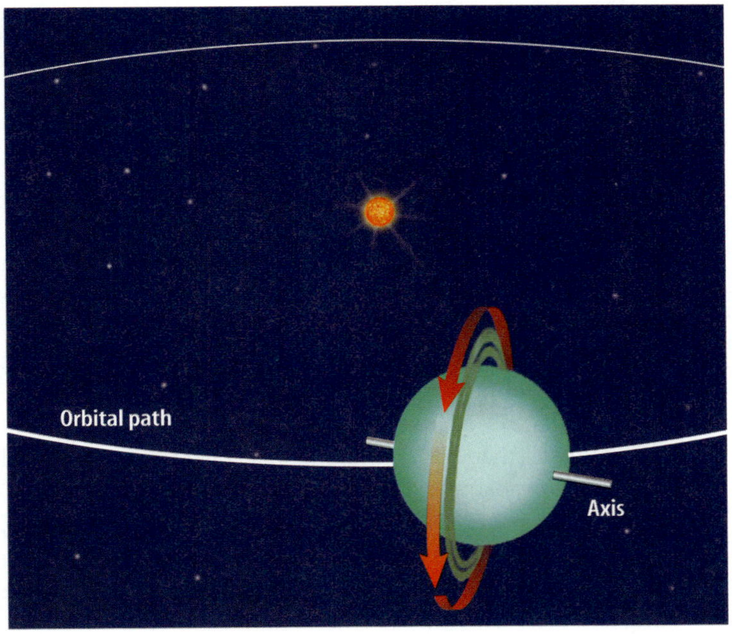

Figure 14 Uranus's axis of rotation is nearly parallel to the plane of its orbit. During its revolution around the Sun, each pole, at different times, points almost directly at the Sun.

Discussion

Outer Planet Moons Remind students that new moons continue to be discovered around the outer planets. Why might astronomers continue to find moons they have not seen before? Students should infer that continued exploration with more powerful telescopes, in addition to images sent back by new probes, allow astronomers to identify small bodies that were previously unseen. L2

Fun Fact

William Herschel discovered a new planet in 1781. Herschel originally named the new planet *Georgium Sidus*. *Sidus* is Latin for "star," and *Georgium* refers to England's King George III. Another astronomer, Johann Bode, suggested it be named after Greek and Roman gods. Uranus, god of the sky and father of Saturn, was chosen as its name.

Identifying Misconceptions

Seeing Planets Many students think that a telescope is required to see planets in the night sky. Make students aware of times when Mercury, Venus, Mars, Jupiter, or Saturn is visible. Invite them to view these planets with the unaided eye.

Differentiated Instruction

English-Language Learners Examine photographs of the outer planets and their moons with the class. Ask native English speakers and English language learners to identify features in the photographs. Students will enjoy learning different words describing the same features. ELL L2

Curriculum Connection

Math Using a scale of 1 mm = 100 km, have students draw the listed moons to scale. Have them use the following diameters: Earth's Moon (3,476 km), Io (3,630 km), Europa (3,138 km), Ganymede (5,262 km), Callisto (4,800 km), Titan (5,150 km), and Triton (2,700 km). How does Earth's Moon compare to the others? It is one of the smaller moons in the group.

INTEGRATE Language Arts

Research Mercury, Roman mythology—messenger god and escort to the underworld; Venus, Roman mythology—goddess of love; Earth, Germanic and Old English origin; Mars, Roman mythology—god of war; Jupiter, Roman mythology—ruler of the gods; Saturn, Roman mythology—god of agriculture; Uranus, Greek mythology—god of the sky; Neptune, Roman mythology—god of the sea; Pluto, Greek mythology—god of the underworld

Fun Fact

Pluto and Neptune's moon Triton share many similarities. Some scientists think that Triton was captured by the gravity of Neptune and could once have been a planet.

Reading Check

Answer methane in its atmosphere

Discussion

Visiting Neptune Is it likely that astronauts will ever visit Neptune? Explain. It isn't likely; Neptune is a gaseous planet, meaning it doesn't have a solid surface like Earth does. It is also very distant and has a hostile environment.

Quick Demo

Orbits of Neptune and Pluto

Materials chalkboard, chalk

Estimated Time 5 minutes

Procedure The reason that Pluto is sometimes closer to the Sun than Neptune is that Pluto has a much more eccentric orbit than Neptune. Make sketches to show how Pluto, with a more eccentric orbit, might sometimes cross Neptune's more circular orbit.

Neptune has a distinctive bluish-green color.

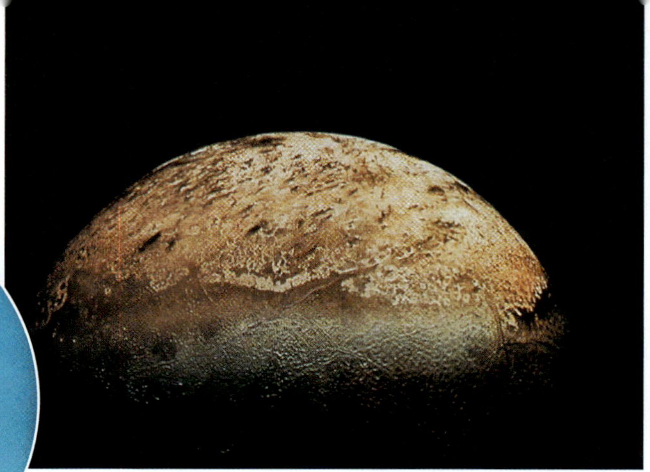

The pinkish hue of Neptune's largest moon, Triton, is thought to come from an evaporating layer of nitrogen and methane ice.

Figure 15 Neptune is the eighth planet from the Sun.

Neptune

Passing Uranus, *Voyager 2* traveled to Neptune, another large, gaseous planet. Discovered in 1846, **Neptune** is usually the eighth planet from the Sun. However, Pluto's orbit crosses inside Neptune's during part of its voyage around the Sun. Between 1979 and 1999, Pluto was closer to the Sun than was Neptune.

Neptune's Characteristics Neptune's atmosphere is similar to Uranus's atmosphere. The methane content gives Neptune, shown in **Figure 15,** its distinctive bluish-green color, just as it does for Uranus.

Reading Check ***What gives Neptune its bluish-green color?***

Neptune has dark-colored storms in its atmosphere that are similar to the Great Red Spot on Jupiter. One discovered by *Voyager 2* in 1989 was called the Great Dark Spot. It was about the size of Earth. However, observations by the *Hubble Space Telescope* in 1994 showed that the Great Dark Spot had disappeared. Bright clouds also form and then disappear. This shows that Neptune's atmosphere is active and changes rapidly.

Under its atmosphere, Neptune is thought to have a layer of liquid water, methane, and ammonia that might change to solid ice. Neptune probably has a rocky core.

Neptune has at least 11 moons and several rings. Triton, shown in **Figure 15,** is Neptune's largest moon. It has a thin atmosphere composed mostly of nitrogen. Neptune's rings are young and probably won't last very long.

INTEGRATE Language Arts

Names of Planets The names of most of the planets in the solar system come from Roman or Greek mythology. For example, Neptune was the Roman god of the sea, and Pluto was the Greek god of the underworld. Research to learn about the names of the other planets. Write a paragraph in your Science Journal that summarizes what you learn.

Science Journal

***Galileo* Probe** Have students write reports in their Science Journals about the information the probe *Galileo* gathered about Jupiter, its moons, and its rings. Ask them to include how the probe was launched on its mission and how NASA used gravity assists to power the space probe during its long flight to Jupiter. L2 LS **Linguistic** P

Differentiated Instruction

Challenge Ask students to research the New Horizons program and plans to send a probe to Pluto and the Kuiper Belt. Funding for this mission was the subject of much debate, and scientists worked hard for approval. Have students research the benefits and costs of the program. Students then could take a position about whether this program should be funded and lobby other students to support their position. L3

Pluto

The smallest planet in the solar system, and the one scientists know the least about, is **Pluto**. During part of its 248-year orbit, Pluto is closer to the Sun than Neptune. However, because Pluto is farther from the Sun than Neptune during most of its orbit, it is considered to be the ninth planet from the Sun. Pluto is vastly different from the other outer planets. It's surrounded by only a thin atmosphere, and it's the only outer planet with a solid, icy-rock surface.

Figure 16 The *Hubble Space Telescope* gave astronomers their first clear view of Pluto and Charon as distinct objects.

Pluto's Moon Pluto's single moon, Charon, has a diameter about half the size of Pluto's. It was discovered in 1978 when a bulge was noticed on a photograph of the planet. Later photographs, taken with improved telescopes, showed that the bulge was a moon. Pluto and Charon are shown in **Figure 16.** Because of their close size and orbit, some scientists consider them to be a double planet.

Data from the *Hubble Space Telescope* indicate the presence of a vast disk of icy objects called the Kuiper Belt beyond Neptune's orbit. Some of the objects are hundreds of kilometers in diameter. One found in 2002 is about 1,300 km in diameter. Many astronomers think that Pluto and Charon are members of the Kuiper Belt.

section 3 review

Summary

Jupiter
- Jupiter is the largest planet in the solar system.
- The Great Red Spot is a huge storm on Jupiter.

Saturn
- Saturn has a complex system of rings.

Uranus
- Uranus has a bluish-green color caused by methane in its atmosphere.

Neptune
- Like Uranus, Neptune has a bluish-green color.
- Neptune's atmosphere has storms and can change rapidly.

Pluto
- Pluto is a small, ice-rock planet.
- Its moon, Charon, is about half as large as Pluto.

Self Check

1. **Describe** the differences between the outer planets and the inner planets.
2. **Describe** what Saturn's rings are made from.
3. **Explain** how Pluto is a unique outer planet.
4. **Explain** how Uranus's axis of rotation differs from those of most other planets.
5. **Think Critically** What would seasons be like on Uranus? Explain.

Applying Skills

6. **Identify a Question** Pluto has never been visited by a space probe, so many questions are unanswered. Think of a question about Pluto that you'd like to have answered. Then, explain how the answer to your question could be discovered.

Science Online bookj.msscience.com/self_check_quiz

3 Assess

DAILY INTERVENTION

Check for Understanding

Logical-Mathematical Provide each student with a list of characteristics describing an outer planet or moon. Ask the students to identify this planet or moon from its characteristics. L2

Reteach

Outer Planet Questions Have each student write down a question about one of the outer planets and then exchange questions with a partner, who will find the answer. Students can then read the questions and answers in class. L1 COOP LEARN IS **Interpersonal**

Assessment

Content Tell students to imagine themselves as NASA scientists who are planning the mission of a new space probe. Direct them to choose which outer planet the probe will visit and write a paragraph that defends the choice. Use **Performance Assessment in the Science Classroom,** p. 89.

section 3 review

1. outer: large and gaseous, except Pluto; inner: small, solid, rocky bodies
2. Saturn's rings are made from pieces of rock and ice.
3. Pluto is smaller, and it is an icy-rock planet. Pluto does not have a thick atmosphere like the gas giant planets.
4. Uranus's axis of rotation is nearly parallel to the plane of its orbit.
5. When the north pole points toward the Sun, the northern hemisphere has summer with perpetual daylight. The southern hemisphere has winter in perpetual darkness. When the south pole points toward the Sun, the southern hemisphere has summer and perpetual daylight. The northern hemisphere has winter and perpetual darkness. When neither pole points toward the Sun, spring/fall seasons occur, and both hemispheres experience day and night.
6. Answers will vary. Assess quality of questions and procedures for answering questions.

Make a Model

Planet Creature Have students choose one planet in this table. Have them make an illustration or a model of an organism that might be able to survive on their chosen planet. In addition, have them describe the conditions to which the organism would have to be adapted. L2 ELL LS **Kinesthetic**

Teacher FYI

Greenhouse Effect Carbon dioxide dissolves in water, therefore, much of the carbon dioxide on Earth is stored in the oceans. If it weren't for the presence of liquid water on Earth, the amount of carbon dioxide in the air would probably be similar to that in Venus's atmosphere, causing a tremendous increase in the greenhouse effect on Earth.

Activity

Planet Table Have students make a table that contains the following information for each planet: name, relative distance from Sun, relative size, type of atmosphere, and number of moons. L2 LS **Logical-Mathematical**

Table 3 Planets

Planet	Image	Characteristics
Mercury		■ closest to the Sun ■ second-smallest planet ■ surface has many craters and high cliffs ■ no atmosphere ■ temperatures range from 425°C during the day to −170°C at night ■ has no moons
Venus		■ similar to Earth in size and mass ■ thick atmosphere made mostly of carbon dioxide ■ droplets of sulfuric acid in atmosphere give clouds a yellowish color ■ surface has craters, faultlike cracks, and volcanoes ■ greenhouse effect causes surface temperatures of 450°C to 475°C ■ has no moons
Earth		■ atmosphere protects life ■ surface temperatures allow water to exist as solid, liquid, and gas ■ only planet where life is known to exist ■ has one large moon
Mars		■ surface appears reddish-yellow because of iron oxide in soil ■ ice caps are made of frozen carbon dioxide and water ■ channels indicate that water had flowed on the surface; has large volcanoes and valleys ■ has a thin atmosphere composed mostly of carbon dioxide ■ surface temperatures range from −125°C to 35°C ■ huge dust storms often blanket the planet ■ has two small moons

LAB DEMONSTRATION

Purpose to show why the planets are not perfect spheres

Materials construction paper, scissors, hole punch, metric ruler, glue, pencil

Preparation Cut two strips of paper. Cross the strips and glue. Bring the ends together, overlap, and glue to form a sphere.

Procedure Punch a hole through the center of the overlapped ends. Push 5 cm of the pencil through the hole. Hold the pencil between your palms, and move your hands back and forth, spinning the sphere.

Expected Outcome The spinning sphere flattens, causing the middle to bulge.

Assessment

What causes the flattening of the sphere? Forces exerted on the spinning sphere cause it to flatten. Why is this bulge hard to observe? Planets don't bulge very much. For example, the difference between Earth's equatorial and polar circumference is about 67 km.

Table 3 Planets

Planet	Image	Characteristics
Jupiter		■ largest planet ■ has faint rings ■ atmosphere is mostly hydrogen and helium; continuous storms swirl on the planet—the largest is the Great Red Spot ■ has four large moons and at least 57 smaller moons; one of its moons, Io, has active volcanoes
Saturn		■ second-largest planet ■ thick atmosphere is mostly hydrogen and helium ■ has a complex ring system ■ has at least 31 moons—the largest, Titan, is larger than Mercury
Uranus		■ large, gaseous planet with thin, dark rings ■ atmosphere is hydrogen, helium, and methane ■ axis of rotation is nearly parallel to plane of orbit ■ has at least 21 moons
Neptune		■ large, gaseous planet with rings that vary in thickness ■ is sometimes farther from the Sun than Pluto is ■ methane atmosphere causes its bluish-green color ■ has dark-colored storms in atmosphere ■ has at least 11 moons
Pluto	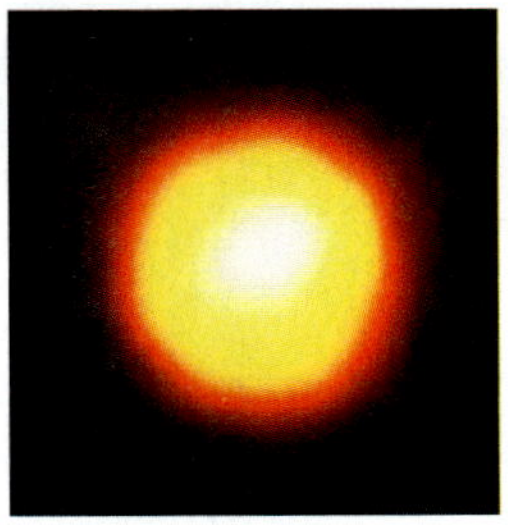	■ small, icy-rock planet with thin atmosphere ■ single moon, Charon, is half the diameter of the planet

Teacher FYI

***Galileo* Missions** The main assignment of the *Galileo* spacecraft ended in December 1997. At that time, NASA extended its mission for two years and renamed it the *Galileo Europa Mission*. During this mission, *Galileo* conducted several fly-bys of Europa. *Galileo* was then repurposed for a third mission, called the *Galileo Millennium Mission*. The *Galileo* spacecraft was crashed into Jupiter in September 2003. NASA was concerned that if Galileo crashed onto Europa, it could contaminate any life that might exist there.

Science Journal

Moons Studies by the *Galileo* spacecraft indicate that three of Jupiter's moons, Europa, Ganymede and Callisto, may have oceans under their crusts. Ask students to speculate about the significance of this discovery in their Science Journals.

Differentiated Instruction

Visually Impaired Help your visually impaired students use this table of planet data by having larger versions of the table produced. Also, try pairing your visually impaired students with mentors who could read aloud the information on the table as it is presented in class. L2

section 4

Other Objects in the Solar System

as you read

What You'll Learn

- **Describe** how comets change when they approach the Sun.
- **Distinguish** among comets, meteoroids, and asteroids.
- **Explain** that objects from space sometimes impact Earth.

Why It's Important

Comets, asteroids, and most meteorites are very old. Scientists can learn about the early solar system by studying them.

Review Vocabulary

crater: a nearly circular depression in a planet, moon, or asteroid that formed when an object from space hit its surface

New Vocabulary

- comet
- meteor
- meteorite
- asteroid

Comets

The planets and their moons are the most noticeable members of the Sun's family, but many other objects also orbit the Sun. Comets, meteoroids, and asteroids are other important objects in the solar system.

You might have heard of Halley's comet. A **comet** is composed of dust and rock particles mixed with frozen water, methane, and ammonia. Halley's comet was last seen from Earth in 1986. English astronomer Edmund Halley realized that comet sightings that had taken place about every 76 years were really sightings of the same comet. This comet, which takes about 76 years to orbit the Sun, was named after him.

Oort Cloud Astronomer Jan Oort proposed the idea that billions of comets surround the solar system. This cloud of comets, called the Oort Cloud, is located beyond the orbit of Pluto. Oort suggested that the gravities of the Sun and nearby stars interact with comets in the Oort Cloud. Comets either escape from the solar system or get captured into smaller orbits.

Comet Hale-Bopp On July 23, 1995, two amateur astronomers made an exciting discovery. A new comet, Comet Hale-Bopp, was headed toward the Sun. Larger than most that approach the Sun, it was the brightest comet visible from Earth in 20 years. Shown in **Figure 17,** Comet Hale-Bopp was at its brightest in March and April 1997.

Figure 17 Comet Hale-Bopp was most visible in March and April 1997.

1 Motivate

INTERACTIVE CHALKBOARD
PowerPoint® Presentations

Bellringer

Section Focus Transparencies also are available on the Interactive Chalkboard CD-ROM. L2 ELL

SECTION 4 Section Focus Transparency

We Don't Even Live There Yet!

Since the start of the space age in 1957, human-generated debris have been accumulating in orbit around Earth. Space junk ranges in size from tiny flecks of paint to dead satellites and burned-out rocket stages.

1. What might happen to junk that falls out of orbit and into Earth's atmosphere?
2. Is debris left on the Moon as great a potential problem as debris in orbit? Explain.
3. How does space junk pose a threat to orbiting satellites? L2

The Solar System

Tie to Prior Knowledge

Meteor Showers Ask students if they've ever seen a meteor shower and, if so, if they know what causes these streaks of light across the night sky. Explain that meteors are meteoroids that enter Earth's atmosphere and that meteoroids are objects in our solar system. L2

Section 4 Resource Manager

Chapter FAST FILE Resources

Transparency Activity, pp. 51, 53–54
Directed Reading for Content Mastery, pp. 21, 22
Enrichment, p. 34
Reinforcement, p. 30
Lab Worksheets, pp. 7–8

Cultural Diversity, p. 31
Lab Management and Safety, p. 69

90 ◆ J CHAPTER 3 The Solar System

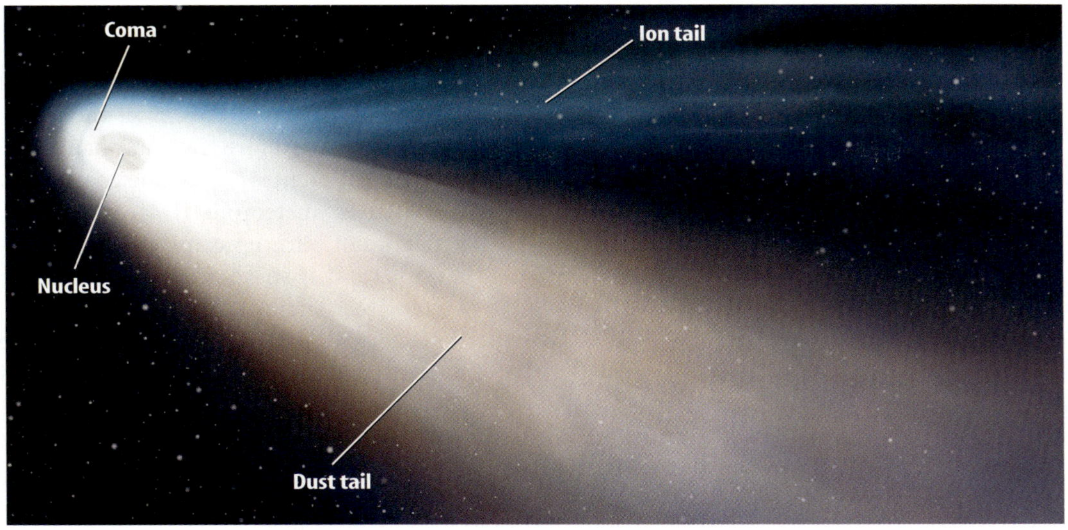

Figure 18 A comet consists of a nucleus, a coma, a dust tail, and an ion tail.

Structure of Comets The *Hubble Space Telescope* and spacecrafts such as the *International Cometary Explorer* have gathered information about comets. In 2006, a spacecraft called *Stardust* will return a capsule to Earth containing samples of dust from a comet's tail. Notice the structure of a comet shown in **Figure 18.** It is a mass of frozen ice and rock similar to a large, dirty snowball.

As a comet approaches the Sun, it changes. Ices of water, methane, and ammonia vaporize because of the heat from the Sun, releasing dust and bits of rock. The gases and released dust form a bright cloud called a coma around the nucleus, or solid part, of the comet. The solar wind pushes on the gases and dust in the coma, causing the particles to form separate tails that point away from the Sun.

After many trips around the Sun, most of the ice in a comet's nucleus has vaporized. All that's left are dust and rock, which are spread throughout the orbit of the original comet.

Meteoroids, Meteors, and Meteorites

You learned that comets vaporize and break up after they have passed close to the Sun many times. The small pieces from the comet's nucleus spread out into a loose group within the original orbit of the comet. These pieces of dust and rock, along with those derived from other sources, are called meteoroids.

Sometimes the path of a meteoroid crosses the position of Earth, and it enters Earth's atmosphere at speeds of 15 km/s to 70 km/s. Most meteoroids are so small that they completely burn up in Earth's atmosphere. A meteoroid that burns up in Earth's atmosphere is called a **meteor,** shown in **Figure 19.**

Figure 19 A meteoroid that burns up in Earth's atmosphere is called a meteor.

2 Teach

Discussion

Halley's Comet Remind students that comets have large orbits that bring them periodically near Earth. Halley's comet returns every 76 years, and it was last visible in 1986. When will Halley's comet again be visible? 2062 Will this pattern of reappearance repeat indefinitely? Explain. No; the frozen ice in a comet's nucleus will eventually be totally vaporized, leaving only debris in its orbit. This will happen to Halley's comet one day.

IDENTIFYING Misconceptions

Shooting Stars Many people refer to meteors as "shooting stars." They were named this long ago when people thought they were bright stars shooting across the sky. Today we know they are chunks of rock that glow as they streak through Earth's atmosphere.

Activity

Meteor Shower Alert students to the occurrence of the next meteor shower. Have students observe the meteor shower and report on the number of meteors they were able to see during a half hour. L1

Science Journal

Comet Studies Have students with Internet access research the mission of one of the five space probes sent by the former Soviet Union, Japan, and Europe to study the last approach of Halley's comet in 1986. They were the Soviet *Vega 1* and *Vega 2*, the Japanese *Suisei* and *Sakigake*, and the European Space Agency's *Giotto*.

Visual Learning

Figure 20 Why would this type of impact be a more serious problem today than it was 50,000 years ago? There are more people and more developed areas on Earth, making it more likely that the meteorite would hit someone or something.

Teacher FYI

Meteorites Most meteoroids burn up in Earth's atmosphere. But some reach the surface as meteorites. In 1954, a small meteorite crashed through the roof of an Alabama house, hitting a woman inside. Fortunately, she wasn't seriously hurt. In 1992, a football-size meteorite crushed the trunk of a parked car in New York.

Reading Check

Answer a meteoroid that strikes Earth

Text Question Answer

Scientists don't know for sure. Jupiter's gravity could have kept a planet from forming there when the solar system was young.

Figure 20 Meteorites occasionally strike Earth's surface. A large meteorite struck Arizona, forming a crater about 1.2 km in diameter and about 200 m deep.

Meteor Showers Each time Earth passes through the loose group of particles within the old orbit of a comet, many small particles of rock and dust enter the atmosphere. Because more meteors than usual are seen, the event is called a meteor shower.

When a meteoroid is large enough, it might not burn up completely in the atmosphere. If it strikes Earth, it is called a **meteorite.** Barringer Crater in Arizona, shown in **Figure 20,** was formed when a large meteorite struck Earth about 50,000 years ago. Most meteorites are probably debris from asteroid collisions or broken-up comets, but some originate from the Moon and Mars.

Reading Check *What is a meteorite?*

Asteroids

An **asteroid** is a piece of rock similar to the material that formed into the planets. Most asteroids are located in an area between the orbits of Mars and Jupiter called the asteroid belt. Find the asteroid belt in **Figure 21.** Why are they located there? The gravity of Jupiter might have kept a planet from forming in the area where the asteroid belt is located now.

Other asteroids are scattered throughout the solar system. They might have been thrown out of the belt by Jupiter's gravity. Some of these asteroids have orbits that cross Earth's orbit. Scientists monitor the positions of these asteroids. However, it is unlikely that an asteroid will hit Earth in the near future.

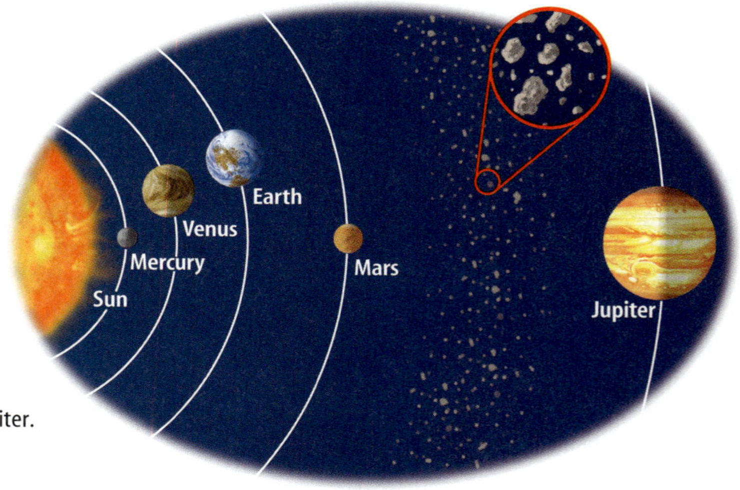

Figure 21 The asteroid belt lies between the orbits of Mars and Jupiter.

Differentiated Instruction

Challenge Have students research the nickel mines of Sudbury Basin in Ontario, Canada. Ask them to write reports telling why so much nickel is found there. It is thought that about 2 billion years ago, an asteroid 6 km in diameter collided with Earth at this location. The collision caused the formation of rich deposits of nickel, cobalt, and platinum. L3 LS **Linguistic** P

Exploring Asteroids The sizes of the asteroids in the asteroid belt range from tiny particles to objects 940 km in diameter. Ceres is the largest and the first one discovered. The next three in order of size are Vesta (530 km), Pallas (522 km), and 10 Hygiea (430 km). The asteroid Gaspra, shown in **Figure 22,** was photographed by *Galileo* on its way to Jupiter.

Figure 22 The asteroid Gaspra is about 20 km long.

NEAR On February 14, 2000, the *Near Earth Asteroid Rendezvous (NEAR)* spacecraft went into orbit around the asteroid 433 Eros and successfully began its one-year mission of data gathering. Data from the spacecraft show that Eros's surface has a large number of craters. Other data indicate that Eros might be similar to the most common type of meteorite that strikes Earth. On February 12, 2001, *NEAR* ended its mission by becoming the first spacecraft to land softly on an asteroid.

Comets, asteroids, and most meteorites formed early in the history of the solar system. Scientists study these space objects to learn what the solar system might have been like long ago. Understanding this could help scientists better understand how Earth formed.

section 4 review

Summary

Comets

- Comets consist of dust, rock, and different types of ice.
- Billions of comets surround the solar system in the Oort Cloud.

Meteoroids, Meteors, Meteorites

- When meteoroids burn up in the atmosphere, they are called meteors.
- Meteor showers occur when Earth crosses the orbital path of a comet.

Asteroids

- Many asteroids occur between the orbits of Mars and Jupiter. This region is called the asteroid belt.

Self Check

1. **Describe** how a comet changes when it comes close to the Sun.
2. **Explain** how craters form.
3. **Summarize** the differences between comets and asteroids.
4. **Describe** the mission of the *NEAR* spacecraft.
5. **Think Critically** A meteorite found in Antarctica is thought to have come from Mars. How could a rock from Mars get to Earth?

Applying Math

6. **Use Proportions** During the 2001 Leonid Meteor Shower, some people saw 20 meteors each minute. Assuming a constant rate, how many meteors did these people see in one hour?

Science Online bookj.msscience.com/self_check_quiz

3 Assess

DAILY INTERVENTION

Check for Understanding

Auditory-Musical Ask students to create a song that distinguishes each of the following objects: comet, asteroid, meteor, meteoroid, and meteorite. L2

Reteach

Write Sentences Have students write three sentences, each containing one of the following terms: *Oort Cloud, Kuiper Belt,* and *asteroid belt between Mars and Jupiter*. Instruct students to take turns reading their sentences aloud. Write a few on the board as you discuss with students whether or not the terms were used correctly. L2 LS **Linguistic**

Assessment

Content Have students draw a picture of a comet at several points in its orbit. The comet's tail, when developed, should always point away from the Sun. Use **Performance Assessment in the Science Classroom,** p. 127. P

section 4 review

1. Some of the ice in the comet's nucleus vaporizes, and dust particles escape into space. The solar wind pushes gas and dust causing tails to form.
2. Craters form when an object from space, such as an asteroid, impacts Earth or some other celestial body.
3. Comets contain ices and some rocky material. Asteroids are all rock and metal. Many other answers are possible.
4. The *NEAR* spacecraft gathered data about asteroid 433 Eros and was the first craft to soft land on an asteroid.
5. An object struck Mars's surface and caused a crater to form. During the impact, a rock from Mars was thrown into space. This rock eventually was caught by Earth's gravity and fell to Earth in Antarctica.
6. 1,200

BENCH TESTED

Real-World Question

Purpose Students design a model that shows the distances between the Sun and the planets. L2 ELL COOP LEARN LS

Logical Mathematical P

Process Skills measure in SI, use numbers, sequence, research, form a hypothesis, separate and control variables, interpret data, make models

Time Required one class period

Materials Photocopy one copy of the Planetary Distances table for each group. Enlarge it slightly if possible.

Safety Precautions Caution students to handle scissors with care.

Make a Model

Possible Procedures Determine a scale for making a solar system model from string. Design a method for placing the planets in proper positions on the model.

Teaching Strategy Have students test their scales by first seeing how far Pluto would be from the Sun. If this distance is too large, they should change the scale.

LAB **Model and Invent**

Solar System Distance Model

Goals

- **Design** a table of scale distances and model the distances between and among the Sun and the planets.

Possible Materials

meterstick
scissors
pencil
string (several meters)
notebook paper (several sheets)

Safety Precautions

Use care when handling scissors.

Real-World Question

Distances between the Sun and the planets of the solar system are large. These large distances can be difficult to visualize. Can you design and create a model that will demonstrate the distances in the solar system?

Make a Model

1. **List** the steps that you need to take to make your model. Describe exactly what you will do at each step.
2. **List** the materials that you will need to complete your model.
3. **Describe** the calculations that you will use to get scale distances from the Sun for all nine planets.
4. **Make** a table of scale distances that you will use in your model. Show your calculations in your table.
5. **Write** a description of how you will build your model. Explain how it will demonstrate relative distances between and among the Sun and planets of the solar system.

Planetary Distances

Planet	Distance to Sun (km)	Distance to Sun (AU)	Scale Distance (1 AU = 10 cm)	Scale Distance (1 AU = 2 cm)
Mercury	5.97×10^7	0.39	3.9 cm	78.00 cm
Venus	1.08×10^8	0.72	7.2 cm	1.44 m
Earth	1.50×10^8	1.00	10.0 cm	2.00 m
Mars	2.28×10^8	1.52	15.2 cm	3.04 m
Jupiter	7.78×10^8	5.20	52.0 cm	10.40 m
Saturn	1.43×10^9	9.54	95.4 cm	19.08 m
Uranus	2.87×10^9	19.19	191.0 cm	38.38 m
Neptune	4.50×10^9	30.07	300.7 cm	60.14 m
Pluto	5.92×10^9	39.48	394.8 cm	78.96 m

Differentiated Instruction

Challenge Have interested students use a computer spreadsheet program to do the calculations for the scale. L3 LS **Logical-Mathematical**

Using Scientific Methods

Test Your Model

1. **Compare** your scale distances with those of other students. Discuss why each of you chose the scale that you did.
2. Make sure your teacher approves your plan before you start.
3. **Construct** the model using your scale distances.
4. While constructing the model, write any observations that you or other members of your group make, and complete the data table in your Science Journal. Calculate the scale distances that would be used in your model if 1 AU = 2 m.

Analyze Your Data

1. **Explain** how a scale distance is determined.
2. Was it possible to work with your scale? Explain why or why not.
3. How much string would be required to construct a model with a scale distance of 1 AU = 2 m?
4. Proxima Centauri, the closest star to the Sun, is about 270,000 AU from the Sun. Based on your scale, how much string would you need to place this star on your model?

Conclude and Apply

1. **Summarize** your observations about distances in the solar system. How are distances between the inner planets different from distances between the outer planets?
2. Using your scale distances, determine which planet orbits closest to Earth. Which planet's orbit is second closest?

Communicating Your Data

Compare your scale model with those of other students. Discuss any differences. **For more help, refer to the Science Skill Handbook.**

Test Your Model

Expected Outcome Students will make scale models that show the distances between bodies in the solar system.

Analyze Your Data

1. by multiplying the AU distance by the scale selected
2. Answers will be subjective and based on the student's individual choices.
3. about 79 m
4. Possible answer: For a scale of 1 AU = 10 cm, Proxima Centauri would be about 27 km away from the model Sun.

Conclude and Apply

1. Distances between the inner planets are smaller than distances between the outer planets.
2. Venus is closest. Mars is second closest.

Assessment

Performance Using the same scales selected for planet distances on their solar system models, have students place the asteroid Ceres in their models. The orbit of Ceres lies 2.77 AU from the Sun. Use **Performance Assessment in the Science Classroom,** p. 123.

Communicating Your Data

Have students compare and contrast their scales and models with representations of the solar system in books and encyclopedias. Did these sources handle the problem of scale in the same way? Ask students to evaluate what they did well and what they could have done better.

Content Background

A large number of meteoroids enter Earth's atmosphere every day. The total mass of this daily onslaught amounts to more than 100 tons of material. However, most of these objects are very small, only a few milligrams each. Only the largest ones ever reach Earth's surface to become meteorites. The largest single meteorite found is the 60–ton Hoba Meteorite, discovered in 1920 in Namibia, in southwestern Africa.

Discussion

Explain to students that, of all the meteorites found on Earth, about a third were witnessed "falls," which means the meteorite was seen by someone as it fell from the sky. The rest are classified as "finds," which means they were found on the ground, but no one witnessed their falling. Would you classify the Burnwell meteorite a fall or a find? Why? It was a fall. Even thought the Peggs did not see it fall, they heard the noise of its fall and its impact when it occurred.

Analyze the Event

Discuss with students how the discovery of the Burnwell meteorite might have been affected if the Peggs had not been home at the time. Possible answer: They might not have realized what caused the damage to their home or realized that it was something that came from outer space. If so, scientists might not have become involved in investigating the incident, and the rare characteristics of the meteorite might never have been discovered.

Oops! Accidents in SCIENCE

SOMETIMES GREAT DISCOVERIES HAPPEN BY ACCIDENT!

IT CAME FROM OUTER SPACE!

On September 4, 1990, Frances Pegg was unloading bags of groceries in her kitchen in Burnwell, Kentucky. Suddenly, she heard a loud crashing sound. Her husband Arthur heard the same sound. The sound frightened the couple's goat and horse. The noise had come from an object that had crashed through the Pegg's roof, their ceiling, and the floor of their porch. They couldn't see what the object was, but the noise sounded like a gunshot, and pieces of wood from their home flew everywhere. The next day the couple looked under their front porch and found the culprit—a chunk of rock from outer space. It was a meteorite.

For seven years, the Peggs kept their "space rock" at home, making them local celebrities. The rock appeared on TV, and the couple was interviewed by newspaper reporters. In 1997, the Peggs sold the meteorite to the National Museum of Natural History in Washington, D.C., which has a collection of more than 9,000 meteorites. Scientists there study meteorites to learn more about the solar system. One astronomer explained, "Meteorites formed at about the same time as the solar system, about 4.6 billion years ago, though some are younger."

Scientists especially are interested in the Burnwell meteorite because its chemical make up is different from other meteorites previously studied. The Burnwell meteorite is richer in metallic iron and nickel than other known meteorites and is less rich in some metals such as cobalt. Scientists are comparing the rare Burnwell rock with other data to find out if there are more meteorites like the one that fell on the Peggs' roof. But so far, it seems the Peggs' visitor from outer space is one-of-a-kind.

The Burnwell meteorite crashed into the Peggs' home and landed in their basement on the right.

Research Do research to learn more about meteorites. How do they give clues to how our solar system formed? Report to the class.

Science Online
For more information, visit bookj.msscience.com/oops

Research Have students discuss why they think it is important for scientists to learn more about how the solar system was formed.

Resources for Teachers and Students

Comets, Asteroids and Meteorites by Cynthia Pratt Nicolson, Kids Can Press, 1999

Collision Earth!: The Threat from Outerspace by Peter Grego, Sterling Publishing, 1998

Reviewing Main Ideas

Section 1 The Solar System

1. Early Greek scientists thought that Earth was at the center of the solar system. They thought that the planets and stars circled Earth.
2. Today, people know that objects in the solar system revolve around the Sun.

Section 2 The Inner Planets

1. The inner planets are Mercury, Venus, Earth, and Mars.
2. The inner planets are small, rocky planets.

Section 3 The Outer Planets

1. The outer planets are Jupiter, Saturn, Uranus, Neptune, and Pluto.
2. Pluto is a small icy planet. The other outer planets are large, gaseous planets.

Section 4 Other Objects in the Solar System

1. Comets are masses of ice and rock. When a comet approaches the Sun, some ice turns to gas and the comet glows brightly.
2. Meteors occur when small pieces of rock enter Earth's atmosphere and burn up.

Visualizing Main Ideas

Copy and complete the following concept map about the solar system.

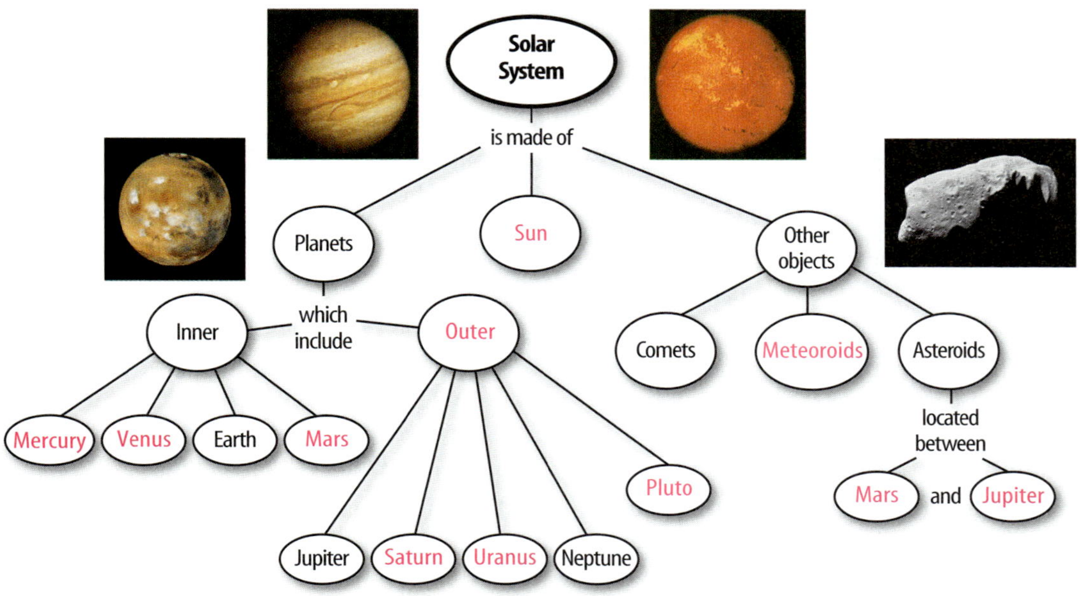

chapter 3 Study Guide

Reviewing Main Ideas

Summary statements can be used by students to review the major concepts of the chapter.

Visualizing Main Ideas

See student page.

Science Online

Visit bookj.msscience.com
/self_check_quiz
/interactive_tutor
/vocabulary_puzzlemaker
/chapter_review
/standardized_test

Identifying Misconceptions

Assess

Use the assessment as follow-up to page F at the beginning of this chapter.

Discussion Use the following table to discuss types of stars. Include the radius, density, and mass of the Sun. Which type of star is the Sun?

Expected Outcome This discussion will assist students in recognizing that the Sun is a star just like other stars in our galaxy.

Star	Class	Radius (solar radii)	Density (solar density)	Mass (solar mass)
Rigel	Supergiant	78	0.00004	20
Deneb	Supergiant	96	0.00002	20
Sirius	Main Sequence	1.9	0.335	2.3
61 Cygni A	Main Sequence	0.7	1.69	0.58
Sun	Main Sequence	1	1	1
40 Erdani B	White Dwarf	0.018	71,000	0.41
Von Maanen's Star	White Dwarf	0.007	47,000	0.14

chapter 3 Review

Using Vocabulary

1. meteor
2. Jupiter
3. Saturn
4. Mars
5. solar system

Checking Concepts

6. B	11. C
7. B	12. C
8. D	13. B
9. C	14. C
10. B	15. B

Assessment Transparency

For additional assessment questions, use the *Assessment Transparency* located in the transparency book.

Assessment

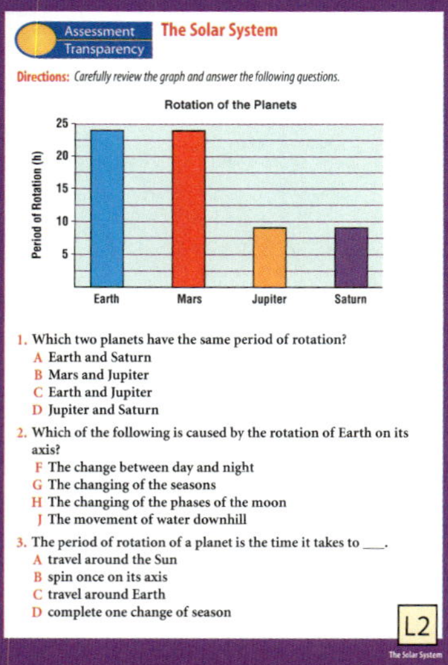
Assessment Transparency — The Solar System

Directions: *Carefully review the graph and answer the following questions.*

1. Which two planets have the same period of rotation?
   A Earth and Saturn
   B Mars and Jupiter
   C Earth and Jupiter
   D Jupiter and Saturn
2. Which of the following is caused by the rotation of Earth on its axis?
   F The change between day and night
   G The changing of the seasons
   H The changing of the phases of the moon
   J The movement of water downhill
3. The period of rotation of a planet is the time it takes to ___.
   A travel around the Sun
   B spin once on its axis
   C travel around Earth
   D complete one change of season

L2

The Solar System

chapter 3 Review

Using Vocabulary

asteroid p. 92	meteorite p. 92
comet p. 90	Neptune p. 86
Earth p. 78	Pluto p. 87
Great Red Spot p. 82	Saturn p. 84
Jupiter p. 82	solar system p. 71
Mars p. 78	Uranus p. 85
Mercury p. 76	Venus p. 77
meteor p. 91	

Fill in the blanks with the correct words.

1. A meteoroid that burns up in Earth's atmosphere is called a(n) ________.
2. The Great Red Spot is a giant storm on ________.
3. ________ is the second largest planet.
4. The *Viking* landers tested for life on ________.
5. The ________ includes the Sun, planets, moons, and other objects.

Checking Concepts

Choose the word or phrase that best answers the question.

6. Who proposed a Sun-centered solar system?
   A) Ptolemy **C)** Galileo
   B) Copernicus **D)** Oort
7. What is the shape of planetary orbits?
   A) circles **C)** squares
   B) ellipses **D)** rectangles
8. Which planet has extreme temperatures because it has no atmosphere?
   A) Earth **C)** Saturn
   B) Jupiter **D)** Mercury
9. Where is the largest volcano in the solar system?
   A) Earth **C)** Mars
   B) Jupiter **D)** Uranus

Use the photo below to answer question 10.

10. Which planet has a complex ring system consisting of thousands of ringlets?
    A) Pluto **C)** Uranus
    B) Saturn **D)** Mars
11. What is a rock from space that strikes Earth's surface?
    A) asteroid **C)** meteorite
    B) meteoroid **D)** meteor
12. By what process does the Sun produce energy?
    A) magnetism
    B) nuclear fission
    C) nuclear fusion
    D) gravity
13. In what direction do comet tails point?
    A) toward the Sun
    B) away from the Sun
    C) toward Earth
    D) away from the Oort Cloud
14. Which planet has abundant surface water and is known to have life?
    A) Mars **C)** Earth
    B) Jupiter **D)** Pluto
15. Which planet has the highest temperatures because of the greenhouse effect?
    A) Mercury **C)** Saturn
    B) Venus **D)** Earth

Science Online bookj.msscience.com/vocabulary_puzzlemaker

Use the Exam*View*® Pro Testmaker CD-ROM to:

- create multiple versions of tests
- create modified tests with one mouse click for inclusion students
- edit existing questions and add your own questions
- build tests aligned with state standards using built-in State Curriculum Tags
- change English tests to Spanish with one mouse click and vice versa

Thinking Critically

16. Infer Why are probe landings on Jupiter not possible?

17. Concept Map Copy and complete the concept map on this page to show how a comet changes as it travels through space.

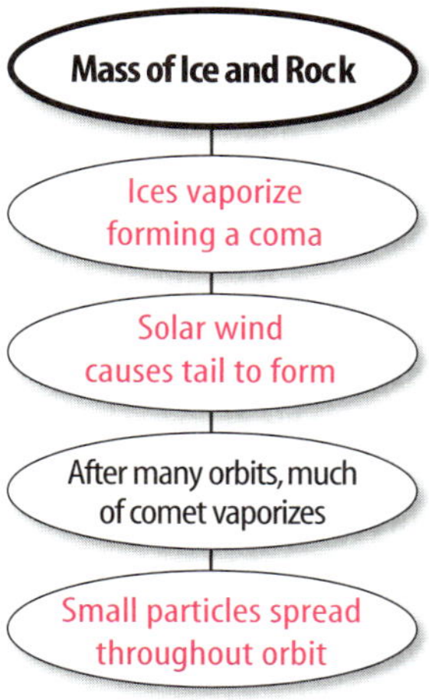

18. Recognize Cause and Effect What evidence suggests that liquid water is or once was present on Mars?

19. Venn Diagram Create a Venn diagram for Earth and Venus. Create a second Venn diagram for Uranus and Neptune. Which two planets do you think are more similar?

20. Recognize Cause and Effect Mercury is closer to the Sun than Venus, yet Venus has higher temperatures. Explain.

21. Make Models Make a model that includes the Sun, Earth, and the Moon. Use your model to demonstrate how the Moon revolves around Earth and how Earth and the Moon revolve around the Sun.

22. Form Hypotheses Why do Mars's two moons look like asteroids?

Performance Activities

23. Display Mercury, Venus, Mars, Jupiter, and Saturn can be observed with the unaided eye. Research when and where in the sky these planets can be observed during the next year. Make a display illustrating your findings. Take some time to observe some of these planets.

24. Short Story Select one of the planets or a moon in the solar system. Write a short story from the planet's or moon's perspective. Include scientifically correct facts and concepts in your story.

Applying Math

25. Saturn's Atmosphere Saturn's atmosphere consists of 96.3% hydrogen and 3.25% helium. What percentage of Saturn's atmosphere is made up of other gases?

26. Length of Day on Pluto A day on Pluto lasts 6.39 times longer than a day on Earth. If an Earth day lasts 24 h, how many hours is a day on Pluto?

Use the graph below to answer question 27.

Weight on Several Planets

Planet	Proportion of Earth's Gravity	Melissa's Weight (lbs)
Mercury	0.378	26.5
Venus	0.903	63.2
Earth	1.000	70
Mars	0.379	26.5
Jupiter	2.54	177
Pluto	0.061	4.27

27. Gravity and Weight Melissa weighs 70 lbs on Earth. Multiply Melissa's weight by the proportion of Earth's gravity for each planet to find out how much Melissa would weigh on each.

Science Online bookj.msscience.com/chapter_review

chapter 3 Review

Thinking Critically

16. Jupiter does not have a solid surface.

17. See student page.

18. Large channels and sedimentary deposits indicate that large amounts of surface water existed on Mars billions of years ago. Some young, erosive features suggest that groundwater might periodically seep out onto the surface of Mars.

19. Students probably will conclude that Uranus and Neptune are more similar than Earth and Venus.

20. The intense greenhouse effect on Venus contributes to the planet's high temperatures.

21. Students should clearly model the Moon revolving around Earth and the Earth-Moon system revolving around the Sun.

22. They probably are captured asteroids.

Performance Activities

23. Data about when planets can be observed are found in many newspapers and through links at bookj.msscience.com. Use **Performance Assessment in the Science Classroom,** p. 141.

24. Stories will vary, but should include correct information about the chosen planet or moon. Use **Performance Assessment in the Science Classroom,** p. 157.

Applying Math

National Math Standards

1,5,9

25. 0.45%

26. 153.4 h

27. See student page.

Assessment Resources

Reproducible Masters

Chapter *Fast File* Resources
Chapter Review, pp. 41–42
Chapter Tests, pp. 43–46
Assessment Transparency Activity, p. 55

Glencoe Science Web site
Chapter Review Test
Standardized Test Practice

Glencoe Technology
Assessment Transparency
Exam*View*® Pro Testmaker
MindJogger Videoquiz
Interactive Chalkboard

FAST FILE

Answer Sheet A practice answer sheet can be found at bookj.msscience.com/answer_sheet.

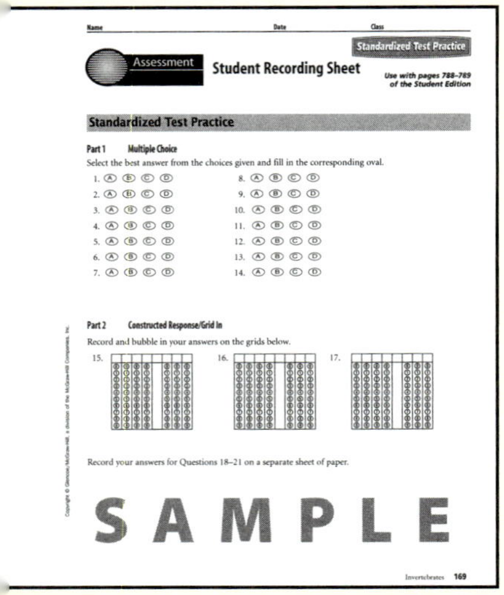
Name Date Class

Assessment Student Recording Sheet

Standardized Test Practice

Use with pages 788–789 of the Student Edition

Standardized Test Practice

Part 1 Multiple Choice

Select the best answer from the choices given and fill in the corresponding oval.

Part 2 Constructed Response/Grid In

Record and bubble in your answers on the grids below.

Record your answers for Questions 18–21 on a separate sheet of paper.

SAMPLE

Part 1 Multiple Choice

1. C
2. D
3. C
4. D
5. A
6. C
7. D
8. B
9. B

Part 1 Multiple Choice

Record your answers on the answer sheet provided by your teacher or on a sheet of paper.

Use the photo below to answer question 1.

1. What is shown in the photo above?
   A. asteroids **C.** meteors
   B. comets **D.** meteorites

2. Which is the ninth planet from the Sun?
   A. Earth **C.** Jupiter
   B. Mars **D.** Pluto

3. What is the name of Pluto's moon?
   A. Ganymede **C.** Charon
   B. Titan **D.** Phobos

4. Which object's gravity holds the planets in their orbits?
   A. Gaspra **C.** Mercury
   B. Earth **D.** the Sun

5. Which of the following occurs in a cycle?
   A. sunspot maxima and minima
   B. condensation of a nebula
   C. formation of a crater
   D. formation of a black hole

Test-Taking Tip

No Peeking During the test, keep your eyes on your own paper. If you need to rest them, close them or look up at the ceiling.

6. Which planet likely will be visited by humans in the future?
   A. Jupiter **C.** Mars
   B. Venus **D.** Neptune

7. Between which two planets' orbits does the asteroid belt occur?
   A. Mercury and Venus
   B. Earth and Mars
   C. Uranus and Neptune
   D. Mars and Jupiter

8. Who discovered that planets have elliptical orbits?
   A. Galileo Galilei
   B. Johannes Kepler
   C. Albert Einstein
   D. Nicholas Copernicus

Use the illustration below to answer question 9.

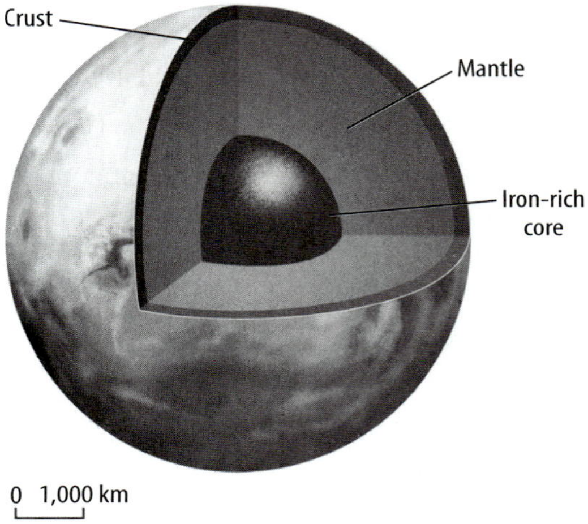

9. Which of the following answers is a good estimate for the diameter of Mars?
   A. 23,122 km **C.** 1,348 km
   B. 6,794 km **D.** 12,583 km

Part 2 Short Response/Grid In

10. The planet's gravity holds the moon in its orbit.
11. The inner planets are small, rocky planets; whereas the outer planets, except for Pluto, are giant, gaseous planets.
12. Pluto has a solid, icy surface.
13. pieces of rock and ice
14. a large, persistent storm in Jupiter's atmosphere
15. Earth is the only planet with abundant surface water. Earth also is the only planet known to support life. Other answers are possible.
16. Each cycle represents one sol, a Martian day.
17. about −25°C
18. about −85°C
19. about 60°C

Part 2 Short Response/Grid In

Record your answers on the answer sheet provided by your teacher or on a sheet of paper.

10. Why does a moon remain in orbit around a planet?

11. Compare and contrast the inner planets and the outer planets.

12. Describe Pluto's surface. How is it different from the other outer planets?

13. Describe Saturn's rings. What are they made of?

14. What is the Great Red Spot?

15. How is Earth different from the other planets in the solar system?

Use the graph below to answfer questions 16–19.

Viking Lander 1 Temperature Data

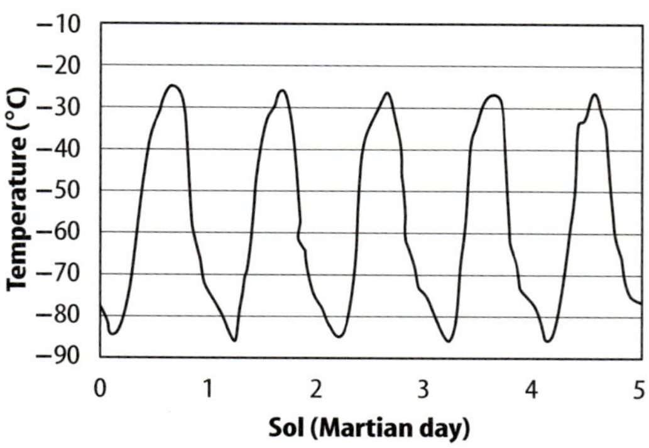

16. Why do the temperatures in the graph vary in a pattern?

17. Approximate the typical high temperature value measured by *Viking I.*

18. Approximate the typical low temperature value measured by *Viking I.*

19. What is the range of these temperature values?

Part 3 Open Ended

Record your answers on a sheet of paper.

20. How might near-Earth-asteroids affect life on Earth? Why do astronomers search for them and monitor their positions?

Use the illustration below to answer question 21.

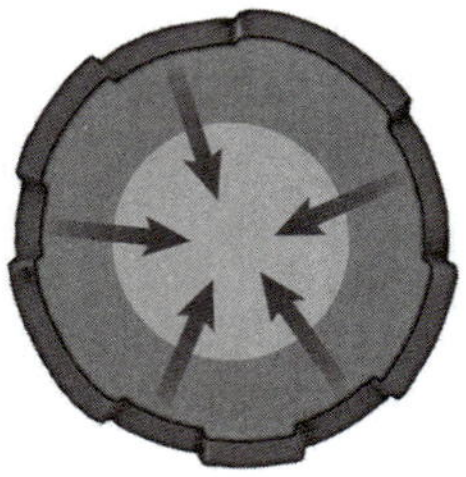

21. Explain how scientists hypothesize that the large cliffs on Mercury formed.

22. Describe the Sun-centered model of the solar system. How is it different from the Earth-centered model?

23. What is an astronomical unit? Why is it useful?

24. Compare and contrast the distances between the planets in the solar system. Which planets are relatively close together? Which planets are relatively far apart?

25. Summarize the current hypothesis about how the solar system formed.

26. Explain how Earth's gravity affects objects that are on or near Earth.

27. Describe the shape of planets' orbits. What is the name of this shape? Where is the Sun located?

28. Describe Jupiter's atmosphere. What characteristics can be observed in images acquired by space probes?

Rubrics

The following rubrics are sample scoring devices for short response and open-ended questions.

Short Response

Points	Description
2	The student demonstrates a thorough understanding of the science of the task. The response may contain minor flaws that do not detract from the demonstration of a thorough understanding.
1	The student has provided a response that is only partially correct.
0	The student has provided a completely incorrect solution or no response at all.

Open Ended

Points	Description
4	The student demonstrates a thorough understanding of the science of the task. The response may contain minor flaws that do not detract from the demonstration of a thorough understanding.
3	The student demonstrates an understanding of the science of the task. The response is essentially correct and demonstrates an essential but less than thorough understanding of the science.
2	The student demonstrates only a partial understanding of the science of the task. Although the student may have used the correct approach to a solution or may have provided a correct solution, the work lacks an essential understanding of the underlying science concepts.
1	The student demonstrates a very limited understanding of the science of the task. The response is incomplete and exhibits many flaws.
0	The student provides a completely incorrect solution or no response at all.

Part 3 Open Ended

20. Some of these asteroids might strike Earth in the future. Astronomers search for and monitor these asteroids to assess the probability of an impact.

21. The cliffs formed when Mercury shrank.

22. The Sun is the central body in the solar system.

23. It is a unit of distance equal to the mean distance between the Sun and Earth. It is useful because distances in the solar system can be measured without using high numbers.

24. Inner planets are relatively close together. Outer planets are far apart.

25. The solar system formed from a portion of a contracting nebula.

26. It pulls them toward Earth's center.

27. ellipses; the Sun is at one focus.

28. thick atmosphere of mostly hydrogen and helium; storms, color bands

chapter 4 Organizer

Section/Objectives	Standards		Labs/Features
	National	State/Local	
Chapter Opener	See pp. 9T–10T for a Key to Standards.		**Launch Lab:** Why do clusters of galaxies move apart?, p. 103 **Foldables,** p. 103
Section 1 Stars 1 session 5 block 1. **Explain** why some constellations are visible only during certain seasons. 2. **Distinguish** between absolute magnitude and apparent magnitude.	National Content Standards: UCP.1–3, UCP.5, A.1, A.2		**MiniLAB:** Observing Star Patterns, p. 105 **Applying Science:** Are distance and brightness related?, p. 106
Section 2 The Sun 2 sessions 1 block 3. **Explain** that the Sun is the closest star to Earth. 4. **Describe** the structure of the Sun. 5. **Describe** sunspots, prominences, and solar flares.	National Content Standards: UCP.1–3, UCP.5, A.1, A.2, D.3		**Science Online,** p. 111 **Lab:** Sunspots, p. 113
Section 3 Evolution of Stars 2 sessions 1 block 6. **Describe** how stars are classified. 7. **Compare** the Sun to other types of stars on the H-R diagram. 8. **Describe** how stars evolve.	National Content Standards: UCP.1–5, A.1, A.2, B.1–3, G.3		**Science Online,** p. 116 **Integrate Chemistry,** p. 117 **Integrate History,** p. 118
Section 4 Galaxies and the Universe 4 sessions 2 blocks 9. **Describe** the Sun's position in the Milky Way Galaxy. 10. **Explain** that the same natural laws that apply to our solar system also apply in other galaxies.	National Content Standards: UCP.1–5, A.1, A.2		**MiniLAB:** Measuring Distance in Space, p. 122 **Visualizing the Big Bang Theory,** p. 124 **Lab:** Measuring Parallax, p. 126 **Science Stats:** Stars and Galaxies, p. 128

Glencoe Exclusive! **TeacherWorks™** All-In-One Planner and Resource Center

Lab Materials	Reproducible Resources	Section Assessment	Technology
Launch Lab: balloon, felt-tip marker, string, metric ruler	**Chapter *FAST FILE* Resources** Foldables Worksheet, p. 17 Directed Reading Overview, p. 19 Note-taking Worksheets, pp. 35–38		**TeacherWorks** includes: • Interactive Teacher Edition • Lesson Planner with calendar • Access to all program blacklines • Correlations to standards • Web links
MiniLAB: paper, drawing materials	**Chapter *FAST FILE* Resources** Transparency Activity, p. 48 MiniLAB, p. 3 Enrichment, p. 31 Reinforcement, p. 27 Directed Reading, p. 20 Lab Activity, pp. 9–12, 13–15 Transparency Activity, pp. 53–54	**Portfolio** Science Journal, p. 105 **Performance** MiniLAB, p. 105 Applying Science, p. 106 Applying Skills, p. 108 **Content** Section Review, p. 108	Section Focus Transparency Teaching Transparency Guided Reading Audio Program Interactive Chalkboard CD-ROM
Lab: several books, piece of cardboard, drawing paper, refracting telescope, clipboard, small tripod, scissors	**Chapter *FAST FILE* Resources** Transparency Activity, p. 49 Lab Worksheets, pp. 5–6 Enrichment, p. 32 Reinforcement, p. 28 Directed Reading, p. 20 **Cultural Diversity,** p. 29	**Portfolio** Differentiated Instruction, p. 110 **Performance** Applying Skills, p. 112 **Content** Section Review, p. 112	Section Focus Transparency Guided Reading Audio Program Interactive Chalkboard CD-ROM
Need materials? Contact Science Kit at 1-800-828-7777 or www.sciencekit.com on the Internet.	**Chapter *FAST FILE* Resources** Transparency Activity, p. 50 Enrichment, p. 33 Reinforcement, p. 29 Directed Reading, p. 21	**Portfolio** Differentiated Instruction, p. 115 **Performance** Applying Math, p. 119 **Content** Section Review, p. 119	Section Focus Transparency Virtual Labs CD-ROM Guided Reading Audio Program Interactive Chalkboard CD-ROM
MiniLAB: large sheet of paper, metric ruler or meterstick, markers **Lab:** meterstick, metric ruler, masking tape, pencil	**Chapter *FAST FILE* Resources** Transparency Activity, p. 51 Lab Worksheets, pp. 7–8 MiniLAB, p. 4 Enrichment, p. 34 Reinforcement, p. 30 Directed Reading, pp. 21, 22 **Cultural Diversity,** p. 31	**Portfolio** Assessment, p. 125 **Performance** MiniLAB, p. 122 Applying Skills, p. 125 **Content** Section Review, p. 125	Guided Reading Audio Program Interactive Chalkboard CD-ROM Video Lab

End of Chapter Assessment

Blackline Masters	Technology	Professional Series
Chapter *FAST FILE* Resources Chapter Review, pp. 41–42 Chapter Tests, pp. 43–46 **Standardized Test Practice,** pp. 19–22	MindJogger Videoquiz Virtual Labs CD-ROM Exam*View*® Pro Testmaker TeacherWorks CD-ROM Interactive Chalkboard CD-ROM	**Performance Assessment in the Science Classroom (PASC)**

Transparencies

Section Focus

Section 1 Section Focus Transparency **A Starry Night**

On a clear night, if you are far away from city lights, you can see hundreds of stars.

1. What constellations can you identify where you live?
2. Make up your own constellations using the stars in this photograph. Name each constellation and explain what it represents.
3. Why might some stars look brighter than others?

L2

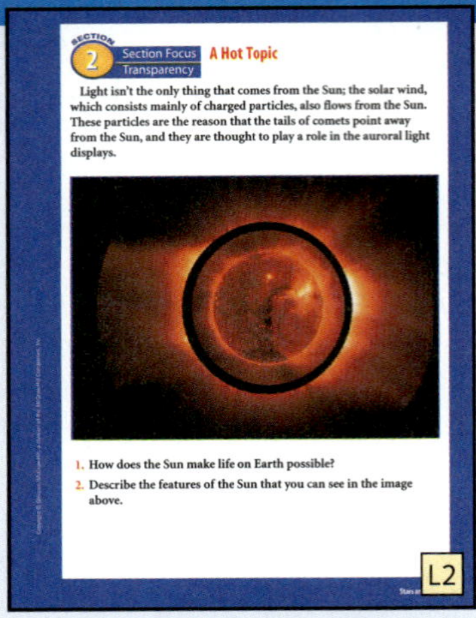
Section 2 Section Focus Transparency **A Hot Topic**

Light isn't the only thing that comes from the Sun; the solar wind, which consists mainly of charged particles, also flows from the Sun. These particles are the reason that the tails of comets point away from the Sun, and they are thought to play a role in the auroral light displays.

1. How does the Sun make life on Earth possible?
2. Describe the features of the Sun that you can see in the image above.

L2

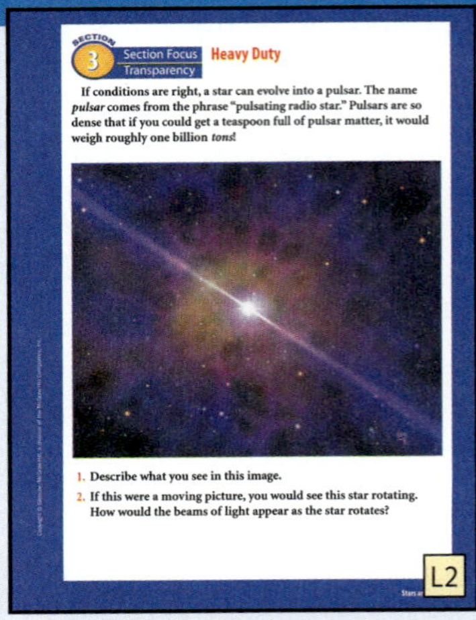
Section 3 Section Focus Transparency **Heavy Duty**

If conditions are right, a star can evolve into a pulsar. The name *pulsar* comes from the phrase "pulsating radio star." Pulsars are so dense that if you could get a teaspoon full of pulsar matter, it would weigh roughly one billion *tons*!

1. Describe what you see in this image.
2. If this were a moving picture, you would see this star rotating. How would the beams of light appear as the star rotates?

L2

This is a representation of key blackline masters available in the Teacher Classroom Resources. See Resource Manager boxes within the chapter for additional information.

Assessment

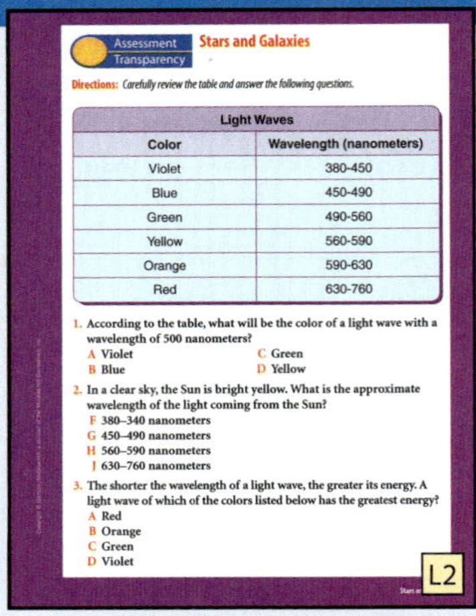
Assessment Transparency **Stars and Galaxies**

Directions: *Carefully review the table and answer the following questions.*

Light Waves	
Color	Wavelength (nanometers)
Violet	380-450
Blue	450-490
Green	490-560
Yellow	560-590
Orange	590-630
Red	630-760

1. According to the table, what will be the color of a light wave with a wavelength of 500 nanometers?
   A Violet C Green
   B Blue D Yellow
2. In a clear sky, the Sun is bright yellow. What is the approximate wavelength of the light coming from the Sun?
   F 380–340 nanometers
   G 450–490 nanometers
   H 560–590 nanometers
   J 630–760 nanometers
3. The shorter the wavelength of a light wave, the greater its energy. A light wave of which of the colors listed below has the greatest energy?
   A Red
   B Orange
   C Green
   D Violet

L2

Teaching

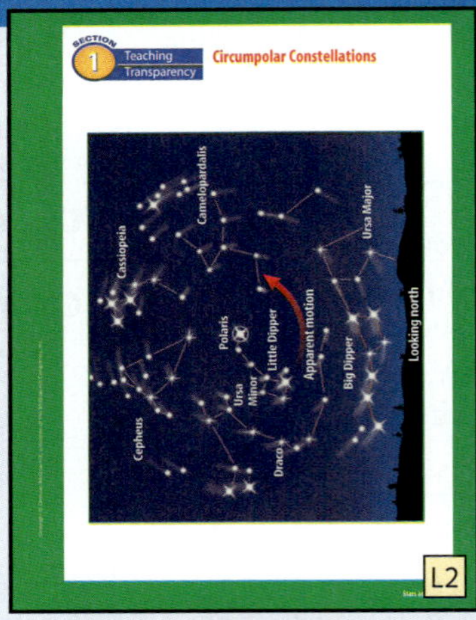
Section 1 Teaching Transparency **Circumpolar Constellations**

L2

Key to Teaching Strategies

The following designations will help you decide which activities are appropriate for your students.

- L1 Level 1 activities should be appropriate for students with learning difficulties.
- L2 Level 2 activities should be within the ability range of all students.
- L3 Level 3 activities are designed for above-average students.
- ELL ELL activities should be within the ability range of English Language Learners.
- COOP LEARN Cooperative Learning activities are designed for small group work.
- LS Multiple Learning Styles logos, as described on page 6T, are used throughout to indicate strategies that address different learning styles.
- P These strategies represent student products that can be placed into a best-work portfolio.
- PBL Problem-Based Learning activities apply real-world situations to learning.

Hands-on Activities

Activity Worksheets

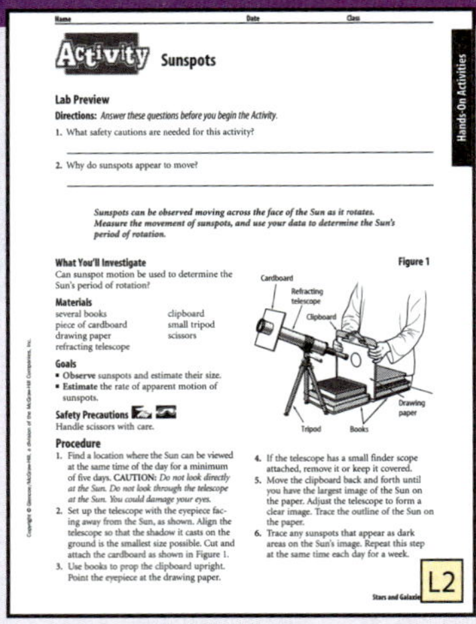
Activity **Sunspots**

Lab Preview

Directions: *Answer these questions before you begin the Activity.*

1. What safety cautions are needed for this activity?
2. Why do sunspots appear to move?

Sunspots can be observed moving across the face of the Sun as it rotates. Measure the movement of sunspots, and use your data to determine the Sun's period of rotation.

L2

Laboratory Activities

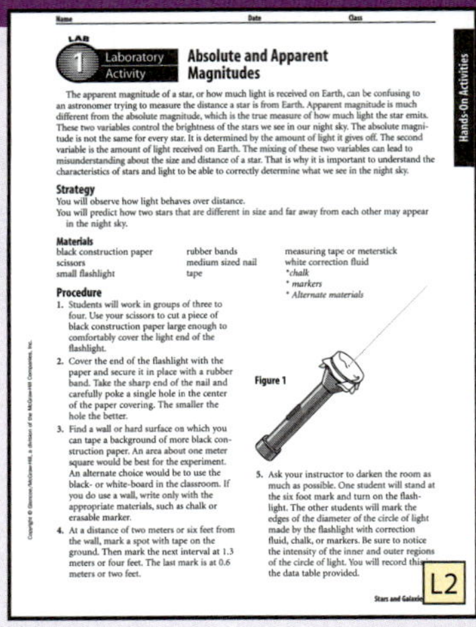
Laboratory Activity 1 **Absolute and Apparent Magnitudes**

L2

Meeting Different Ability Levels

Content Outline

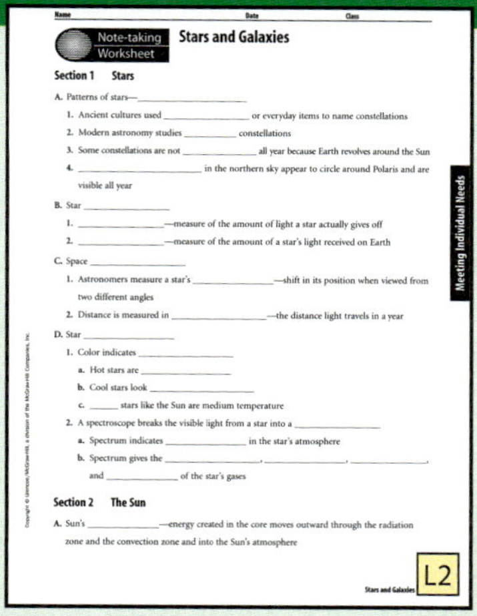

Reinforcement

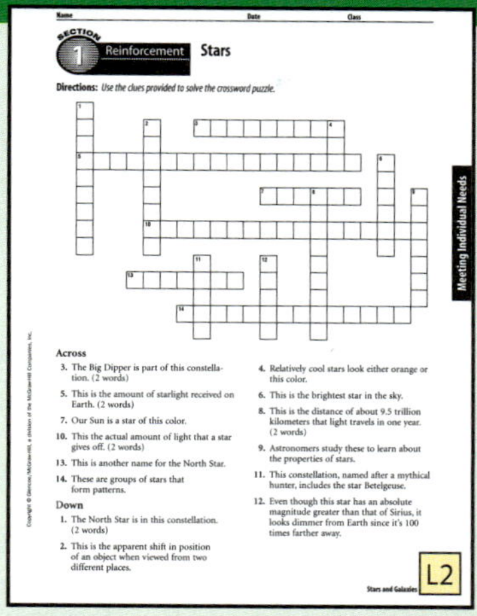

Enrichment

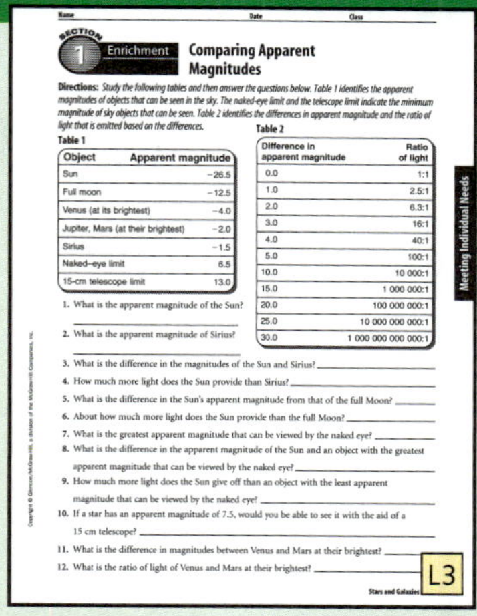

Directed Reading (English/Spanish)

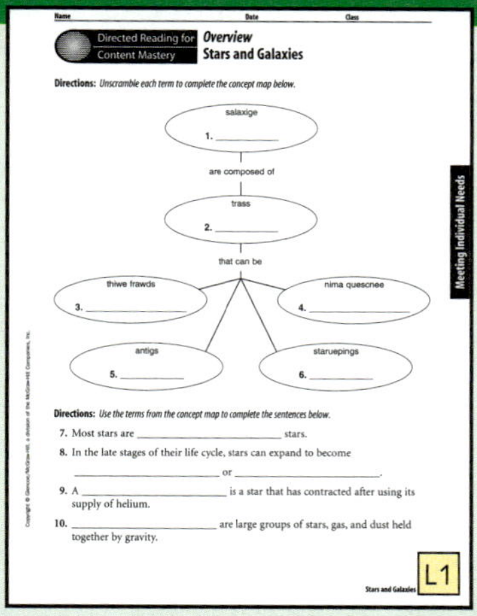

Study Guide

Reading Essentials

Assessment

Test Practice Workbook

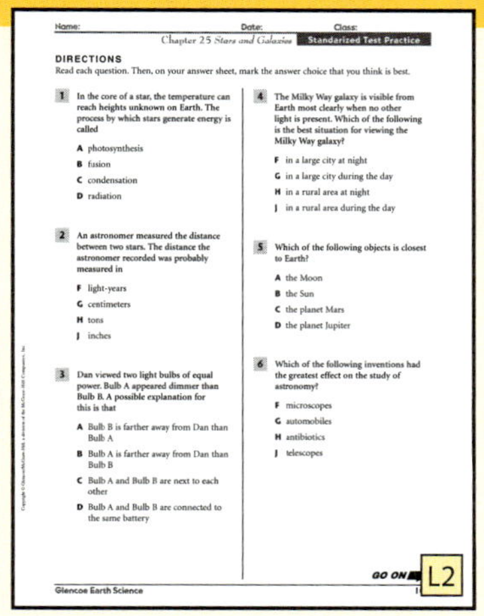

Chapter Review

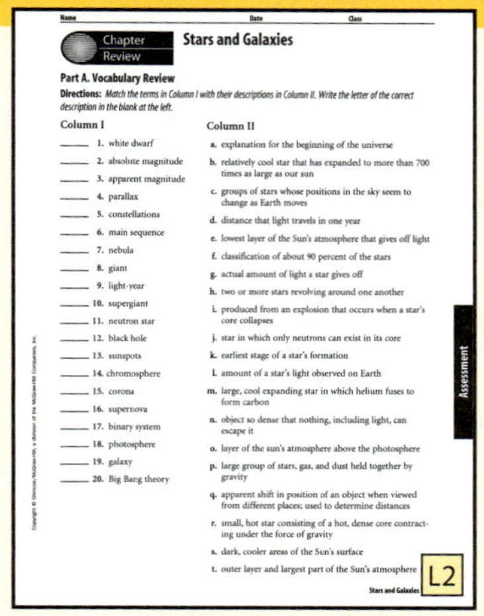

Chapter Tests

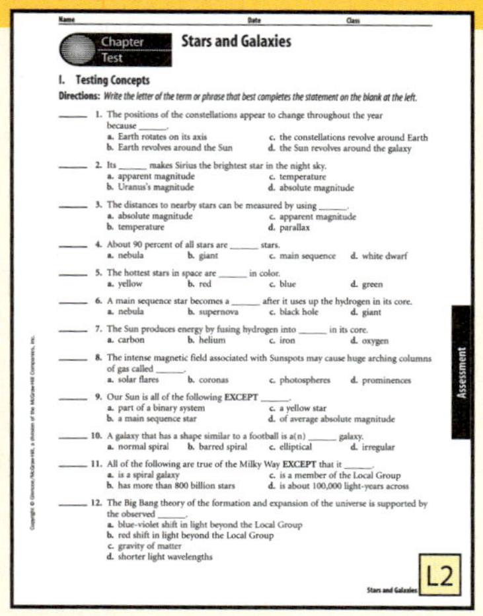

Science Content Background

section 1 Stars

Absolute and Apparent Magnitudes

The size and temperature of a star affect its absolute magnitude. The size and temperature of a star, along with its distance from Earth, affect its apparent magnitude. If the absolute and apparent magnitudes of a star are known, the distance from the star to Earth can be determined.

Teacher to Teacher
Dennis L. Stockdale
Asheville High School
Asheville, N.C.

"While studying the stars, my students build a planetarium from a 50-foot square of plastic film. The final product will be 30 feet in diameter and 16 feet tall. Plans are available in *The Science Teacher,* vol. 64, no. 7, October 1997."

Dennis L. Stockdale

Clouds of gas and dust in space, called nebulae, can obscure starlight, thus reducing a star's apparent magnitude. Interstellar nebulae reduce the apparent magnitude of stars in many parts of our galaxy to the point where they cannot be seen. The gas and dust from these interstellar nebulae can also make a star appear red in color. Interstellar material tends to be opaque to light of shorter wavelengths, allowing longer wavelengths of red light to pass through.

section 2 The Sun

Surface Features

The phenomenon called granulation seen on the surface of the Sun was first observed by Galileo. He thought the patterns he observed resembled grains of wheat and thus called the features granules. Granules on the Sun's surface are the tops of large convection cells that extend deep into the Sun. Their presence is evidence that convection currents do exist inside the Sun, just under the photosphere. These granules form when hot gases are forced upward by surrounding denser gases. The hot gases flow toward the surface, emit energy, cool, and sink back toward the Sun's interior. Each granule measures about 1,000 km across.

Sunspots

Sunspots occur in a periodic pattern. Magnetic reversals occur every 11 years; therefore, a sunspot cycle is complete after 22 years. Solar activity affects circulation within Earth's atmosphere. Some scientists see a correlation between the Sun's 22-year activity cycle and droughts on Earth. Solar activity might affect Earth's climate over longer periods of time. The Little Ice Age that gripped northern Europe for over 300 years has been associated with the Maunder minimum, a time during the late 1600s when sunspot occurrences were very low.

chapter content resources

Internet Resources
For additional content background, visit bookj.msscience.com to:

- access your book onlineonline
- find references to related articles in popular science magazines
- access Web links with related content background
- access current events with science journal topics

Print Resources

Discovering the Universe, by Neil F. Comins, and William J. Kaufmann, W.H. Freeman, 2003

A Dictionary of Astronomy, Ian Ridpath (ed.), Oxford University Press, 1998

Star Chart Map, American Map Corporation, 1995

section 3 Evolution of Stars

Classifying Stars

Not all stars shine with a steady light. Stars that change brightness in a cyclic manner are called variable stars. Stars may vary because the outer layers of the star expand and contract, causing a change in the temperature and the absolute magnitude of the star.

One class of variable stars, the Cepheid variable, provides important information related to the study of the universe. These stars vary regularly, and their period of variation is an indication of their absolute magnitudes. Because of this, they can be used to determine distances to faraway clusters and galaxies. If two Cepheid variables have the same period of pulsation, they are the same average size and have the same average absolute magnitude. Any difference noted in the apparent magnitudes of the two stars is caused by a difference in the distance to each of the stars. Polaris is an example of a Cepheid variable.

Black Holes

New infrared studies of the center of the Milky Way provide more evidence for a supermassive black hole at the core of our galaxy. The total output of energy from this object at the Milky Way's core is more than 1 million times that of the Sun, and it appears to contain more than 2 million solar masses (1 solar mass = mass of the Sun = 2×10^{30} kg). Most other galaxies also contain supermassive black holes.

Quasars are supermassive black holes that existed billions of years ago but still can be seen today at great distances. These objects emitted much more energy than modern supermassive black holes emit. Some quasars are thousands of times brighter than entire galaxies. Some have suggested that today's supermassive black holes might be "fossil" quasars.

section 4 Galaxies and the Universe

Galaxies

The Milky Way and Andromeda Galaxies both have a number of companion dwarf spheroidal galaxies. A dwarf spheroidal galaxy called the Sagittarius Dwarf Galaxy is about 50,000 light-years from the nucleus of the Milky Way. It lies in the direction of the constellation Sagittarius on the opposite side of the Milky Way from Earth. Researchers hypothesize that the Sagittarius Dwarf Galaxy has orbited the Milky Way many times. The stars from this small galaxy eventually will be incorporated into the Milky Way. This supports a theory that large galaxies, such as the Milky Way, are formed, in part, by the incorporation of smaller galaxies. Larger galaxies, therefore, are thought to have been built up over time from smaller galaxies.

Luke Dodd/Science Library/Photo Researchers, Inc.

chapter 4

Chapter Vocabulary

constellation, p. 104
absolute magnitude, p. 106
apparent magnitude, p. 106
light-year, p. 107
photosphere, p. 109
chromosphere, p. 109
corona, p. 109
sunspot, p. 110
nebula, p. 116
giant, p. 117
white dwarf, p. 117
supergiant, p. 118
neutron star, p. 118
black hole, p. 118
galaxy, p. 120
big bang theory, p. 125

Science Journal Students' descriptions should include spiral arms and a bright galactic core.

INTERACTIVE CHALKBOARD with Image Bank

PowerPoint® Presentations

This CD-ROM is an editable Microsoft® PowerPoint® presentation that includes:

- a pre-made presentation for every chapter
- interactive graphics
- animations
- audio clips
- image bank
- all new section and chapter questions
- Standardized Test Practice
- transparencies
- pre-lab questions for all labs
- Foldables directions
- links to bookj.msscience.com

chapter 4

Stars and Galaxies

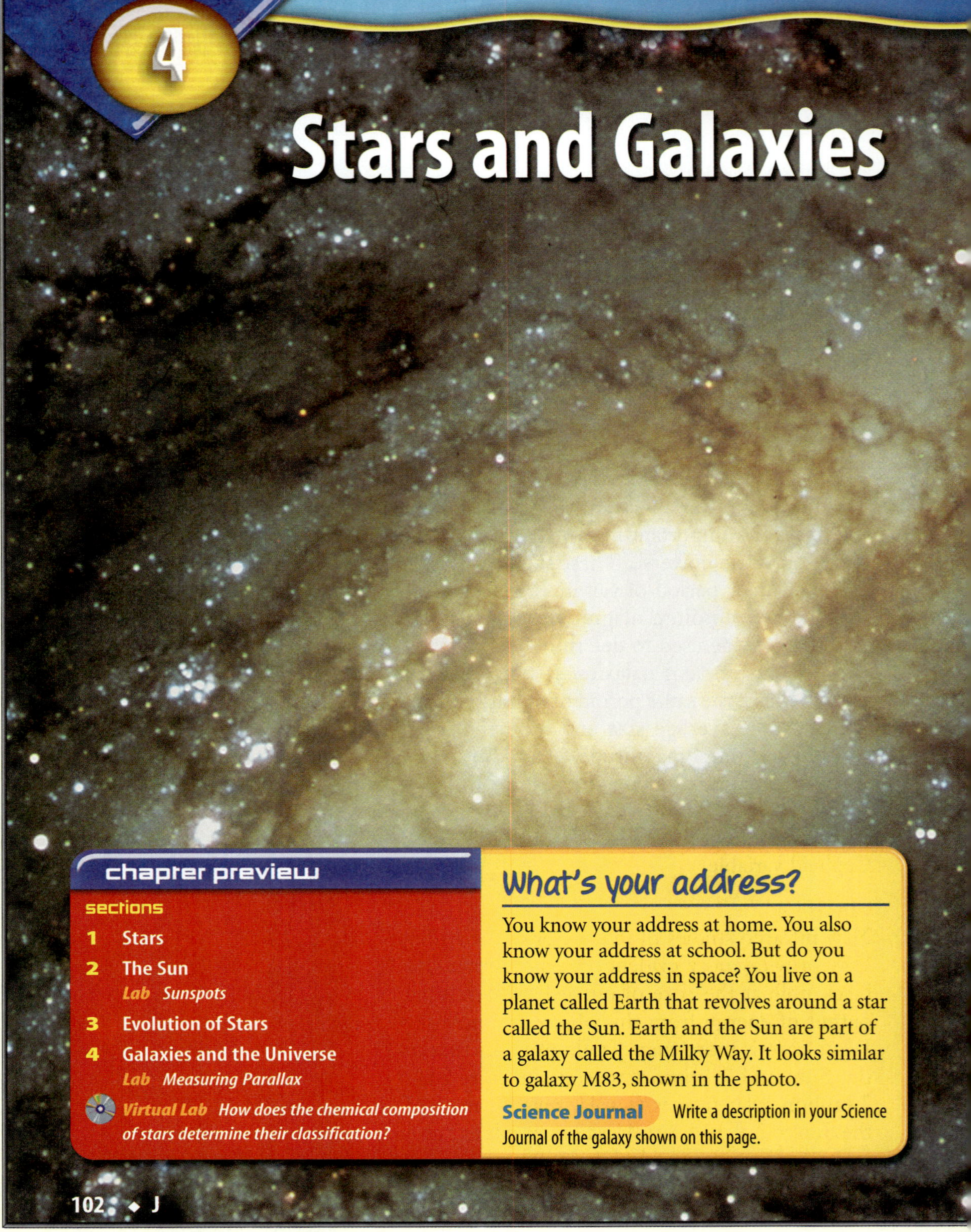

chapter preview

sections

1 Stars
2 The Sun
Lab Sunspots
3 Evolution of Stars
4 Galaxies and the Universe
Lab Measuring Parallax
Virtual Lab How does the chemical composition of stars determine their classification?

What's your address?

You know your address at home. You also know your address at school. But do you know your address in space? You live on a planet called Earth that revolves around a star called the Sun. Earth and the Sun are part of a galaxy called the Milky Way. It looks similar to galaxy M83, shown in the photo.

Science Journal Write a description in your Science Journal of the galaxy shown on this page.

102 ◆ J

Theme Connection

Scale and Structure and **Stability and Change** The major themes of this chapter deal with the vast scale and structure of the universe and how the universe, and the matter that it contains, changes.

About the Photo

Galaxy M83 The photo shows galaxy M83 in the constellation Hydra. M83 is a barred spiral galaxy that probably is similar to the Milky Way.

Start-Up Activities

Launch LAB

Why do clusters of galaxies move apart?

Astronomers know that most galaxies occur in groups of galaxies called clusters. These clusters are moving away from each other in space. The fabric of space is stretching like an inflating balloon.

1. Partially inflate a balloon. Use a piece of string to seal the neck.
2. Draw six evenly spaced dots on the balloon with a felt-tipped marker. Label the dots A through F.
3. Use string and a ruler to measure the distance, in millimeters, from dot A to each of the other dots.
4. Inflate the balloon more.
5. Measure the distances from dot A again.
6. Inflate the balloon again and make new measurements.
7. **Think Critically** Imagine that each dot represents a cluster of galaxies and that the balloon represents the universe. Describe the motion of the clusters in your Science Journal.

Science Online — Preview this chapter's content and activities at bookj.msscience.com

FOLDABLES™ Study Organizer

Stars, Galaxies, and the Universe Make the following Foldable to show what you know about stars, galaxies, and the universe.

STEP 1 **Fold** a sheet of paper from side to side. Make the front edge about 1.25 cm shorter than the back edge.

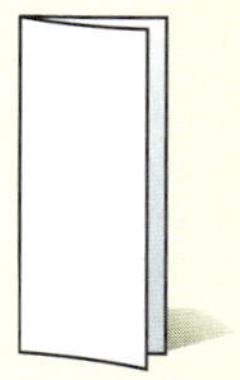

STEP 2 **Turn** lengthwise and **fold** into thirds.

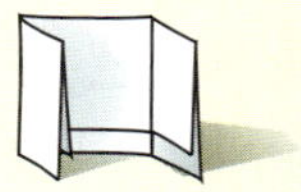

STEP 3 **Unfold and cut** only the top layer along both folds to make three tabs.

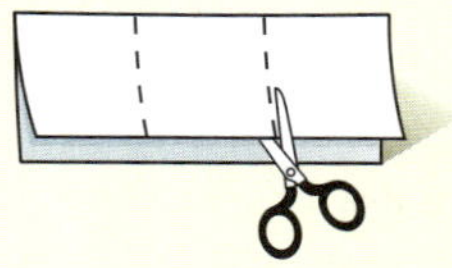

STEP 4 **Label** the tabs *Stars, Galaxies,* and *Universe.*

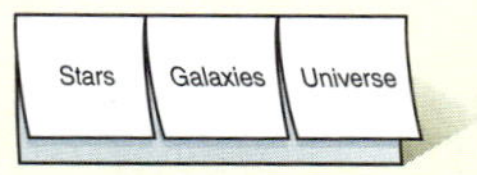

Read and Write Before you read the chapter, write what you already know about stars, galaxies, and the universe. As you read the chapter, add to or correct what you have written under the tabs.

Launch LAB

Purpose Students are introduced to the concept of an expanding universe. L1 ELL COOP LEARN

LS **Kinesthetic**

Materials balloon, felt-tipped marker, string, metric ruler

Teaching Strategies

- Divide the class into groups of four students. Assign roles to students. One student should blow up the balloon and tie it shut. Another should draw the dots on the balloon. The remaining students should measure the distance between the dots.
- Have students compare their results with those of classmates. Discuss differences and similarities.

Safety Precautions Caution students not to blow up the balloons to the breaking point.

Think Critically

The galaxy clusters are moving away from each other. This type of motion implies that the universe is expanding.

Assessment

Process Have students make illustrations that depict the motion of galaxy clusters. Drawings should indicate that clusters are moving away from each other. Use **Performance Assessment in the Science Classroom,** p. 127.

FOLDABLES™ Study Organizer **Dinah Zike Study Fold**

Student preparation materials for this Foldable are available in the Chapter *FAST FILE* Resources.

1 Motivate

INTERACTIVE CHALKBOARD PowerPoint® Presentations

Bellringer

Section Focus Transparencies also are available on the Interactive Chalkboard CD-ROM. L2 ELL

Tie to Prior Knowledge

Big Dipper Ask students if they have observed the Big Dipper, part of the constellation known as Ursa Major. Invite a volunteer to draw the stars of the Big Dipper on the chalkboard to illustrate why it's called a dipper.

section 1

Stars

as you read

What You'll Learn

- **Explain** why some constellations are visible only during certain seasons.
- **Distinguish** between absolute magnitude and apparent magnitude.

Why It's Important

The Sun is a typical star.

Review Vocabulary

star: a large, spherical mass of gas that gives off light and other types of radiation

New Vocabulary

- constellation
- absolute magnitude
- apparent magnitude
- light-year

Constellations

It's fun to look at clouds and find ones that remind you of animals, people, or objects that you recognize. It takes more imagination to play this game with stars. Ancient Greeks, Romans, and other early cultures observed patterns of stars in the night sky called **constellations.** They imagined that the constellations represented mythological characters, animals, or familiar objects.

From Earth, a constellation looks like spots of light arranged in a particular shape against the dark night sky. **Figure 1** shows how the constellation of the mythological Greek hunter Orion appears from Earth. It also shows that the stars in a constellation often have no relationship to each other in space.

Stars in the sky can be found at specific locations within a constellation. For example, you can find the star Betelgeuse (BEE tul jooz) in the shoulder of the mighty hunter Orion. Orion's faithful companion is his dog, Canis Major. Sirius, the brightest star that is visible from the northern hemisphere, is in Canis Major.

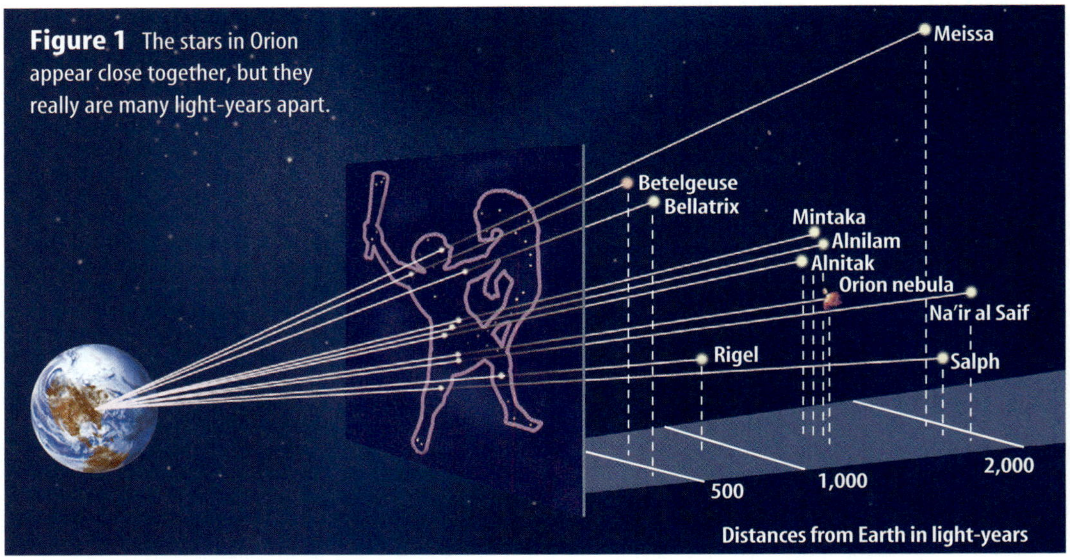

Figure 1 The stars in Orion appear close together, but they really are many light-years apart.

Section 1 Resource Manager

Chapter *FAST FILE* Resources

- Transparency Activity, pp. 48, 53–54
- Directed Reading for Content Mastery, pp. 19, 20
- MiniLAB, p. 3
- Note-taking Worksheets, pp. 35–38
- Enrichment, p. 31
- Reinforcement, p. 27
- Lab Activity, pp. 9–12, 13–15

104 ◆ J CHAPTER 4 Stars and Galaxies

Figure 2 The Big Dipper, in red, is part of the constellation Ursa Major. It is visible year-round in the northern hemisphere. Constellations close to Polaris rotate around Polaris, which is almost directly over the north pole.

Modern Constellations Modern astronomy divides the sky into 88 constellations, many of which were named by early astronomers. You probably know some of them. Can you recognize the Big Dipper? It's part of the constellation Ursa Major, shown in **Figure 2.** Notice how the front two stars of the Big Dipper point almost directly at Polaris, which often is called the North Star. Polaris is located at the end of the Little Dipper in the constellation Ursa Minor. It is positioned almost directly over Earth's north pole.

Circumpolar Constellations As Earth rotates, Ursa Major, Ursa Minor, and other constellations in the northern sky circle around Polaris. Because of this, they are called circumpolar constellations. The constellations appear to move, as shown in **Figure 2,** because Earth is in motion. The stars appear to complete one full circle in the sky in about 24 h as Earth rotates on its axis. One circumpolar constellation that's easy to find is Cassiopeia (ka see uh PEE uh). You can look for five bright stars that form a big W or a big M in the northern sky, depending on the season.

As Earth orbits the Sun, different constellations come into view while others disappear. Because of their unique position, circumpolar constellations are visible all year long. Other constellations are not. Orion, which is visible in the winter in the northern hemisphere, can't be seen there in the summer because the daytime side of Earth is facing it.

Mini LAB

Observing Star Patterns

Procedure

1. On a clear night, go outside after dark and study the stars. Take an adult with you.
2. Let your imagination flow to find patterns of stars that look like something familiar.
3. Draw the stars you see, note their positions, and include a drawing of what you think each star pattern resembles.

Analysis

1. Which of your constellations match those observed by your classmates?
2. How can recognizing star patterns be useful?

Try at Home

2 Teach

Mini LAB

Purpose Students identify constellations. L2 ELL LS **Visual-Spatial**

Materials paper, pencil

Teaching Strategy Provide star charts for students to refer to.

Analysis

1. Many students will likely include familiar constellations; other constellations will be unique.
2. Possible answer: Constellations can be used to identify directions.

Assessment

Performance Have students write a myth about one of the drawings they made. Use **Performance Assessment in the Science Classroom,** p. 155.

Try at Home

IDENTIFYING Misconceptions

Constellations Students might think that stars within a constellation are close to one another in space. Explain that two stars that appear to be side by side can be separated by hundreds of light-years.

Teacher FYI

Asterisms A recognizable pattern of stars within a constellation or among stars in different constellations is called an asterism. The Big Dipper and Little Dipper are asterisms in the constellations of Ursa Major and Ursa Minor.

Science Journal

Constellations from Mythology Have students research a constellation of their choice and write a report in their Science Journals that identifies the stars in the constellation and describes the mythological character or story for which the constellation is named. L2 LS **Linguistic** P

SECTION 1 Stars J ◆ 105

Quick Demo

Bright Lights

Materials identical flashlights (2)

Estimated Time five minutes

Procedure Move students into a hallway. Give two volunteers identical flashlights, have them walk to different locations, and have them shine the flashlights toward the class. Which has a brighter apparent magnitude? the closer flashlight Which has the brighter absolute magnitude? Neither; they have equal absolute magnitudes. L2 LS **Visual-Spatial**

Text Question Answer

The brighter star may be bigger or hotter.

Reading Check

Answer absolute—a measure of the amount of light given off by a star; apparent—a measure of the amount of light received on Earth from a star

Fun Fact

Luminosity, *L*, is a measure of how many times brighter or dimmer a star is compared with the Sun.

Applying Science

Answers

1. As distance is doubled (20 cm to 40 cm), light intensity decreases to 1/4 the original. As distance is tripled (20 cm to 60 cm), light intensity decreases to 1/9 the original.
2. The relationship can be expressed as 1 divided by the square of how many times distance is increased. At 100 cm, the distance will have increased five fold (to five times the original), and the light intensity will be decreased to 1/25 of its original intensity, or 166 luxes.

Absolute and Apparent Magnitudes

When you look at constellations, you'll notice that some stars are brighter than others. For example, Sirius looks much brighter than Rigel. Is Sirius a brighter star, or is it just closer to Earth, making it appear to be brighter? As it turns out, Sirius is 100 times closer to Earth than Rigel is. If Sirius and Rigel were the same distance from Earth, Rigel would appear much brighter in the night sky than Sirius would.

When you refer to the brightness of a star, you can refer to its absolute magnitude or its apparent magnitude. The **absolute magnitude** of a star is a measure of the amount of light it gives off. A measure of the amount of light received on Earth is the **apparent magnitude.** A star that's dim can appear bright in the sky if it's close to Earth, and a star that's bright can appear dim if it's far away. If two stars are the same distance away, what might cause one of them to be brighter than the other?

Reading Check *What is the difference between absolute and apparent magnitude?*

Applying Science

Are distance and brightness related?

The apparent magnitude of a star is affected by its distance from Earth. This activity will help you determine the relationship between distance and brightness.

Identifying the Problem

Luisa conducted an experiment to determine the relationship between distance and the brightness of stars. She used a meterstick, a light meter, and a lightbulb. She placed the bulb at the zero end of the meterstick, then placed the light meter at the 20-cm mark and recorded the distance and the light-meter reading in her data table. Readings are in luxes, which are units for measuring light intensity. Luisa then increased the distance from the bulb to the light meter and took more readings. By examining the data in the table, can you see a relationship between the two variables?

Effect of Distance on Light

Distance (cm)	Meter Reading (luxes)
20	4150.0
40	1037.5
60	461.1
80	259.4

Solving the Problem

1. What happened to the amount of light recorded when the distance was increased from 20 cm to 40 cm? When the distance was increased from 20 cm to 60 cm?
2. What does this indicate about the relationship between light intensity and distance? What would the light intensity be at 100 cm? Would making a graph help you visualize the relationship?

Differentiated Instruction

Challenge Have students measure the distance between their eyes in centimeters. Then have them do calculations using the following equation to determine the parallax angle they observe at various distances from an object: $P° = (360°)(B)/2\pi r$, where $P°$ is the parallax angle, B is the student's baseline (distance between eyes), and r is the distance to the object. Students should plug in different values for r in centimeters. L3

Learning Disabled Pair each learning disabled student with another student. Have pairs cut a 4-cm square from black construction paper and punch pinholes in the square to show a constellation's outline. Then have students put the square at one end of a cardboard tube and point the tube toward a light source to view the constellation. L2 ELL LS **Visual-Spatial**

Measurement in Space

How do scientists determine the distance from Earth to nearby stars? One way is to measure parallax—the apparent shift in the position of an object when viewed from two different positions. Extend your arm and look at your thumb first with your left eye closed and then with your right eye closed, as the girl in **Figure 3A** is doing. Your thumb appears to change position with respect to the background. Now do the same experiment with your thumb closer to your face, as shown in **Figure 3B.** What do you observe? The nearer an object is to the observer, the greater its parallax is.

Astronomers can measure the parallax of relatively close stars to determine their distances from Earth. **Figure 4** shows how a close star's position appears to change. Knowing the angle that the star's position changes and the size of Earth's orbit, astronomers can calculate the distance of the star from Earth.

Because space is so vast, a special unit of measure is needed to record distances. Distances between stars and galaxies are measured in light-years. A **light-year** is the distance that light travels in one year. Light travels at 300,000 km/s, or about 9.5 trillion km in one year. The nearest star to Earth, other than the Sun, is Proxima Centauri. Proxima Centauri is a mere 4.3 light-years away, or about 40 trillion km.

Figure 3 A Your thumb appears to move less against the background when it is farther away from your eyes. B It appears to move more when it is closer to your eyes.

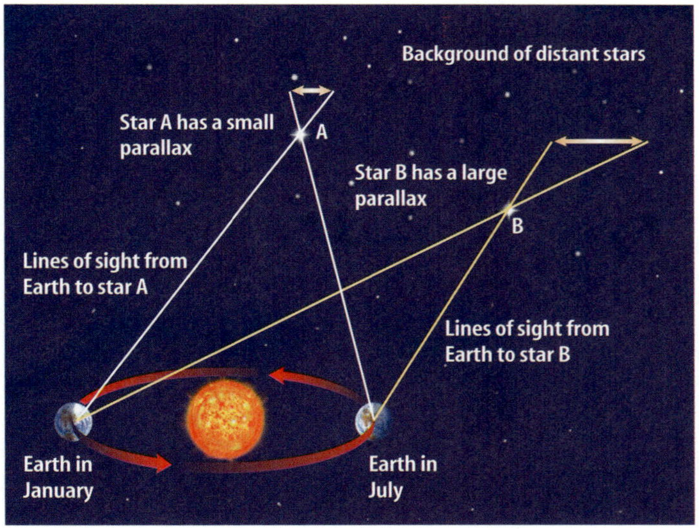

Figure 4 Parallax is determined by observing the same star when Earth is at two different points in its orbit around the Sun. The star's position relative to more distant background stars will appear to change.
Infer *whether star* **A** *or* **B** *is farther from Earth.*

Activity

Parallax Angle Arrange a numbered scale in the front of the room. Place two objects at different distances from the scale. Have students observe the objects relative to the scale from three different classroom locations. Ask them to explain differences in their observations. Students should note that as the distance to the viewed object decreases, the shift in the position of the object in relation to the scale increases. This occurs because the parallax angle of the closer object is greater than that of the farther object. **LS Visual-Spatial**

Visual Learning

Figure 3 Have students repeat the activity shown in the figure. They should notice that the closer their thumb is to their face, the more the position of the thumb appears to shift and the greater the parallax.

Caption Answer

Figure 4 Star A is more distant.

Discussion

Parallax Why can the parallax of close stars be measured against the background of distant stars? The distant stars have no measurable parallax. Why are the two positions in Earth's orbit that are chosen for parallax measurements usually far apart? A longer baseline produces higher parallax angles, which are easier to measure.

Curriculum Connection

Math Use the following equation to calculate the distance in light-years to two stars:

$d = 3.26/p$

In this equation, d represents light-years and p represents parallax angle. Star 1 has a parallax angle of 0.77″, and Star 2 has a parallax angle of 0.04″. Star 1 = 4.23 light-years; Star 2 = 81.5 light-years

Caption Answer

Figure 5 absorption of light by specific elements in the star's atmosphere

3 Assess

DAILY INTERVENTION

Check for Understanding

Visual-Spatial Provide students with unlabeled star charts and have them make and explain their own constellations. L2 ELL LS

Reteach

Compass Headings Take students to an open area to demonstrate parallax. Have two volunteers stand approximately 50 m apart. Give one student a compass with a sighting mirror. Have the student determine the exact compass heading to the other student. Now have the student holding the compass move sideways 20 m, and determine the new compass heading to the other student. Repeat this activity, varying the distances between students. L2 ELL

COOP LEARN LS **Kinesthetic**

Assessment

Performance Have students write paragraphs explaining why Deneb appears so bright and yet is more than 1,400 light-years away. Deneb is either very large, very hot, or both. Use **Performance Assessment in the Science Classroom,** p. 159.

Figure 5 This star spectrum was made by placing a diffraction grating over a telescope's objective lens. A diffraction grating produces a spectrum by causing interference of light waves.
Explain *what causes the lines in spectra.*

Properties of Stars

The color of a star indicates its temperature. For example, hot stars are a blue-white color. A relatively cool star looks orange or red. Stars that have the same temperature as the Sun have a yellow color.

Astronomers study the composition of stars by observing their spectra. When fitted into a telescope, a spectroscope acts like a prism. It spreads light out in the rainbow band called a spectrum. When light from a star passes through a spectroscope, it breaks into its component colors. Look at the spectrum of a star in **Figure 5.** Notice the dark lines caused by elements in the star's atmosphere. Light radiated from a star passes through the star's atmosphere. As it does, elements in the atmosphere absorb some of this light.The wavelengths of visible light that are absorbed appear as dark lines in the spectrum. Each element absorbs certain wavelengths, producing a unique pattern of dark lines. Like a fingerprint, the patterns of lines can be used to identify the elements in a star's atmosphere.

section 1 review

Summary

Constellations

- Constellations are patterns of stars in the night sky.
- The stars in a constellation often have no relationship to each other in space.

Absolute and Apparent Magnitudes

- Absolute magnitude is a measure of how much light is given off by a star.
- Apparent magnitude is a measure of how much light from a star is received on Earth.

Measurement in Space

- Distances between stars are measured in light-years.

Properties of Stars

- Astronomers study the composition of stars by observing their spectra.

Self Check

1. **Describe** circumpolar constellations.
2. **Explain** why some constellations are visible only during certain seasons.
3. **Infer** how two stars could have the same apparent magnitude but different absolute magnitudes.
4. **Explain** how a star is similar to the Sun if it has the same absorption lines in its spectrum that occur in the Sun's spectrum.
5. **Think Critically** If a star's parallax angle is too small to measure, what can you conclude about the star's distance from Earth?

Applying Skills

6. **Recognize Cause and Effect** Suppose you viewed Proxima Centauri, which is 4.3 light-years from Earth, through a telescope. How old were you when the light that you see left this star?

Science Online bookj.msscience.com/self_check_quiz

section 1 review

1. Circumpolar constellations are constellations that occur high above or below Earth's celestial equator. In the northern hemisphere, these constellations appear to rotate around Polaris.
2. As Earth revolves around the Sun, its nighttime side faces different directions. As a result, different constellations are visible during the year.
3. The star with the greater absolute magnitude is farther away.
4. The star's composition is similar to the Sun's composition.
5. The star is not close to Earth.
6. student's current age minus 4.3 years

108 ◆ J CHAPTER 4 Stars and Galaxies

section 2

The Sun

The Sun's Layers

The Sun is an ordinary star, but it's important to you. The Sun is the center of the solar system, and the closest star to Earth. Almost all of the life on Earth depends on energy from the Sun.

Notice the different layers of the Sun, shown in **Figure 6,** as you read about them. Like other stars, the Sun is an enormous ball of gas that produces energy by fusing hydrogen into helium in its core. This energy travels outward through the radiation zone and the convection zone. In the convection zone, gases circulate in giant swirls. Finally, energy passes into the Sun's atmosphere.

The Sun's Atmosphere

The lowest layer of the Sun's atmosphere and the layer from which light is given off is the **photosphere.** The photosphere often is called the surface of the Sun, although the surface is not a smooth feature. Temperatures there are about 6,000 K. Above the photosphere is the **chromosphere.** This layer extends upward about 2,000 km above the photosphere. A transition zone occurs between 2,000 km and 10,000 km above the photosphere. Above the transition zone is the **corona.** This is the largest layer of the Sun's atmosphere and extends millions of kilometers into space. Temperatures in the corona are as high as 2 million K. Charged particles continually escape from the corona and move through space as solar wind.

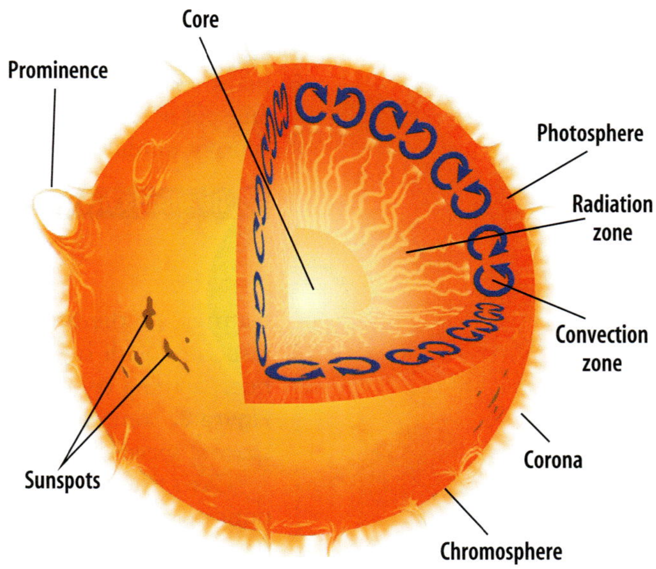

Figure 6 Energy produced in the Sun's core by fusion travels outward by radiation and convection. The Sun's atmosphere shines by the energy produced in the core.

as you read

What You'll Learn

- **Explain** that the Sun is the closest star to Earth.
- **Describe** the structure of the Sun.
- **Describe** sunspots, prominences, and solar flares.

Why It's Important

The Sun is the source of most energy on Earth.

Review Vocabulary

cycle: a repeating sequence of events, such as the sunspot cycle

New Vocabulary

- photosphere
- chromosphere
- corona
- sunspot

1 Motivate

INTERACTIVE CHALKBOARD PowerPoint® Presentations

Bellringer

Section Focus Transparencies also are available on the Interactive Chalkboard CD-ROM. L2 ELL

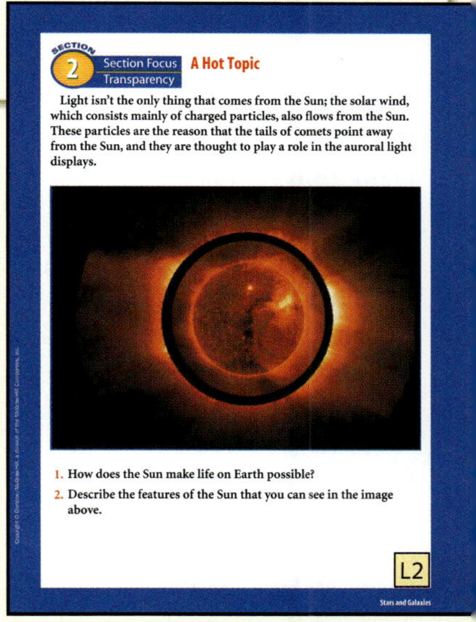
Section 2 Section Focus Transparency **A Hot Topic**

Light isn't the only thing that comes from the Sun; the solar wind, which consists mainly of charged particles, also flows from the Sun. These particles are the reason that the tails of comets point away from the Sun, and they are thought to play a role in the auroral light displays.

1. How does the Sun make life on Earth possible?
2. Describe the features of the Sun that you can see in the image above.

L2

Stars and Galaxies

Tie to Prior Knowledge

The Sun's Warmth Have students recall the warmth of a sunny summer day. Remind them that the Sun's energy keeps Earth's temperatures appropriate for life.

Section 2 Resource Manager

Chapter *FAST FILE* Resources

Transparency Activity, p. 49

Directed Reading for Content Mastery, p. 20

Enrichment, p. 32

Lab Worksheets, pp. 5–6

Reinforcement, p. 28

Cultural Diversity, p. 29

2 Teach

Caption Answer

Figure 7 counterclockwise

Fun Fact

Temperatures rise from 6,000 K to more than 1,000,000 K in a transition zone between the chromosphere and corona.

Quick Demo

Sunspots

Materials overhead projector, blank transparency

Estimated Time 10 minutes

Procedure Use this demo to help students understand why sunspots look dark even though they are quite bright. Turn on an overhead projector, and have students observe the light that falls on the screen. Then partially cover the surface of the projector with a blank transparency. Have students compare both sides of the screen. They should note a difference in brightness. This is analogous to what occurs on the Sun's surface. Next to the brighter surrounding areas of the Sun's surface, the sunspots look dark.

Reading Check

Answer the regular pattern of increasing and decreasing sunspot activity on the Sun

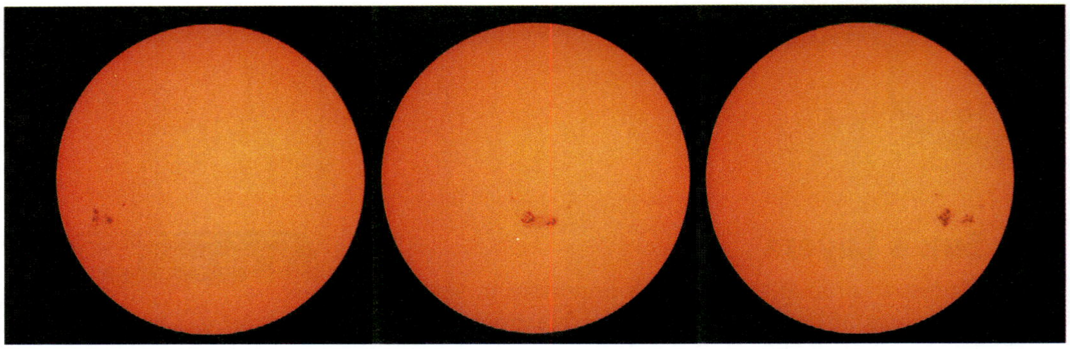

Figure 7 Sunspots are bright, but when viewed against the rest of the photosphere, they appear dark. Notice how these sunspots move as the Sun rotates.
Describe *the Sun's direction of rotation.*

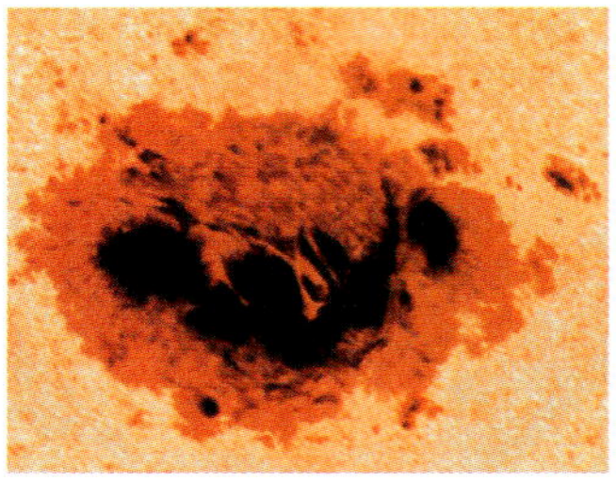

This is a close-up photo of a large sunspot.

Surface Features

From the viewpoint that you observe the Sun, its surface appears to be a smooth layer. But the Sun's surface has many features, including sunspots, prominences, flares, and CMEs.

Sunspots Areas of the Sun's surface that appear dark because they are cooler than surrounding areas are called **sunspots.** Ever since Galileo Galilei made drawings of sunspots, scientists have been studying them. Because scientists could observe the movement of individual sunspots, shown in **Figure 7,** they concluded that the Sun rotates. However, the Sun doesn't rotate as a solid body, as Earth does. It rotates faster at its equator than at its poles. Sunspots at the equator take about 25 days to complete one rotation. Near the poles, they take about 35 days.

Sunspots aren't permanent features on the Sun. They appear and disappear over a period of several days, weeks, or months. The number of sunspots increases and decreases in a fairly regular pattern called the sunspot, or solar activity, cycle. Times when many large sunspots occur are called sunspot maximums. Sunspot maximums occur about every 10 to 11 years. Periods of sunspot minimum occur in between.

Reading Check *What is a sunspot cycle?*

Prominences and Flares Sunspots are related to several features on the Sun's surface. The intense magnetic fields associated with sunspots might cause prominences, which are huge, arching columns of gas. Notice the huge prominence in **Figure 8.** Some prominences blast material from the Sun into space at speeds ranging from 600 km/s to more than 1,000 km/s.

Gases near a sunspot sometimes brighten suddenly, shooting outward at high speed. These violent eruptions are called solar flares. You can see a solar flare in **Figure 8.**

Differentiated Instruction

Challenge Have students research the relationship between sunspot activity and Earth's climate. Have students research and write a report about the Maunder Minimum, a period from 1645 through 1715 when few sunspots were observed. Possible answer: During this time, Europe experienced record low temperatures, and severe droughts occurred in the western United States. L3 P

Teacher FYI

Sunspots Sunspots are caused by intense magnetic storms on the Sun. As the Sun rotates, lines of magnetic force wrap around it. These lines dip into the Sun's interior and then back out, forming sunspots. This explains why many sunspots occur in pairs and why the members of the pair have opposite magnetic polarity.

110 ◆ J CHAPTER 4 Stars and Galaxies

CMEs Coronal mass ejections (CMEs) occur when large amounts of electrically-charged gas are ejected suddenly from the Sun's corona. CMEs can occur as often as two or three times each day during a sunspot maximum.

CMEs present little danger to life on Earth, but they do have some effects. CMEs can damage satellites in orbit around Earth. They also can interfere with radio and power distribution equipment. CMEs often cause auroras. High-energy particles contained in CMEs and the solar wind are carried past Earth's magnetic field. This generates electric currents that flow toward Earth's poles. These electric currents ionize gases in Earth's atmosphere. When these ions recombine with electrons, they produce the light of an aurora, shown in **Figure 8.**

Science Online

Topic: Space Weather

Visit bookj.msscience.com for Web links to information about space weather and its effects.

Activity Record space weather conditions for several weeks. How does space weather affect Earth?

Figure 8 Features such as solar prominences and solar flares can reach hundreds of thousands of kilometers into space. CMEs are generated as magnetic fields above sunspot groups rearrange. CMEs can trigger events that produce auroras.

Solar prominence

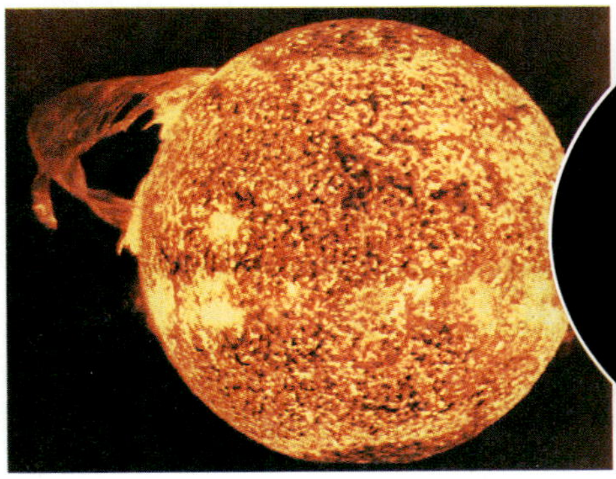

Solar flare

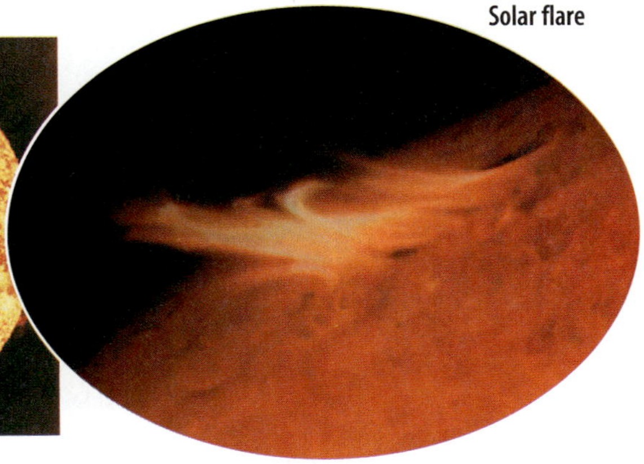

Aurora borealis, or northern lights

Discussion

Coronagraph Ask the class to suggest ways to view a CME around the bright disk of the Sun. Encourage students to be inventive and brainstorm the topic for several minutes. After the discussion, tell students that CMEs are viewed through telescopes called coronagraphs. These telescopes use a disk to block light from the Sun's surface. **WARNING:** *Students should not attempt to view the Sun with the unaided eye.*

Visual Learning

Figure 8 Explain that CMEs can send as much as ten billion tons of plasma away from the Sun at speeds up to 2,000 km/s. It takes from 2 to 4 days for the shock wave caused by a CME to reach Earth. If a CME is observed, additional sightings of auroras can be expected within a few days as the shock wave passes Earth. Additional geomagnetic activity lasts for 1 to 2 days after the shock wave passes. Have students make an events-chain concept map using the information from Figure 8 and the text detailing the events from the production of a CME to the sighting of auroras on Earth.

Use Science Words

Word Meaning Have students find out why a study of sunspots, solar flares, CMEs, and magnetic disturbances such as auroras on Earth are part of a study called *space weather.* Space weather deals with any factors affecting the environment of space in the vicinity of Earth and the Sun.

Curriculum Connection

Math Using the data provided, have student groups graph sunspot activity. Have each group describe any observed trends. Sunspot activity seems to increase and decrease in a cycle. L2 COOP LEARN

Year	Sunspots	Year	Sunspots	Year	Sunspots	Year	Sunspots
1985	18	1989	159	1993	54	1997	22
1986	14	1990	147	1994	31	1998	65
1987	29	1991	145	1995	18	1999	94
1988	100	1992	94	1996	9	2000	127

SECTION 2 The Sun J ◆ 111

Caption Answer

Figure 9 Student sketches should include both star clusters.

3 Assess

DAILY INTERVENTION

Check for Understanding

Intrapersonal Provide students with six circles that represent the Sun. Label the circles Day 1, Day 5, Day 10, Day 15, Day 20, and Day 25. Draw a group of sunspots on the equator at Day 1. Ask students to predict and draw the appearance of the Sun for each of the other days. Assume that the Sun rotates once in about 25 days at its equator and that the sunspots persist through this time.

Reteach

Light and Heat Have students draw illustrations that show how light and heat from the Sun are produced. L2 ELL LS **Visual-Spatial**

Assessment

Oral To further assess students' abilities to interpret scientific illustrations, have them hypothesize what causes the glow around stars in some photographs. Radiation from the stars causes surrounding gas and dust to glow. Use **Performance Assessment in the Science Classroom,** p. 93.

Figure 9 Most stars originally formed in large clusters containing hundreds, or even thousands, of stars.
Draw and label *a sketch of the double cluster.*

The Sun—An Average Star

The Sun is an average star. It is middle-aged, and its absolute magnitude is about average. It shines with a yellow light. Although the Sun is an average star, it is much closer to Earth than other stars. Light from the Sun reaches Earth in about eight minutes. Light from other stars takes from many years to many millions of years to reach Earth.

The Sun is unusual in one way. It is not close to any other stars. Most stars are part of a system in which two or more stars orbit each other. When two stars orbit each other, it is called a binary system. When three stars orbit each other, it is called a triple star system. The closest star system to the Sun—the Alpha Centauri system, including Proxima Centauri—is a triple star.

Stars also can move through space as a cluster. In a star cluster, many stars are relatively close, so the gravitational attraction among the stars is strong. Most star clusters are far from the solar system. They sometimes appear as a fuzzy patch in the night sky. The double cluster in the northern part of the constellation Perseus is shown in **Figure 9.** On a dark night in autumn, you can see the double cluster with binoculars, but you can't see its individual stars. The Pleiades star cluster can be seen in the constellation of Taurus in the winter sky. On a clear, dark night, you might be able to see seven of the stars in this cluster.

section 2 review

Summary

The Sun's Layers

- The Sun's interior has layers that include the core, radiation zone, and convection zone.

The Sun's Atmosphere

- The Sun's atmosphere includes the photosphere, chromosphere, and corona.

Surface Features

- The number of sunspots on the Sun varies in a 10- to 11-year cycle.
- Auroras occur when charged particles from the Sun interact with Earth's magnetic field.

The Sun—An Average Star

- The Sun is an average star, but it is much closer to Earth than any other star.

Self Check

1. **Explain** why the Sun is important for life on Earth.
2. **Describe** the sunspot cycle.
3. **Explain** why sunspots appear dark.
4. **Explain** why the Sun, which is an average star, appears so much brighter from Earth than other stars do.
5. **Think Critically** When a CME occurs on the Sun, it takes a couple of days for effects to be noticed on Earth. Explain.

Applying Skills

6. **Communicate** Make a sketch that shows the Sun's layers in your Science Journal. Write a short description of each layer.

Science Online bookj.msscience.com/self_check_quiz

section 2 review

1. The Sun provides warmth, and it provides energy for photosynthesis.
2. The sunspot cycle is a regular pattern of increase and decrease in the number of sunspots.
3. Sunspots appear dark because they are cooler than surrounding regions of the photosphere.
4. It is much closer to Earth.
5. The charged particles take this much time to travel to Earth.
6. Students' sketches should include the core, radiation zone, and convection zone, and also the photosphere, chromosphere, and corona.

LAB

Sunspots

Sunspots can be observed moving across the face of the Sun as it rotates. Measure the movement of sunspots, and use your data to determine the Sun's period of rotation.

Real-World Question

Can sunspot motion be used to determine the Sun's period of rotation?

Goals

- **Observe** sunspots and estimate their size.
- **Estimate** the rate at which sunspots move across the face of the Sun.

Materials

several books
piece of cardboard
drawing paper
refracting telescope
clipboard
small tripod
scissors

Safety Precautions

WARNING: *Handle scissors with care.*

Procedure

1. Find a location where the Sun can be viewed at the same time of day for a minimum of five days. **WARNING:** *Do not look directly at the Sun. Do not look through the telescope at the Sun. You could damage your eyes.*
2. If the telescope has a small finder scope attached, remove it or keep it covered.
3. Set up the telescope with the eyepiece facing away from the Sun, as shown. Align the telescope so that the shadow it casts on the ground is the smallest size possible. Cut and attach the cardboard as shown in the photo.
4. Use books to prop the clipboard upright. Point the eyepiece at the drawing paper.
5. Move the clipboard back and forth until you have the largest image of the Sun on the paper. Adjust the telescope to form a clear image. Trace the outline of the Sun on the paper.
6. Trace any sunspots that appear as dark areas on the Sun's image. Repeat this step at the same time each day for a week.
7. Using the Sun's diameter (approximately 1,390,000 km), estimate the size of the largest sunspots that you observed.
8. **Calculate** how many kilometers the sunspots move each day.
9. **Predict** how many days it will take for the same group of sunspots to return to the same position in which they appeared on day 1.

Conclude and Apply

1. What was the estimated size and rate of motion of the largest sunspots?
2. **Infer** how sunspots can be used to determine that the Sun's surface is not solid like Earth's surface.

Communicating Your Data

Compare your conclusions with those of other students in your class. **For more help, refer to the Science Skill Handbook.**

BENCH TESTED

Real-World Question

Purpose Students estimate the size and rate of movement of sunspots. L2 ELL IN **Visual-Spatial**

Process Skills observe and infer, interpret data, use numbers

Time Required 10 minutes each day for five days

Procedure

Alternate Materials binoculars

Safety Precautions Caution students not to look directly at the Sun. Do not allow them to look at the Sun through the telescope or binoculars.

Troubleshooting If no sunspots are visible, check to see whether it is currently a sunspot minimum.

Conclude and Apply

1. Answers will vary. Possible answer: size—3 to 4 times the size of Earth; rate of motion—about 16 days to cross the Sun's face
2. Sunspots near the Sun's equator move at different rates than those near the poles.

Assessment

Content To further assess students' understanding of sunspots, ask them to hypothesize about sunspot movement on the hemisphere they did not observe. How might sunspots on the opposite hemisphere move compared to those plotted in this activity? Possible answer: Sunspots in the opposite hemisphere move in a similar manner, except that they move in the opposite direction from Earth's frame of reference (right to left, not left to right) Use **PASC**, p. 93.

Communicating Your Data

Use a computer to produce a table to record calculations and predictions. Explain possible causes of differences. Possible answer: Students' estimates of sunspot movement might vary.

1 Motivate

Bellringer

INTERACTIVE CHALKBOARD
PowerPoint® Presentations

Section Focus Transparencies also are available on the Interactive Chalkboard CD-ROM. L2 ELL

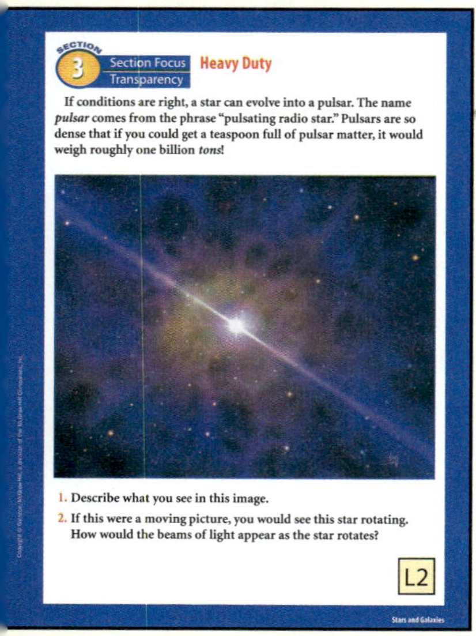
Section 3 Section Focus Transparency **Heavy Duty**

If conditions are right, a star can evolve into a pulsar. The name *pulsar* comes from the phrase "pulsating radio star." Pulsars are so dense that if you could get a teaspoon full of pulsar matter, it would weigh roughly one billion *tons*!

1. Describe what you see in this image.
2. If this were a moving picture, you would see this star rotating. How would the beams of light appear as the star rotates?

L2

Stars and Galaxies

Tie to Prior Knowledge

Plasma Have students recall the four states of matter: solid, liquid, gas, and plasma. Explain that stars are composed of matter in the plasma state. Ask students to describe traits of matter in this state. Plasma is the state of matter in which there are roughly equal numbers of positively and negatively charged particles. It is produced when the atoms in a gas become ionized.

Caption Answer

Figure 10 white dwarf

section 3

Evolution of Stars

as you read

What You'll Learn

- **Describe** how stars are classified.
- **Compare** the Sun to other types of stars on the H-R diagram.
- **Describe** how stars evolve.

Why It's Important

Earth and your body contain elements that were made in stars.

Review Vocabulary
gravity: an attractive force between objects that have mass

New Vocabulary
- nebula
- giant
- white dwarf
- supergiant
- neutron star
- black hole

Classifying Stars

When you look at the night sky, all stars might appear to be similar, but they are quite different. Like people, they vary in age and size, but stars also vary in temperature and brightness.

In the early 1900s, Ejnar Hertzsprung and Henry Russell made some important observations. They noticed that, in general, stars with higher temperatures also have brighter absolute magnitudes.

Hertzsprung and Russell developed a graph, shown in **Figure 10,** to show this relationship. They placed temperatures across the bottom and absolute magnitudes up one side. A graph that shows the relationship of a star's temperature to its absolute magnitude is called a Hertzsprung-Russell (H-R) diagram.

The Main Sequence As you can see, stars seem to fit into specific areas of the graph. Most stars fit into a diagonal band that runs from the upper left to the lower right of the graph. This band, called the main sequence, contains hot, blue, bright stars in the upper left and cool, red, dim stars in the lower right. Yellow main sequence stars, like the Sun, fall in between.

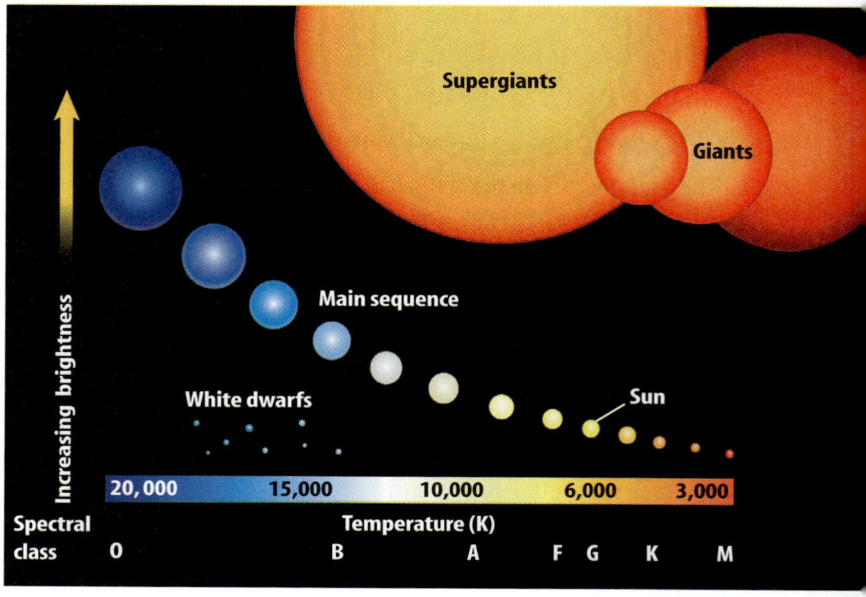

Figure 10 The relationships among a star's color, temperature, and brightness are shown in this H-R diagram. Stars in the upper left are hot, bright stars, and stars in the lower right are cool, dim stars.
Classify *Which type of star shown in the diagram is the hottest, dimmest star?*

Section 3 Resource Manager

Chapter FAST FILE Resources

Transparency Activity, p. 50
Directed Reading for Content Mastery, p. 21
Enrichment, p. 33
Reinforcement, p. 29

Dwarfs and Giants About 90 percent of all stars are main sequence stars. Most of these are small, red stars found in the lower right of the H-R diagram. Among main sequence stars, the hottest stars generate the most light and the coolest ones generate the least. What about the ten percent of stars that are not part of the main sequence? Some of these stars are hot but not bright. These small stars are located on the lower left of the H-R diagram and are called white dwarfs. Other stars are extremely bright but not hot. These large stars on the upper right of the H-R diagram are called giants, or red giants, because they are usually red in color. The largest giants are called supergiants. **Figure 11** shows the supergiant, Antares—a star 300 times the Sun's diameter—in the constellation Scorpius. It is more than 11,000 times as bright as the Sun.

Reading Check *What kinds of stars are on the main sequence?*

Figure 11 Antares is a bright supergiant located 400 light-years from Earth. Although its temperature is only about 3,500 K, it is the 16th brightest star in the sky.

How do stars shine?

For centuries, people were puzzled by the questions of what stars were made of and how they produced light. Many people had estimated that Earth was only a few thousand years old. The Sun could have been made of coal and shined for that long. However, when people realized that Earth was much older, they wondered what material possibly could burn for so many years. Early in the twentieth century, scientists began to understand the process that keeps stars shining for billions of years.

Generating Energy In the 1930s, scientists discovered reactions between the nuclei of atoms. They hypothesized that temperatures in the center of the Sun must be high enough to cause hydrogen to fuse to make helium. This reaction releases tremendous amounts of energy. Much of this energy is emitted as different wavelengths of light, including visible, infrared, and ultraviolet light. Only a tiny fraction of this light comes to Earth. During the fusion reaction, four hydrogen nuclei combine to create one helium nucleus. The mass of one helium nucleus is less than the mass of four hydrogen nuclei, so some mass is lost in the reaction.

Years earlier, in 1905, Albert Einstein had proposed a theory stating that mass can be converted into energy. This was stated as the famous equation $E = mc^2$. In this equation, E is the energy produced, m is the mass, and c is the speed of light. The small amount of mass "lost" when hydrogen atoms fuse to form a helium atom is converted to a large amount of energy.

2 Teach

Reading Check

Answer Most of the stars on the main sequence are small, red stars. Hot, blue, bright stars and medium-sized, medium-temperature yellow stars also are on the main sequence.

Quick Demo

E = mc²

Materials calculator

Estimated Time 10 minutes

Procedure Plug values into the right side of this equation to show students how much energy can come from the complete conversion of a relatively small amount of matter. For example, using 1 kg, $E = (1\text{ kg})(3.0 \times 10^8\text{ m/s})^2 = 9.0 \times 10^{16}$ Joules. This is enough energy to supply power to about 2.8 million U.S. homes for one year.

Virtual Labs

Star Classification *How does the chemical composition of stars determine their classification?*

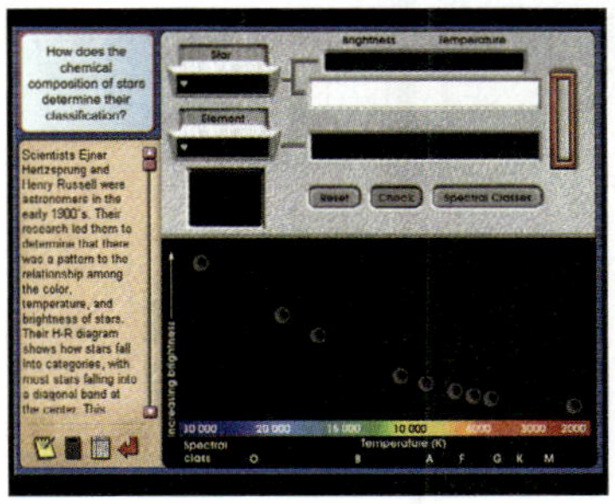

Differentiated Instruction

Challenge Have students research the work of A. S. Eddington, Robert Atkinson, and Hans Bethe, the scientists whose discoveries led to an understanding of energy production in stars. Have them write a brief report on their findings. Eddington hypothesized that temperatures in the center of the Sun must be high, Robert Atkinson suggested that hydrogen could fuse into helium. Bethe explained energy production in hotter stars. L2 LS **Linguistics** P

English-Language Learners Have students prepare a multilingual bulletin board display illustrating the Hertzsprung-Russell (H-R) diagram. Encourage them to use **Figure 10** as a guide. L2 COOP LEARN LS **Kinesthetic** ELL

Caption Answer

Figure 12 It is converted into energy; $E = mc^2$.

Make a Model

Fusion Ask students to make a model of the fusion of hydrogen into helium. Have them use medium-sized and small polystyrene balls to represent neutrons, protons, neutrinos, and positrons. Ask students to use the model to demonstrate how the fusion reaction progresses. Two protons fuse to form hydrogen-2 and release a positron and a neutrino. Hydrogen-2 fuses with a proton to produce helium-3. Two helium-3 nuclei fuse to form helium-4 and two protons. L3
LS Visual-Spatial

Use Science Words

Word Meaning Have students compare and contrast the words *fusion* and *fission*. Both are nuclear processes. Fusion is the combining of nuclei. Fission is the splitting apart of nuclei.

Teacher FYI

Stellar Class L Astronomers have devised a new class of stars called stellar class L. Stars of this class are relatively cool—1,500 K to 2,000 K, and have spectra that show the presence of iron hydride and chromium hydride. As with M-class stars, L-class stars are dwarfs.

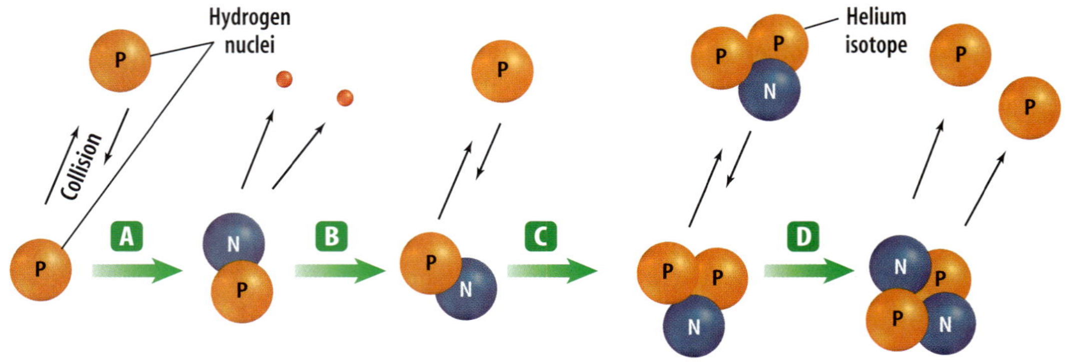

A Two protons (hydrogen nuclei) collide. One proton decays to a neutron, releasing subatomic particles and some energy.

B Another proton fuses with a proton and neutron to form an isotope of helium. Energy is given off again.

C Two helium isotopes collide with enough energy to fuse.

D A helium nucleus (two protons and two neutrons) forms as two protons break away. During the process, still more energy is released.

Figure 12 Fusion of hydrogen into helium occurs in a star's core. **Infer** *what happens to the "lost" mass during this process.*

Fusion Shown in **Figure 12,** fusion occurs in the cores of stars. Only in the core are temperatures high enough to cause atoms to fuse. Normally, they would repel each other, but in the core of a star where temperatures can exceed 15,000,000 K, atoms can move so fast that some of them fuse upon colliding.

Evolution of Stars

INTEGRATE Physics The H-R diagram explained a lot about stars. However, it also led to more questions. Many wondered why some stars didn't fit in the main sequence group and what happened when a star depleted its supply of hydrogen fuel. Today, scientists have theories about how stars evolve, what makes them different from one another, and what happens when they die. **Figure 13** illustrates the lives of different types of stars.

When hydrogen fuel is depleted, a star loses its main sequence status. This can take less than 1 million years for the brightest stars to many billions of years for the dimmest stars. The Sun has a main sequence life span of about 10 billion years. Half of its life span is still in the future.

Nebula Stars begin as a large cloud of gas and dust called a **nebula.** As the particles of gas and dust exert a gravitational force on each other, the nebula begins to contract. Gravitational forces cause instability within the nebula. The nebula can break apart into smaller and smaller pieces. Each piece eventually might collapse to form a star.

Science Online

Topic: Evolution of Stars
Visit bookj.msscience.com for Web links to information about the evolution of stars.

Activity Make a three-circle Venn diagram to compare and contrast white dwarfs, neutron stars, and black holes.

LAB DEMONSTRATION

Purpose to demonstrate how color changes as a material is heated

Materials Bunsen burner, wire, tongs, insulated pad

Preparation Set up the Bunsen burner and obtain safety goggles and gloves.

Procedure Using tongs, hold a wire over a burner flame until it begins to glow. Have students record the color of the glowing wire. Continue heating the wire as students record changes and summarize their observations.

Expected Outcome The wire changed from red to reddish-orange as it became hotter.

Assessment

How do the color changes you observed relate to star colors? Hotter stars shine at different colors than cooler stars.

116 ◆ J CHAPTER 4 Stars and Galaxies

A Star Is Born As the particles in the smaller pieces of nebula move closer together, the temperatures in each nebula piece increase. When the temperature inside the core of a nebula piece reaches 10 million K, fusion begins. The energy released radiates outward through the condensing ball of gas. As the energy radiates into space, stars are born.

Reading Check *How are stars born?*

Main Sequence to Giant Stars In the newly formed star, the heat from fusion causes pressure to increase. This pressure balances the attraction due to gravity. The star becomes a main sequence star. It continues to use its hydrogen fuel.

When hydrogen in the core of the star is depleted, a balance no longer exists between pressure and gravity. The core contracts, and temperatures inside the star increase. This causes the outer layers of the star to expand and cool. In this late stage of its life cycle, a star is called a **giant.**

After the core temperature reaches 100 million K, helium nuclei fuse to form carbon in the giant's core. By this time, the star has expanded to an enormous size, and its outer layers are much cooler than they were when it was a main sequence star. In about 5 billion years, the Sun will become a giant.

White Dwarfs After the star's core uses much of its helium, it contracts even more and its outer layers escape into space. This leaves behind the hot, dense core. At this stage in a star's evolution, it becomes a **white dwarf.** A white dwarf is about the size of Earth. Eventually, the white dwarf will cool and stop giving off light.

INTEGRATE Chemistry

White Dwarf Matter The matter in white dwarf stars is more than 500,000 times as dense as the matter in Earth. In white dwarf matter, there are free electrons and atomic nuclei. The resistance of the electrons to pack together more provides pressure that keeps the star from collapsing. This state of matter is called electron degeneracy.

Figure 13 The life of a star depends on its mass. Massive stars eventually become neutron stars or black holes.
Explain *what happens to stars that are the size of the Sun.*

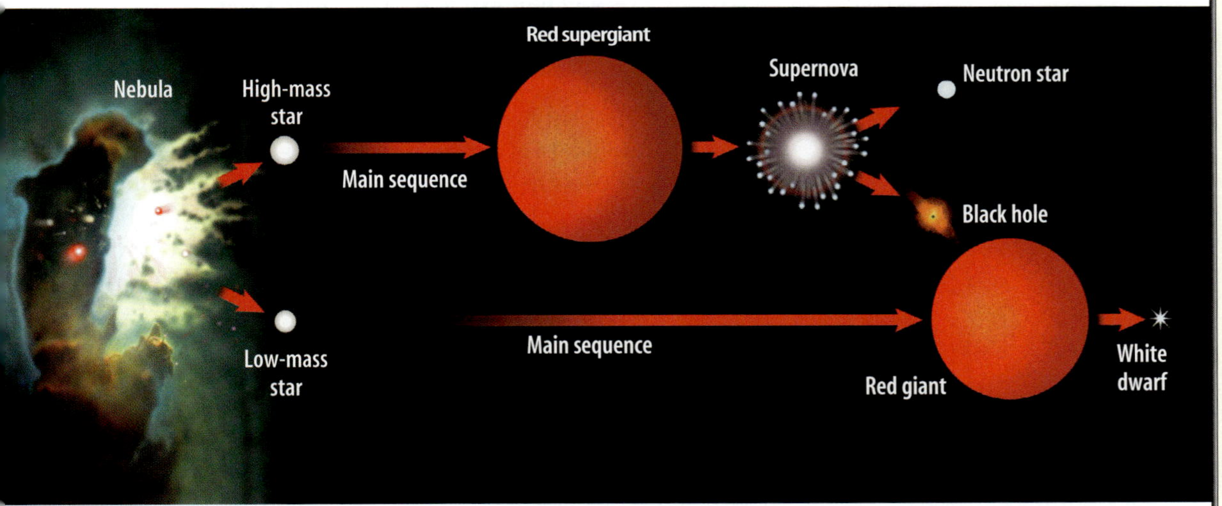

INTEGRATE Chemistry

White Dwarf Matter White dwarf matter is electron degenerate. Electron degeneracy is effective to 1.4 solar masses, a value called the Chandrasekhar Limit. If a white dwarf acquires more than 1.4 solar masses, it will collapse and explode as a type IA supernova.

Reading Check

Answer temperatures in nebulae increase, fusion begins; energy radiates

Activity

Nebulae Have students research the Horsehead and Orion nebulae, located in the constellation of Orion. Ask them to describe each nebula. The Horsehead Nebula is composed of a dust cloud that blocks light from behind it, whereas the Orion Nebula is composed of young stars and hot, glowing gases. L3 LS **Linguistic**

Caption Answer

Figure 13 Stars the size of the Sun will become giants, white dwarfs, and finally black dwarfs.

Visual Learning

Figure 13 Have students construct sequence diagrams showing the evolution of a massive star. After students read about black holes, ask the following questions: After a supernova, what type of object forms if the remaining dense core collapses? a black hole If no light is given off, how are black holes located? X-rays are emitted by matter as it is pulled into the black hole.

Teacher FYI

Orion Nebula A nebula may produce a steady stream of stars over a given period. In the Orion Nebula (M42), a star nursery 1,500 light-years away, hundreds of stars have already formed and more will follow. The brightest star seen in the Orion Nebula by the *Hubble Space Telescope* is the Becklin-Neugebauer object.

INTEGRATE History

Research Historic accounts like the description of the 386 supernova are very important to astronomers. Knowing the exact age of this pulsar provides an opportunity to test and calibrate models of how pulsars evolve.

Reading Check

Answer the remaining dense core from a massive supernova, collapsed down into a point, where gravity is so strong nothing can escape, not even light

Fun Fact

In a star with a mass similar to that of the Sun, outer layers become unstable and begin escaping into space when the core runs out of hydrogen fuel. The escaping gases form a shell around the star, producing a nebula. The Ring Nebula in the constellation Lyra is a result of this process.

Teacher FYI

Brown Dwarfs Brown dwarfs are stars that never generated fusion in their cores. These dim stars are found by observing the movement of a companion star. A single star's path through space is a nearly straight line. The path of a star orbited by a brown dwarf has a distinct wobble. This wobble is used to locate brown dwarfs as well as planets beyond the solar system.

INTEGRATE History

386 Supernova In 386 A.D., Chinese observers described a new star—a supernova—in the night sky. More recently, astronomers using the *Chandra X-ray Observatory* found evidence of a spinning neutron star, called a pulsar, in exactly the same location. Because of the Chinese account, astronomers better understand how neutron stars form and evolve.

Supergiants and Supernovas In stars that are more than about eight times more massive than the Sun, the stages of evolution occur more quickly and more violently. Look back at **Figure 13.** In massive stars, the core heats up to much higher temperatures. Heavier and heavier elements form by fusion, and the star expands into a **supergiant.** Eventually, iron forms in the core. Because of iron's atomic structure, it cannot release energy through fusion. The core collapses violently, and a shock wave travels outward through the star. The outer portion of the star explodes, producing a supernova. A supernova can be millions of times brighter than the original star was.

Neutron Stars If the collapsed core of a supernova is between about 1.4 and 3 times as massive as the Sun, it will shrink to approximately 20 km in diameter. Only neutrons can exist in the dense core, and it becomes a **neutron star.** Neutron stars are so dense that a teaspoonful would weigh more than 600 million metric tons in Earth's gravity. As dense as neutron stars are, they can contract only so far because the neutrons resist the inward pull of gravity.

Black Holes If the remaining dense core from a supernova is more than about three times more massive than the Sun, probably nothing can stop the core's collapse. Under these conditions, all of the core's mass collapses to a point. The gravity near this mass is so strong that nothing can escape from it, not even light. Because light cannot escape, the region is called a **black hole.** If you could shine a flashlight on a black hole, the light simply would disappear into it.

Reading Check *What is a black hole?*

Black holes, however, are not like giant vacuum cleaners, sucking in distant objects. A black hole has an event horizon, which is a region inside of which nothing can escape. If something—including light—crosses the event horizon, it will be pulled into the black hole. Beyond the event horizon, the black hole's gravity pulls on objects just as it would if the mass had not collapsed. Stars and planets can orbit around a black hole.

The photograph in **Figure 14** was taken by the *Hubble Space Telescope.* It shows a jet of gas streaming out of the center of galaxy M87. This jet of gas formed as matter flowed toward a black hole, and some of the gas was ejected along the polar axis.

Figure 14 The black hole at the center of galaxy M87 pulls matter into it at extremely high velocities. Some matter is ejected to produce a jet of gas that streams away from the center of the galaxy at nearly light speed.

Active Reading

Learning Journal Students should draw a vertical line down each page of their Learning Journals. The left-column entries can be research notes, lecture notes, or vocabulary terms. The right-column entries are the student's response to, interpretation of, questions about, or analysis of the left-column entries. Have students write a Learning Journal related to stars and galaxies.

118 ◆ J CHAPTER 4 Stars and Galaxies

Recycling Matter A star begins its life as a nebula, such as the one shown in **Figure 15.** Where does the matter in a nebula come from? Nebulas form partly from the matter that was once in other stars. A star ejects enormous amounts of matter during its lifetime. Some of this matter is incorporated into nebulas, which can evolve to form new stars. The matter in stars is recycled many times.

What about the matter created in the cores of stars and during supernova explosions? Are elements such as carbon and iron also recycled? These elements can become parts of new stars. In fact, spectrographs have shown that the Sun contains some carbon, iron, and other heavier elements. Because the Sun is an average, main sequence star, it is too young and its mass is too small to have formed these elements itself. The Sun condensed from material that was created in stars that died many billions of years ago.

Some elements condense to form planets and other bodies rather than stars. In fact, your body contains many atoms that were fused in the cores of ancient stars. Evidence suggests that the first stars formed from hydrogen and helium and that all the other elements have formed in the cores of stars or as stars explode.

Figure 15 Stars are forming in the Orion Nebula and other similar nebulae.
Describe *a star-forming nebula.*

section 3 review

Summary

Classifying Stars

- Most stars plot on the main sequence of an H-R diagram.
- As stars near the end of their lives, they move off of the main sequence.

How do stars shine?

- Stars shine because of a process called fusion.
- During fusion, nuclei of a lighter element merge to form a heavier element.

Evolution of Stars

- Stars form in regions of gas and dust called nebulae.
- Stars evolve differently depending on how massive they are.

Self Check

1. **Explain** how the Sun is different from other stars on the main sequence. How is it different from a giant star? How is it different from a white dwarf?
2. **Describe** how stars release energy.
3. **Outline** the past and probable future of the Sun.
4. **Define** a black hole.
5. **Think Critically** How can white dwarf stars be both hot and dim?

Applying Math

6. **Convert Units** A neutron star has a diameter of 20 km. One kilometer equals 0.62 miles. What is the neutron star's diameter in miles?

Caption Answer

Figure 15 a cloud of gas, dust, and ice in which stars are forming

Discussion

The Sun's Elements What is the origin of the carbon and iron found in the Sun? These and other elements that did not form in the Sun were in the nebula from which the Sun formed. The nebula obtained its carbon and iron from stars that evolved earlier.

3 Assess

DAILY INTERVENTION

Check for Understanding

Logical-Mathematical Refer students to **Figure 10.** Why do supergiant stars have such a high absolute magnitude even though they have a relatively low temperature? Supergiant stars are very large compared with other stars. Why do supergiant stars have such low temperatures? Supergiant stars form because their outer shells have expanded. This expansion causes the outer layers to cool. LS

Reteach

Stellar Evolution Make note cards for the different stages of stellar evolution. Have students place the cards in the correct order. You might want to have student groups compete for the fastest time.

Assessment

Content Have students explain where elements in their bodies, such as carbon, may have come from. They were most likely produced in a supernova explosion. Use **Performance Assessment in the Science Classroom,** p. 93.

section 3 review

1. The Sun is smaller and its surface is hotter than a giant star. The Sun is larger and cooler than a white dwarf.
2. Stars release energy by fusing a lighter element, such as hydrogen, into a heavier element, such as helium.
3. The Sun formed when a solar nebula contracted. It has been a main sequence star for about 4.6 billion years. In another 5 billion years, the Sun will become a giant star. Eventually, its core will shrink to a white dwarf, and it will cool gradually.
4. A black hole is the completely collapsed core of a massive star.
5. White dwarfs are very small, so they can be both hot and dim.
6. 12.4 miles

1 Motivate

INTERACTIVE CHALKBOARD
PowerPoint® Presentations

Bellringer

Section Focus Transparencies also are available on the Interactive Chalkboard CD-ROM. L2 ELL

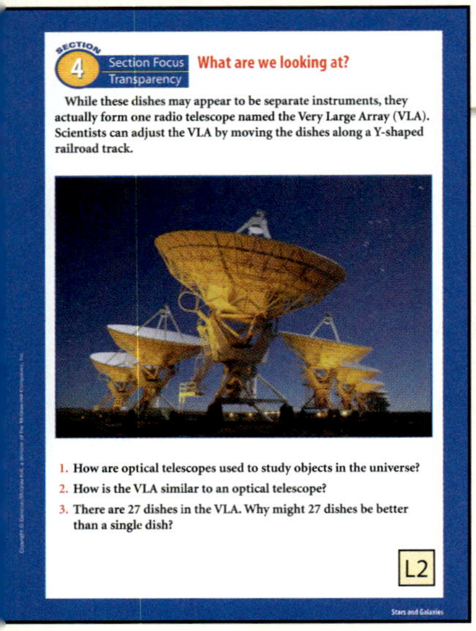
Section 4 Section Focus Transparency **What are we looking at?**

While these dishes may appear to be separate instruments, they actually form one radio telescope named the Very Large Array (VLA). Scientists can adjust the VLA by moving the dishes along a Y-shaped railroad track.

1. How are optical telescopes used to study objects in the universe?
2. How is the VLA similar to an optical telescope?
3. There are 27 dishes in the VLA. Why might 27 dishes be better than a single dish?

L2

Stars and Galaxies

Tie to Prior Knowledge

Milky Way Ask students if they have ever observed the night sky far from city lights. If so, they may have seen the band of stars that stretches across the sky—the Milky Way, our home galaxy.

section 4

Galaxies and the Universe

as you read

What You'll Learn

- **Describe** the Sun's position in the Milky Way Galaxy.
- **Explain** that the same natural laws that apply to our solar system also apply in other galaxies.

Why It's Important

Studying the universe could determine whether life exists elsewhere.

Review Vocabulary
universe: the space that contains all known matter and energy

New Vocabulary
- galaxy
- big bang theory

Galaxies

If you enjoy science fiction, you might have read about explorers traveling through the galaxy. On their way, they visit planets around other stars and encounter strange alien beings. Although this type of space exploration is futuristic, it is possible to explore galaxies today. Using a variety of telescopes, much is being learned about the Milky Way and other galaxies.

A **galaxy** is a large group of stars, gas, and dust held together by gravity. Earth and the solar system are in a galaxy called the Milky Way. It might contain as many as one trillion stars. Countless other galaxies also exist. Each of these galaxies contains the same elements, forces, and types of energy that occur in Earth's solar system. Galaxies are separated by huge distances—often millions of light-years.

In the same way that stars are grouped together within galaxies, galaxies are grouped into clusters. The cluster that the Milky Way belongs to is called the Local Group. It contains about 45 galaxies of various sizes and types. The three major types of galaxies are spiral, elliptical, and irregular.

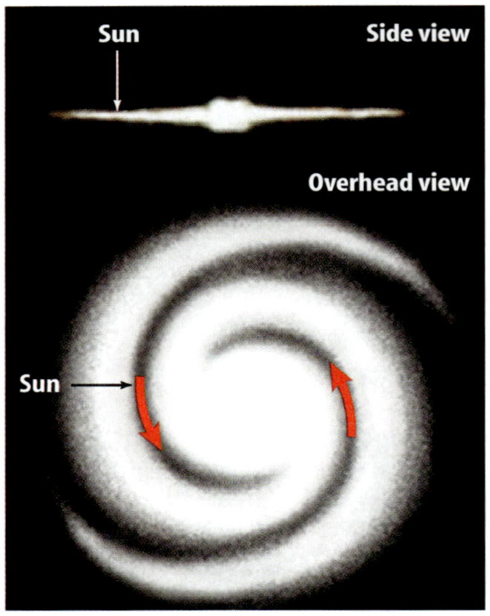

Spiral Galaxies Spiral galaxies are galaxies that have spiral arms that wind outward from the center. The arms consist of bright stars, dust, and gas. The Milky Way Galaxy, shown in **Figure 16,** is a spiral galaxy. The Sun and the rest of the solar system are located near the outer edge of the Milky Way Galaxy.

Spiral galaxies can be normal or barred. Arms in a normal spiral start close to the center of the galaxy. Barred spirals have spiral arms extending from a large bar of stars and gas that passes through the center of the galaxy.

Figure 16 This illustration shows a side view and an overhead view of the Milky Way.
Describe *where the Sun is in the Milky Way.*

Section 4 Resource Manager

Chapter FAST FILE Resources
- Transparency Activity, pp. 51, 53–54
- Directed Reading for Content Mastery, pp. 21, 22
- Enrichment, p. 34
- Reinforcement, p. 30
- MiniLAB, p. 4
- Lab Worksheets, pp. 7–8

Cultural Diversity, p. 31
Reading and Writing Skill Activities, p. 47
Lab Management and Safety, p. 65

Elliptical Galaxies A common type of galaxy is the elliptical galaxy. **Figure 17** shows an elliptical galaxy in the constellation Andromeda. These galaxies are shaped like large, three-dimensional ellipses. Many are football shaped, but others are round. Some elliptical galaxies are small, while others are so large that several galaxies the size of the Milky Way would fit inside one of them.

Irregular Galaxies The third type—an irregular galaxy—includes most of those galaxies that don't fit into the other categories. Irregular galaxies have many different shapes. They are smaller than the other types of galaxies. Two irregular galaxies called the Clouds of Magellan orbit the Milky Way. The Large Magellanic Cloud is shown in **Figure 18.**

Reading Check *How do the three different types of galaxies differ?*

Figure 17 This photo shows an example of an elliptical galaxy. **Identify** *the two other types of galaxies.*

The Milky Way Galaxy

The Milky Way might contain one trillion stars. The visible disk of stars shown in **Figure 16** is about 100,000 light-years across. Find the location of the Sun. Notice that it is located about 26,000 light-years from the galaxy's center in one of the spiral arms. In the galaxy, all stars orbit around a central region, or core. It takes about 225 million years for the Sun to orbit the center of the Milky Way.

The Milky Way often is classified as a normal spiral galaxy. However, recent evidence suggests that it might be a barred spiral. It is difficult to know for sure because astronomers have limited data about how the galaxy looks from the outside.

You can't see the shape of the Milky Way because you are located within one of its spiral arms. You can, however, see the Milky Way stretching across the sky as a misty band of faint light. You can see the brightest part of the Milky Way if you look low in the southern sky on a moonless summer night. All the stars you can see in the night sky belong to the Milky Way.

Like many other galaxies, the Milky Way has a supermassive black hole at its center. This black hole might be more than 2.5 million times as massive as the Sun. Evidence for the existence of the black hole comes from observing the orbit of a star near the galaxy's center. Additional evidence includes X-ray emissions detected by the *Chandra X-ray Observatory.* X rays are produced when matter spirals into a black hole.

Figure 18 The Large Magellanic Cloud is an irregular galaxy. It's a member of the Local Group, and it orbits the Milky Way.

2 Teach

Caption Answers

Figure 16 in one of the spiral arms

Figure 17 spiral and irregular

Reading Check

Answer by their shape

Visual Learning

Figures 16, 17, and 18 Have student pairs make flash cards illustrating the types of galaxies. Pairs should sketch a type of galaxy on one side of the card and write its name on the reverse side. Partners can then test each other using the flash cards. L2 ELL COOP LEARN **Interpersonal**

Teacher FYI

Irregular Galaxy The Large Magellanic Cloud is an irregular galaxy located about 169,000 light-years from the Milky Way. Supernova 1987A was observed there in 1987. It would have actually occurred about 169,000 years ago.

Cultural Diversity

The Milky Way The names various cultures have assigned to the Milky Way show that most observers are struck by its fluid form. Akkadians called it the River of the Snake. In Judea, Armenia, and Syria, it was a long bandage. In China and Japan it was called the Celestial River. Polynesians called it the Long, Blue, Cloud-Eating Shark. In India, it was known as the Bed of the Ganges.

Mini LAB

Purpose Students develop a scale illustrating the vastness of space. L2 ELL IS **Logical-Mathematical**

Materials large sheet of paper, metric ruler or meterstick

Teaching Strategies Help students conceptualize the vast scale of their model. If our Sun were represented by a dot, the circle representing the distance to Proxima Centauri would be nearly within that same dot.

Analysis

1. Answers will vary depending on the scale chosen.
2. The distance to the Andromeda Galaxy should be 29 times the diameter of the student's Milky Way model. This can be determined by dividing the distance to the Andromeda Galaxy (2.9 million light-years) by the diameter of the Milky Way (100,000 light-years).

Assessment

Content Have students determine the scale distance to the Large Magellanic Cloud. Its true distance is about 169,000 light-years. The distance to the Large Magellanic Cloud should be about 1.69 times the diameter of the student's Milky Way model. Use **Performance Assessment in the Science Classroom,** p. 123.

Mini LAB

Measuring Distance in Space

Procedure

1. On a large sheet of **paper,** draw an overhead view of the Milky Way. If necessary, refer to **Figure 16.** Choose a scale to show distance in light-years.
2. Mark the approximate location of the solar system, which is about two-thirds of the way out on one of the spiral arms.
3. Now, draw a side view of the Milky Way Galaxy. Mark the position of the solar system.

Analysis

1. What scale did you use to represent distance on your model of the Milky Way?
2. The Andromeda Galaxy is about 2.9 million light-years from Earth. What scale distance would this represent?

Origin of the Universe

People long have wondered how the universe formed. Several models of its origin have been proposed. One model is the steady state theory. It suggests that the universe always has been the same as it is now. The universe always existed and always will. As the universe expands, new matter is created to keep the overall density of the universe the same or in a steady state. However, evidence indicates that the universe was much different in the past.

A second idea is called the oscillating model. In this model, the universe began with expansion. Over time, the expansion slowed and the universe contracted. Then the process began again, oscillating back and forth. Some scientists still hypothesize that the universe expands and contracts in a cycle.

A third model of how the universe formed is called the big bang theory. The universe started with a big bang and has been expanding ever since. This theory will be described later.

Expansion of the Universe

What does it sound like when a train is blowing its whistle while it travels past you? The whistle has a higher pitch as the train approaches you. Then the whistle seems to drop in pitch as the train moves away. This effect is called the Doppler shift. The Doppler shift occurs with light as well as with sound. **Figure 19** shows how the Doppler shift causes changes in the light coming from distant stars and galaxies. If a star is moving toward Earth, its wavelengths of light are compressed. If a star is moving away from Earth, its wavelengths of light are stretched.

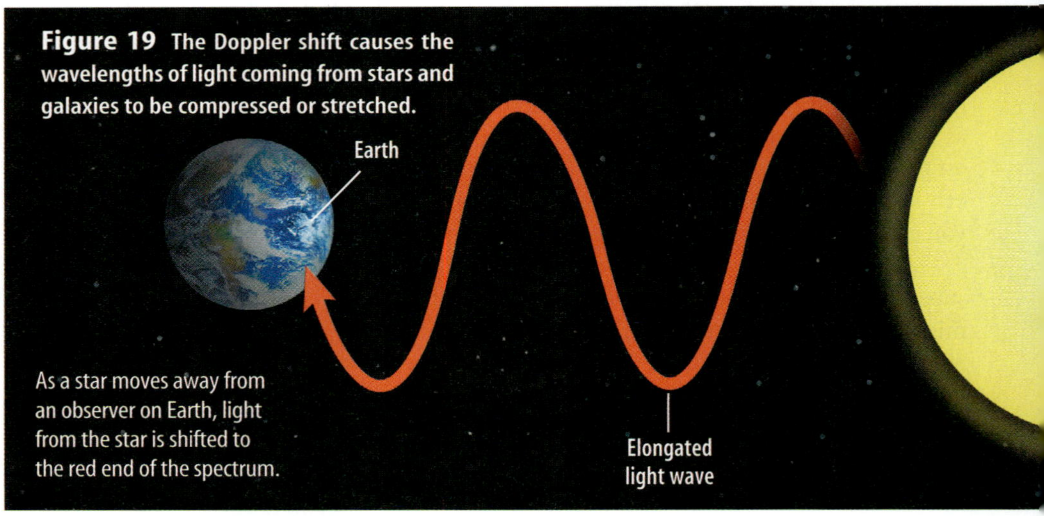

Figure 19 The Doppler shift causes the wavelengths of light coming from stars and galaxies to be compressed or stretched.

Curriculum Connection

Art Suggest that a group of students research and draw a diagram of the *Hubble Space Telescope,* which is used to gather data about the origin of the universe. When they present their diagram, have students explain the advantage of having a telescope in space. Images are clearer because there is no interference from the atmosphere. L2 COOP LEARN IS **Visual-Spatial**

Differentiated Instruction

Learning Disabled Use a balloon to demonstrate the oscillating model of the origin of the universe. Blow up the balloon, and explain that this is like the expansion phase of this model. Then slowly let the air out of the balloon, and explain that this is like the contraction phase of this model. Have students describe what would happen next. The balloon (universe) would again begin to expand. L1 ELL IS **Visual-Spatial**

The Doppler Shift Look at the spectrum of a star in **Figure 20A.** Note the position of the dark lines. How do they compare with the lines in **Figures 20B** and **20C?** They have shifted in position. What caused this shift? As you just read, when a star is moving toward Earth, its wavelengths of light are compressed, just as the sound waves from the train's whistle are. This causes the dark lines in the spectrum to shift toward the blue-violet end of the spectrum. A red shift in the spectrum occurs when a star is moving away from Earth. In a red shift, the dark lines shift toward the red end of the spectrum.

Figure 20 **A** This spectrum shows dark absorption lines. **B** The dark lines shift toward the blue-violet end for a star moving toward Earth. **C** The lines shift toward the red end for a star moving away from Earth.

Red Shift In 1929, Edwin Hubble published an interesting fact about the light coming from most galaxies. When a spectrograph is used to study light from galaxies beyond the Local Group, a red shift occurs in the light. What does this red shift tell you about the universe?

Because all galaxies beyond the Local Group show a red shift in their spectra, they must be moving away from Earth. If all galaxies outside the Local Group are moving away from Earth, then the entire universe must be expanding. Remember the Launch Lab at the beginning of the chapter? The dots on the balloon moved apart as the model universe expanded. Regardless of which dot you picked, all the other dots moved away from it. In a similar way, galaxies beyond the Local Group are moving away from Earth.

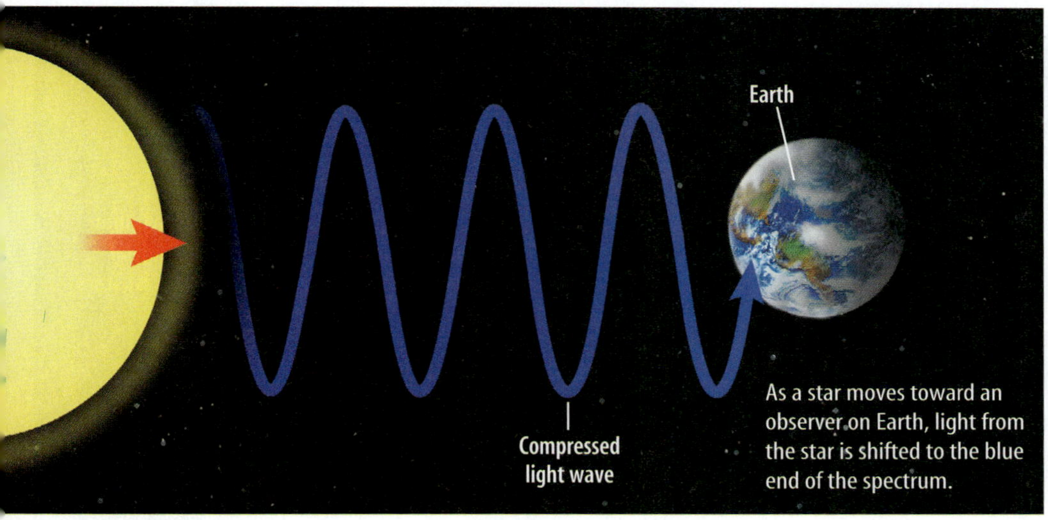

As a star moves toward an observer on Earth, light from the star is shifted to the blue end of the spectrum.

Text Question Answer

The universe is expanding.

Quick Demo

Red Shift

Materials large balloon that has a long, slender shape; black marker

Estimated Time five minutes

Procedure Partially inflate the balloon and draw a waveform along its length with a marker. Continue inflating the balloon to show students how an expanding universe causes a red shift in light coming from distant galaxies.

Inquiry Lab

Modeling a Red Shift

Question How can sound waves be red-shifted?

Possible Materials horn, whistle

Estimated Time 1 class period

Teaching Strategies

- Challenge students to develop a plan to produce a red shift in sound waves.
- Evaluate student plans for safety and feasibility.
- One possible procedure is to place a horn or whistle at a fixed location. Students then bike past the fixed location. They listen to the pitch of the sound while approaching and retreating from the source of the sound.

For additional inquiry activities, see *Science Inquiry Labs.*

Differentiated Instruction

Challenge Have students learn about mapping the universe by researching the work of Margaret Geller and John Huchra of the Harvard-Smithsonian Center for Astrophysics. Invite volunteers to share their findings in an oral report to the class. Reports should note that great voids and large concentrations of galaxies, such as the Local Group, alternate throughout the universe. L3 LS **Linguistic**

SECTION 4 Galaxies and the Universe J ◆ 123

NATIONAL GEOGRAPHIC

Visualizing the Big Bang Theory

Have students examine the pictures and read the captions. Then ask the following questions.

If a galaxy were discovered at a distance of 24 billion light years from Earth, how would this change the estimated age of the universe? According to the big bang theory, it would mean that the universe had been expanding for at least 24 billion years, making it much older that its present estimated age.

When you look at distant galaxies, why can you say that you are looking back in time? Because light that is now reaching Earth left those galaxies millions or even billions of years ago.

Activity

Penzias and Wilson Have small groups of students research the work of Arno A. Penzias and Robert W. Wilson that resulted in their winning a Nobel Prize. Ask each group to make a comic book that illustrates the story of their work and explains how it relates to the big bang theory. L2 COOP LEARN LS **Visual-Spatial and Linguistic**

NATIONAL GEOGRAPHIC

VISUALIZING THE BIG BANG THEORY

Figure 21

The big bang theory states that the universe probably began about 13.7 billion years ago with an enormous explosion. Even today, galaxies are rushing apart from this explosion.

A Within fractions of a second of the initial explosion, the universe grew from the size of a pinhead to 2,000 times the size of the Sun.

B By the time the universe was one second old, it was a dense, opaque, swirling mass of elementary particles.

C Matter began collecting in clumps. As matter cooled, hydrogen and helium gases formed.

D More than a billion years after the initial explosion, the first stars were born.

Science Journal

The Big Bang Theory Ask students to write a short newspaper article about the big bang theory in their Science Journals. Students should copy the style of articles in the local paper. They also should assume that they are writing for an audience that is not familiar with the theory.

124 ◆ J CHAPTER 4 Stars and Galaxies

The Big Bang Theory

When scientists determined that the universe was expanding, they developed a theory to explain their observations. The leading theory about the formation of the universe is called the **big bang theory**. **Figure 21** illustrates the big bang theory. According to this theory, approximately 13.7 billion years ago, the universe began with an enormous explosion. The entire universe began to expand everywhere at the same time.

Figure 22 The light from the galaxies in this photo mosaic took billions of years to reach Earth.

Looking Back in Time The time-exposure photograph shown in **Figure 22** was taken by the *Hubble Space Telescope.* It shows more than 1,500 galaxies at distances of more than 10 billion light-years. These galaxies could date back to when the universe was no more than 1 billion years old. The galaxies are in various stages of development. One astronomer says that humans might be looking back to a time when the Milky Way was forming.

Whether the universe will expand forever or stop expanding is still unknown. If enough matter exists, gravity might halt the expansion, and the universe will contract until everything comes to a single point. However, studies of distant supernovae indicate that an energy, called dark energy, is causing the universe to expand faster. Scientists are trying to understand how dark energy might affect the fate of the universe.

section 4 review

Summary

Galaxies

- The three main types of galaxies are spiral, elliptical, and irregular.

The Milky Way Galaxy

- The Milky Way is a spiral galaxy and the Sun is about 26,000 light-years from its center.

Origin of the Universe

- Theories about how the universe formed include the steady state theory, the oscillating universe theory, and the big bang theory.

The Big Bang Theory

- This theory states that the universe began with an explosion about 13.7 billion years ago.

Self Check

1. **Describe** elliptical galaxies. How are they different from spiral galaxies?
2. **Identify** the galaxy that you live in.
3. **Explain** the Doppler shift.
4. **Explain** how all galaxies are similar.
5. **Think Critically** All galaxies outside the Local Group show a red shift. Within the Local Group, some show a red shift and some show a blue shift. What does this tell you about the galaxies in the Local Group?

Applying Skills

6. **Compare and contrast** the theories about the origin of the universe.

Science Online bookj.msscience.com/self_check_quiz

Use an Analogy

Raisin Bread Ask students to think of a loaf of raisin bread. When the bread is unbaked, the raisins are relatively close together. As the bread bakes, all the raisins move apart from one another. Have students compare this to the way scientists hypothesize galaxies are moving in the universe.

3 Assess

DAILY INTERVENTION

Check for Understanding

Logical-Mathematical Have students write brief paragraphs in their Science Journals explaining why the spiral shape of the Milky Way Galaxy is not visible from Earth. To see the spiral shape, an observer would have to be above or below the plane of the galaxy. Earth is located in one of the spiral arms. LS

Reteach

Light Waves Have a student use a spring toy to demonstrate the compression and expansion of light waves described by the Doppler shift. L1 LS **Visual-Spatial**

Assessment

Content Have students write a paragraph about galaxies. Encourage them to illustrate their summaries. Use **Performance Assessment in the Science Classroom**, p. 159. P

section 4 review

1. Elliptical galaxies have ellipsoidal shapes. Spiral galaxies have flattened, pinwheel shapes.
2. the Milky Way
3. The Doppler shift occurs when waves of light or sound are stretched or compressed because of relative motion between the source and the receiver.
4. Answers will vary. Possible answer: All galaxies are made of similar chemical elements and have the same forms of energy.
5. Some galaxies are moving toward the Milky Way, and some are moving away from the Milky Way.
6. Student answers will vary.

BENCH TESTED

Real-World Question

Purpose Students conduct an experiment that shows how distance from an object affects the object's parallax shift. L2 ELL COOP LEARN LS **Logical-Mathematical**

Process Skills measure in SI, use numbers, compare and contrast, recognize cause and effect, formulate models, communicate

Time Required one class period

Materials meterstick, metric ruler, masking tape, pencil

Safety Precautions Caution students to wear goggles during the experiment.

Form a Hypothesis

Possible Hypothesis Students may hypothesize that the amount of observed parallax increases as an object is moved closer and that a simple demonstration can illustrate this effect.

Test Your Hypothesis

Possible Procedures Choose and mark three different positions on the meterstick (20 cm, 40 cm, and 60 cm). Attach a metric ruler to one end, forming a T shape, and place it on a table. Hold a pencil upright at one of the marked positions. Observe the pencil's apparent movement against the meterstick with one eye closed and then the other. Repeat for all positions.

LAB Design Your Own

Measuring Parallax

Goals

- **Design** a model to show how the distance from an observer to an object affects the object's parallax shift.
- **Describe** how parallax can be used to determine the distance to a star.

Possible Materials

meterstick
masking tape
metric ruler
pencil

Safety Precautions

WARNING: *Be sure to wear goggles to protect your eyes.*

Real-World Question

Parallax is the apparent shift in the position of an object when viewed from two locations. How can you build a model to show the relationship between distance and parallax?

Form a Hypothesis

State a hypothesis about how parallax varies with distance.

Test Your Hypothesis

Make a Plan

1. As a group, agree upon and write your hypothesis statement.
2. **List** the steps you need to take to build your model. Be specific, describing exactly what you will do at each step.
3. **Devise** a method to test how distance from an observer to an object, such as a pencil, affects the parallax of the object.
4. **List** the steps you will take to test your hypothesis. Be specific, describing exactly what you will do at each step.
5. Read over your plan for the model to be used in this experiment.

Alternative Inquiry Lab

Extend the Activity Ask students to develop demonstrations that show how people observe parallax as part of their everyday experience. For example, students might demonstrate that a moving person loses sight of nearby objects very quickly but can see a distant object for a long time. One notices this effect when driving—trees along the roadside pass quickly, but one can observe the Moon during the entire drive.

6. How will you determine changes in observed parallax? Remember, these changes should occur when the distance from the observer to the object is changed.
7. You should measure shifts in parallax from several different positions. How will these positions differ?
8. How will you measure distances accurately and compare relative position shift?

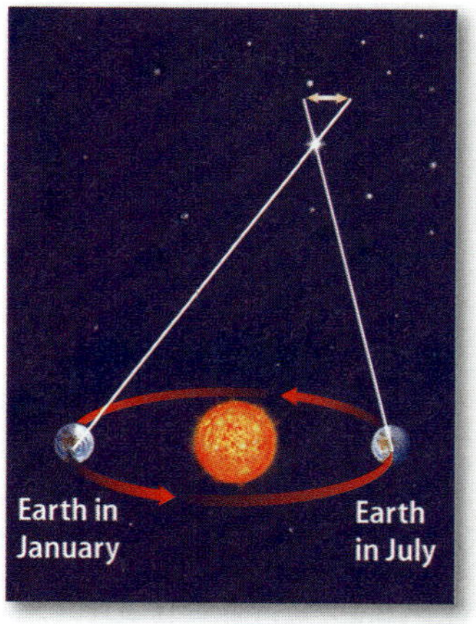

Follow Your Plan

1. Make sure your teacher approves your plan before you start.
2. **Construct** the model your team has planned.
3. Carry out the experiment as planned.
4. While conducting the experiment, record any observations that you or other members of your group make in your Science Journal.

Analyze Your Data

1. **Compare** what happened to the object when it was viewed with one eye closed, then the other.
2. At what distance from the observer did the object appear to shift the most?
3. At what distance did it appear to shift the least?

Conclude and Apply

1. **Infer** what happened to the apparent shift of the object's location as the distance from the observer was increased or decreased.
2. **Describe** how astronomers might use parallax to study stars.

Communicating Your Data

Prepare a chart showing the results of your experiment. Share the chart with members of your class. **For more help, refer to the** Science Skill Handbook.

Teaching Strategy Pair students and have one act as holder while the other student views the pencil. Students should then switch roles.

Analyze Your Data

Expected Outcome If students position themselves so that the meterstick is lined up with their noses and distances of 20 cm, 40 cm, and 60 cm are used, the following results can be expected: Observed parallax is about 20 cm at the 20-cm mark, 8.5 cm at the 40-cm mark, and 4.9 cm at the 60-cm mark.

Answers to Questions

1. The position of the pencil seemed to shift, but this motion was only apparent.
2. the closest distance used
3. the farthest distance

Error Analysis It important that students not move while viewing the pencil. Movement could cause erroneous results.

Conclude and Apply

1. As distance from the observer increased, the pencil's apparent shift decreased. As the distance from the observer decreased, the pencil's apparent shift increased.
2. Astronomers use parallax angles or shifts in the apparent position of astronomical objects to determine distances to the objects.

Assessment

Oral How can astronomers use parallax to determine distances to stars relatively close to Earth? Parallax angle decreases with distance from observer. Very distant stars will show no parallax.

Communicating Your Data

Have students use a computer to prepare a table on which they can enter the data from their experiment.

Content Background

Supernova explosions often are hypothesized to initiate the contraction of a gaseous nebula and hence star formation. The shock wave of a supernova can compress gas in a nebula. The increased density can be enough to initiate contraction under the influence of gravity.

Discussion

Betelgeuse Tell students that Betelgeuse is 520 light-years from Earth. If Betelgeuse is so much larger than the Sun, why isn't it brighter than the Sun? It is much farther away from Earth, so it looks much smaller and dimmer than the Sun.

Activity

Scale Models Have students select a scale to use and make scale models that compare the diameters of the Sun and Betelgeuse. Then have students research the sizes of other stars and make additional scale models showing how their diameters compare to those of the Sun and Betelgeuse.

Applying Math

Answer four hundred fifty million five hundred twenty thousand

Write About It

Student reports will vary.

SCIENCE Stats

Stars and Galaxies

Did you know...

. . . A star in Earth's galaxy explodes as a supernova about once a century. The most famous supernova of this galaxy occurred in 1054 and was recorded by the ancient Chinese and Koreans. The explosion was so powerful that it could be seen during the day, and its brightness lasted for weeks. Other major supernovas in the Milky Way that were observed from Earth occurred in 185, 386, 1006, 1181, 1572, and 1604.

. . . The large loops of material called solar prominences can extend more than 320,000 km above the Sun's surface. This is so high that two Jupiters and three Earths could fit under the arch.

. . . The red giant star Betelgeuse has a diameter larger than that of Earth's Sun. This gigantic star measures 450,520,000 km in diameter, while the Sun's diameter is a mere 1,390,176 km.

Applying Math Use words to express the number 450,520,000.

Write About It

Visit bookj.msscience.com/science_stats to learn whether it might be possible for Earth astronauts to travel to the nearest stars. How long would such a trip take? What problems would have to be overcome? Write a brief report about what you find.

Visual Learning

Solar Prominence Art Make certain that students notice the relative sizes of Earth and Jupiter. Have them estimate how many Earths would need to be stacked to equal the diameter of Jupiter. about 11

Reviewing Main Ideas

Section 1 Stars

1. Constellations are patterns of stars in the night sky. Some constellations can be seen all year. Other constellations are visible only during certain seasons.
2. Parallax is the apparent shift in the position of an object when viewed from two different positions. Parallax is used to find the distance to nearby stars.

Section 2 The Sun

1. The Sun is the closest star to Earth.
2. Sunspots are areas on the Sun's surface that are cooler and less bright than surrounding areas.

Section 3 Evolution of Stars

1. Stars are classified according to their position on the H-R diagram.
2. Low-mass stars end their lives as white dwarfs. High-mass stars become neutron stars or black holes.

Section 4 Galaxies and the Universe

1. A galaxy consists of stars, gas, and dust held together by gravity.
2. Earth's solar system is in the Milky Way, a spiral galaxy.
3. The universe is expanding. Scientists don't know whether the universe will expand forever or contract to a single point.

Visualizing Main Ideas

Copy and complete the following concept map that shows the evolution of a main sequence star with a mass similar to that of the Sun.

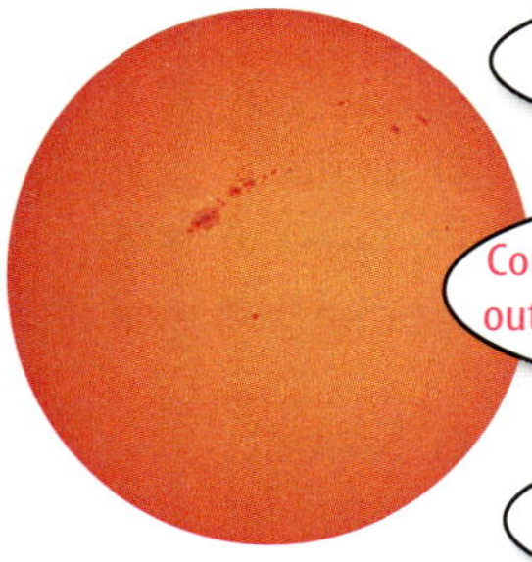

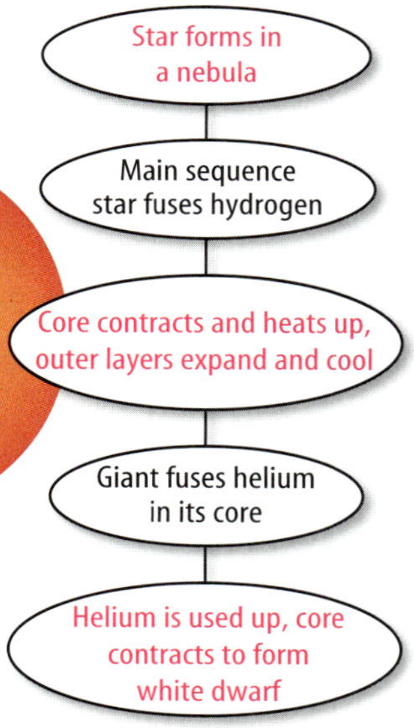

Science Online bookj.msscience.com/interactive_tutor

chapter 4 Study Guide

Reviewing Main Ideas

Summary statements can be used by students to review the major concepts of the chapter.

Visualizing Main Ideas

See student page.

Science Online

Visit bookj.msscience.com
/self_check_quiz
/interactive_tutor
/vocabulary_puzzlemaker
/chapter_review
/standardized_test

Assessment Transparency

For additional assessment questions, use the *Assessment Transparency* located in the transparency book.

Assessment

Assessment Transparency **Stars and Galaxies**

Directions: *Carefully review the table and answer the following questions.*

Light Waves	
Color	**Wavelength (nanometers)**
Violet	380-450
Blue	450-490
Green	490-560
Yellow	560-590
Orange	590-630
Red	630-760

1. According to the table, what will be the color of a light wave with a wavelength of 500 nanometers?
   A Violet
   B Blue
   C Green
   D Yellow
2. In a clear sky, the Sun is bright yellow. What is the approximate wavelength of the light coming from the Sun?
   F 380–340 nanometers
   G 450–490 nanometers
   H 560–590 nanometers
   J 630–760 nanometers
3. The shorter the wavelength of a light wave, the greater its energy. A light wave of which of the colors listed below has the greatest energy?
   A Red
   B Orange
   C Green
   D Violet

L2

Stars and Galaxies

chapter 4 Review

Using Vocabulary

1. Absolute magnitude is a measure of the light given off by a star. Apparent magnitude is a measure of the light received on Earth.
2. A constellation is a pattern of stars in the night sky. A galaxy is a group of stars held together by gravity.
3. A giant is the late stage of a relatively low mass star. A supergiant is the late stage of a very massive star.
4. The photosphere is the lowest layer of the Sun's atmosphere. The chromosphere is directly above it.
5. Black holes form when the cores of very massive stars collapse to a point. Neutron stars form when slightly less massive stars collapse to a small core consisting of neutrons.

Checking Concepts

6. B	11. D
7. C	12. D
8. C	13. A
9. C	14. B
10. D	15. A

chapter 4 Review

Using Vocabulary

absolute magnitude p. 106	giant p. 117
apparent magnitude p. 106	light-year p. 107
big bang theory p. 125	nebula p. 116
black hole p. 118	neutron star p. 118
chromosphere p. 109	photosphere p. 109
constellation p. 104	sunspot p. 110
corona p. 109	supergiant p. 118
galaxy p. 120	white dwarf p. 117

Explain the difference between the terms in each of the following sets.

1. absolute magnitude—apparent magnitude
2. galaxy—constellation
3. giant—supergiant
4. chromosphere—photosphere
5. black hole—neutron star

Checking Concepts

Choose the word or phrase that best answers the question.

6. What is a measure of the amount of a star's light that is received on Earth?
   A) absolute magnitude
   B) apparent magnitude
   C) fusion
   D) parallax
7. What is higher for closer stars?
   A) absolute magnitude
   B) red shift
   C) parallax
   D) blue shift
8. What happens after a nebula contracts and its temperature increases to 10 million K?
   A) a black hole forms
   B) a supernova occurs
   C) fusion begins
   D) a white dwarf forms
9. Which of these has an event horizon?
   A) giant **C)** black hole
   B) white dwarf **D)** neutron star
10. What forms when the Sun fuses hydrogen?
    A) carbon **C)** iron
    B) oxygen **D)** helium

Use the illustration below to answer question 11.

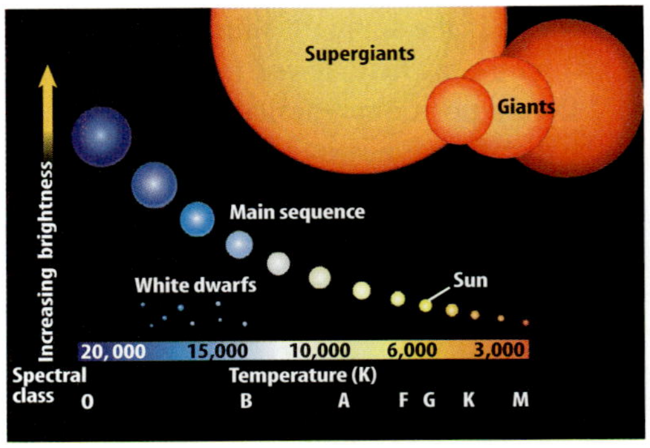

11. Which of the following best describes giant stars?
    A) hot, dim stars
    B) cool, dim stars
    C) hot, bright stars
    D) cool, bright stars
12. Which of the following are loops of matter flowing from the Sun?
    A) sunspots **C)** coronas
    B) auroras **D)** prominences
13. What are groups of galaxies called?
    A) clusters **C)** giants
    B) supergiants **D)** binary systems
14. Which galaxies are sometimes shaped like footballs?
    A) spiral **C)** barred
    B) elliptical **D)** irregular
15. What do scientists study to determine shifts in wavelengths of light?
    A) spectrum **C)** corona
    B) parallax **D)** nebula

Science Online bookj.msscience.com/vocabulary_puzzlemaker

Use the Exam*View*® Pro Testmaker CD-ROM to:

- create multiple versions of tests
- create modified tests with one mouse click for inclusion students
- edit existing questions and add your own questions
- build tests aligned with state standards using built-in State Curriculum Tags
- change English tests to Spanish with one mouse click and vice versa

130 ◆ J CHAPTER REVIEW

Thinking Critically

Use the table below to answer question 16.

Magnitude and Distance of Stars

Star	Apparent Magnitude	Absolute Magnitude	Distance in Light-Years
A	−26	4.8	0.00002
B	−1.5	1.4	8.7
C	0.1	4.4	4.3
D	0.1	−7.0	815
E	0.4	−5.9	520
F	1.0	−0.6	45

16. Interpret Data Use the table above to answer the following questions. *Hint: lower magnitude values are brighter than higher magnitude values.*

a. Which star appears brightest from Earth?

b. Which star would appear brightest from a distance of 10 light-years?

c. Infer which star in the table above is the Sun.

17. Infer How do scientists know that black holes exist if these objects don't emit visible light?

18. Recognize Cause and Effect Why can parallax only be used to measure distances to stars that are relatively close to Earth?

19. Compare and contrast the Sun with other stars on the H-R diagram.

20. Concept Map Make a concept map showing the life history of a very large star.

21. Make Models Make a model of the Sun. Include all of the Sun's layers in your model.

Performance Activities

22. Story Write a short science-fiction story about an astronaut traveling through the universe. In your story, describe what the astronaut observes. Use as many vocabulary words as you can.

23. Photomontage Gather photographs of the aurora borealis from magazines and other sources. Use the photographs to create a photomontage. Write a caption for each photo.

Applying Math

24. Travel to Vega Vega is a star that is 26 light-years away. If a spaceship could travel at one-tenth the speed of light, how long would it take to reach this star?

Use the illustration below to answer question 25.

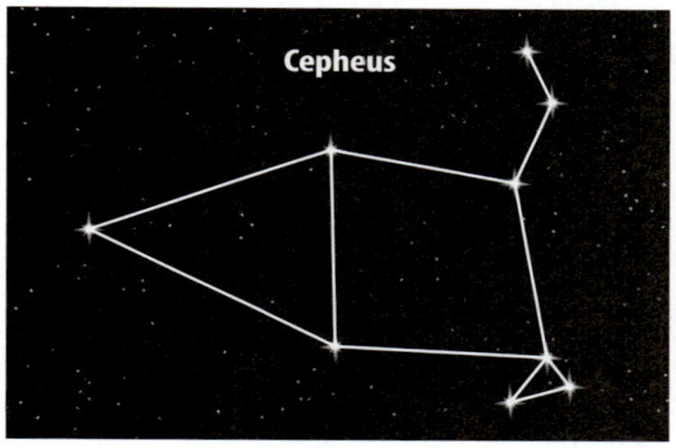

25. Constellation Cepheus The illustration above shows the constellation Cepheus. Answer the following questions about this contellation.

a. Which of the labeled line segments are nearly parallel?

b. Which line segments are nearly perpendicular?

c. Which line segments are oblique?

d. What geometric shape do the three stars at the left side of the drawing form?

Science Online bookj.msscience.com/chapter_review

chapter Review 4

Thinking Critically

16. a. A

b. D

c. A

17. Matter emits radiation as it is drawn into black holes. A black hole's gravity can affect nearby objects.

18. Stars that are far from Earth have parallax angles that are too small to measure accurately.

19. The Sun is smaller and hotter than giants and supergiants. It is larger and cooler than white dwarf stars.

20. Student maps should parallel the upper branch of **Figure 13.**

21. Make certain that student models are accurate and complete.

Performance Activities

22. Stories will be futuristic. However, they should be based on sound science. Use **PASC,** p. 155

23. Students also might create a photomontage for the aurora australis. Use **PASC,** p. 149

Applying Math

National Math Standards

1, 3, 9

24. 260 years

25. a. $\overline{AE}$ and $\overline{CD}$; $\overline{BC}$ and $\overline{DE}$

b. $\overline{BE}$ and $\overline{ED}$

c. ∠AED; ∠ABC; ∠BCD

d. triangle

Assessment Resources

Reproducible Masters

Chapter *Fast File* Resources

Chapter Review, pp. 41–42

Chapter Tests, pp. 43–46

Assessment Transparency Activity, p. 55

Glencoe Science Web site

Chapter Review Test

Standardized Test Practice

Glencoe Technology

Assessment Transparency

Exam*View*® Pro Testmaker

MindJogger Videoquiz

Interactive Chalkboard

CHAPTER REVIEW J ◆ 131

chapter 4 Standardized Test Practice

FAST FILE

Answer Sheet A practice answer sheet can be found at bookj.msscience.com/answer_sheet.

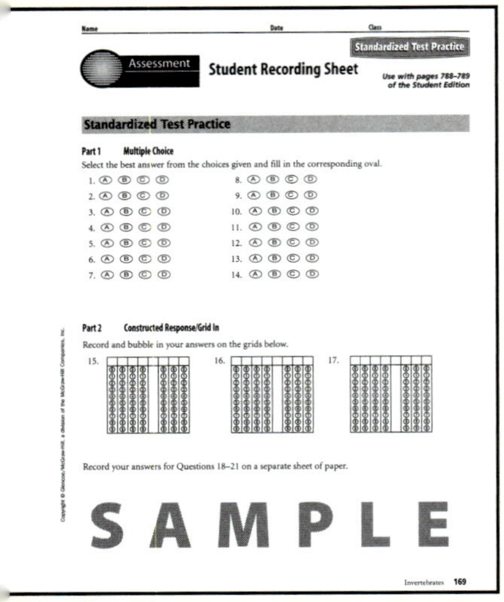
Name Date Class

Assessment Student Recording Sheet

Standardized Test Practice

Use with pages 788–789 of the Student Edition

Standardized Test Practice

Part 1 Multiple Choice

Select the best answer from the choices given and fill in the corresponding oval.

1.–14. Ⓐ Ⓑ Ⓒ Ⓓ

Part 2 Constructed Response/Grid In

Record and bubble in your answers on the grids below.

15. 16. 17.

Record your answers for Questions 18–21 on a separate sheet of paper.

SAMPLE

Part 1 Multiple Choice

1. C
2. D
3. A
4. C
5. B
6. A
7. B
8. A

Part 2 Short Response

9. Charged particles from the Sun can travel to Earth after CMEs or solar flares. These particles can cause communication and power transmission problems. They also can trigger the formation of aurorae.
10. A red shift occurs when the wavelength of light is increased because the source is moving away from the receiver. A blue shift occurs when the wavelength of light is decreased because the source is moving toward the receiver.
11. Light from galaxies outside of the local group displays a red shift.

Part 1 Multiple Choice

Record your answers on the answer sheet provided by your teacher or on a sheet of paper.

Use the illustration below to answer question 1.

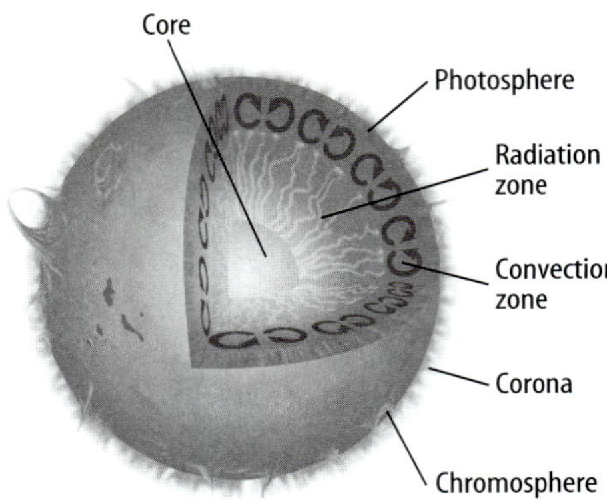

1. The illustration above shows the interior of which object?
A. Earth **C.** the Sun
B. Saturn **D.** the Moon

2. Which is a group of stars, gas, and dust held together by gravity?
A. constellation **C.** black hole
B. supergiant **D.** galaxy

3. The most massive stars end their lives as which type of object?
A. black hole **C.** neutron star
B. white dwarf **D.** black dwarf

4. In which galaxy does the Sun exist?
A. Arp's Galaxy **C.** Milky Way Galaxy
B. Barnard's Galaxy **D.** Andromeda Galaxy

Test-Taking Tip

Process of Elimination If you don't know the answer to a multiple-choice question, eliminate as many incorrect choices as possible. Mark your best guess from the remaining answers before moving on to the next question.

5. Which is the closest star to Earth?
A. Sirius **C.** Betelgeuse
B. the Sun **D.** the Moon

6. In which of the following choices are the objects ordered from smallest to largest?
A. stars, galaxies, galaxy clusters, universe
B. galaxy clusters, galaxies, stars, universe
C. universe, galaxy clusters, galaxies, stars
D. universe, stars, galaxies, galaxy clusters

Use the graph below to answer questions 7 and 8.

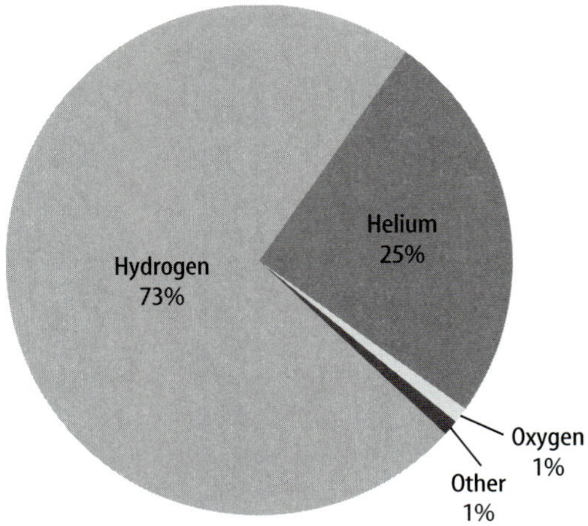

7. Which is the most abundant element in the Sun?
A. helium
B. hydrogen
C. oxygen
D. carbon

8. How will this circle graph change as the Sun ages?
A. The hydrogen slice will get smaller.
B. The hydrogen slice will get larger.
C. The helium slice will get smaller.
D. The circle graph will not change.

Part 2 Short Response/Grid In

12. The main sequence is a diagonal band on the H-R diagram. Most stars plot on the main sequence. It represents the longest stage of stellar evolution during which stars shine by hydrogen fusion.
13. A constellation is a pattern of stars in the night sky.
14. 4.3
15. 4.3
16. 100
17. A star's color provides general information about temperature. Blue-white stars are hotter than yellow stars, which are hotter than red stars.
18. Light from the Sun reaches Earth in just a few minutes. However, it takes years, and in some cases millions of years, for light to reach Earth from other stars.
19. When compared to other stars in the Milky Way Galaxy, the Sun is a typical star in terms of temperature, age, and brightness. In general, it is of average size when compared to most stars, but is a lot smaller when compared to giant stars.

Part 2 Short Response/Grid In

Record your answers on the answer sheet provided by your teacher or on a sheet of paper.

9. How can events on the Sun affect Earth? Give one example.
10. How does a red shift differ from a blue shift?
11. How do astronomers know that the universe is expanding?
12. What is the main sequence?
13. What is a constellation?

Use the illustration below to answer questions 14–16.

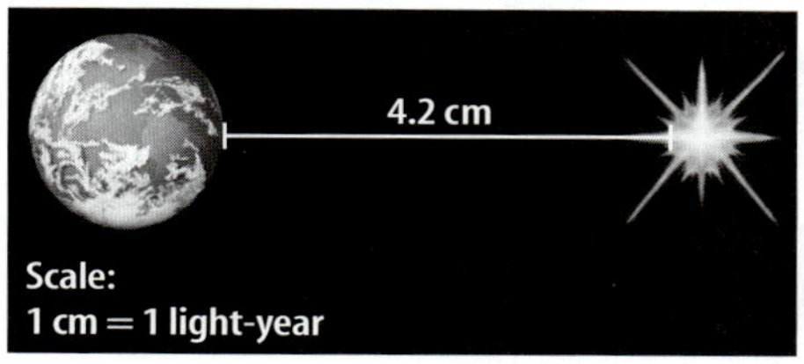

14. According to the illustration, how many light-years from Earth is Proxima Centauri?
15. How many years would it take for light from Proxima Centauri to get to Earth?
16. At this scale, how many centimeters would represent the distance to a star that is 100 light-years from Earth?
17. How can a star's color provide information about its temperature?
18. Approximately how long does it take light from the Sun to reach Earth? In general, how does this compare to the amount of time it takes light from all other stars to reach Earth?
19. How does the size, temperature, age, and brightness of the Sun compare to other stars in the Milky Way Galaxy?

Part 3 Open Ended

Record your answers on a sheet of paper.

Use the graph below to answer question 20.

Visual Light Curve of Supernova

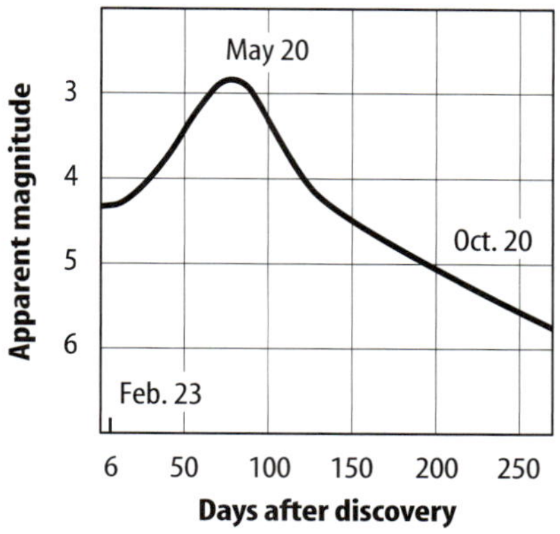

20. The graph above shows the brightness of a supernova that was observed from Earth in 1987. Describe how the brightness of this supernova changed through time. When was it brightest? What happened before May 20? What happened after May 20? How much did the brightness change?
21. Compare and contrast the different types of galaxies.
22. Write a detailed description of the Sun. What is it? What is it like?
23. Explain how parallax is used to measure the distance to nearby stars.
24. Why are some constellations visible all year? Why are other constellations only visible during certain seasons?
25. What are black holes? How do they form?
26. Explain the big bang theory.
27. What can be learned by studying the dark lines in a star's spectrum?

The parallax angle decreases as the distance to a star increases, and the angle can be used to calculate the distance to the star.

24. Circumpolar constellations are visible all year. The stars in these constellations are well above or below the projection of Earth's equator into space. Other constellations that are closer to the celestial equator are only visible when the night side of Earth faces in the correct direction.
25. the cores of massive stars that have collapsed to a point
26. The universe formed in an explosion and has been expanding since.
27. the composition of the star's atmosphere

Rubrics

For more help evaluating open-ended assessment questions, see the rubric on p. 10T.

Part 3 Open Ended

20. brightest on about May 20th, 1987; before May 20th, the brightness increased; after May 20th, the brightness decreased; apparent magnitude ranged from about 3 to about 5.5 or 6.
21. All galaxies consist of gas, dust, and stars held together by gravity. Spiral galaxies have spiral arms. Barred spirals have a bar-shaped band of stars near their center. Elliptical galaxies have ellipsoidal shapes, and irregular galaxies have irregular shapes.
22. The Sun is a typical star. It is middle-aged and shines with a yellow light.
23. A star is sighted at two different positions in Earth's orbit. The angular difference against the background stars is the parallax.

Student Resources

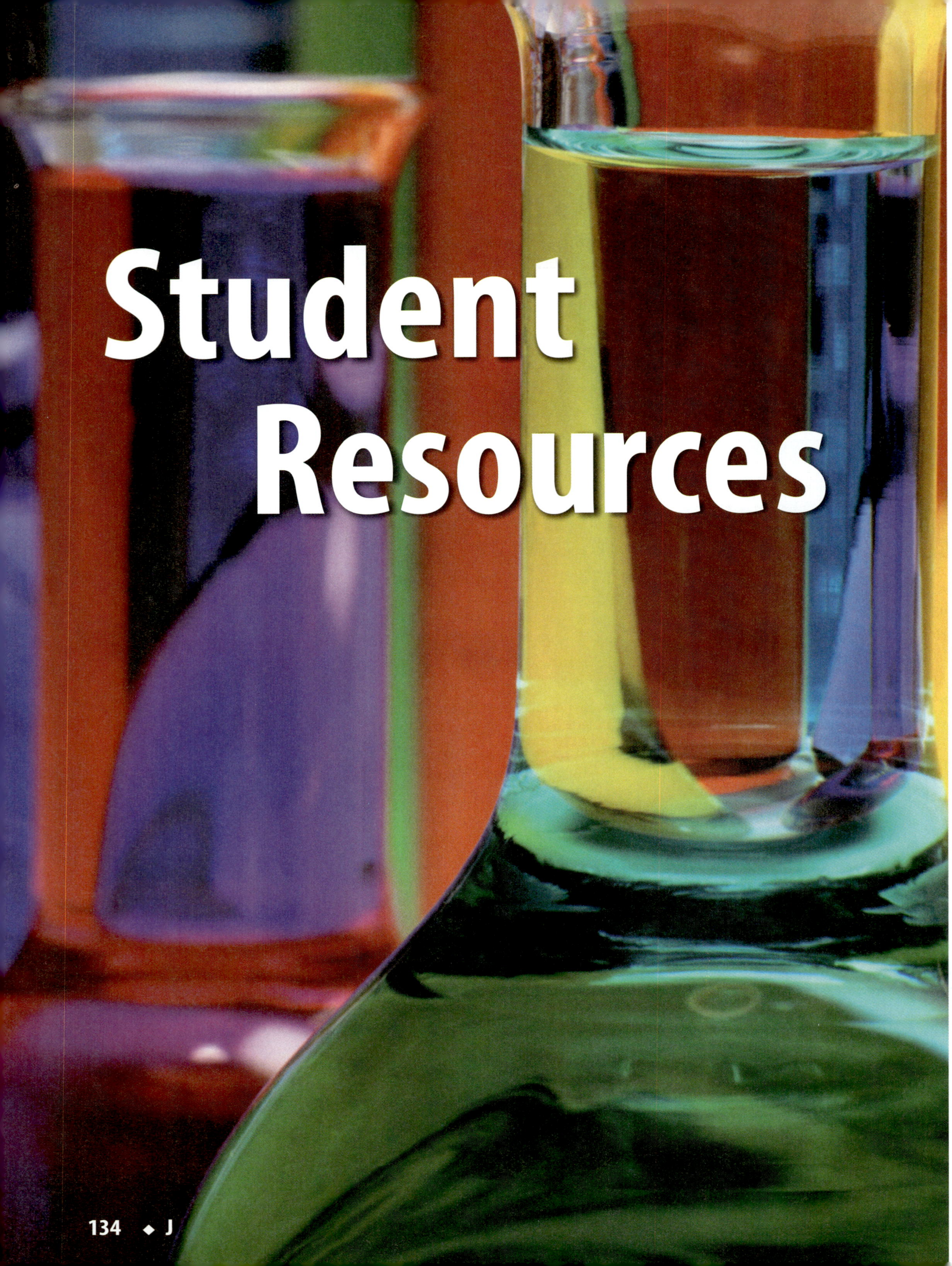

CONTENTS

Scientific Methods

Scientists use an orderly approach called scientific methods to solve problems. These include organizing and recording data so others can understand them. Scientists use many variations in these methods when they solve problems.

Identify a Question

The first step in a scientific investigation or experiment is to identify a question to be answered or a problem to be solved. For example, you might ask which gasoline is the most efficient.

Gather and Organize Information

After you have identified your question, begin gathering and organizing information. There are many ways to gather information, such as researching in a library, interviewing those knowledgeable about the subject, testing and working in the laboratory and field. Fieldwork is investigations and observations done outside of a laboratory.

Researching Information Before moving in a new direction, it is important to gather the information that already is known about the subject. Start by asking yourself questions to determine exactly what you need to know. Then you will look for the information in various reference sources, like the student is doing in **Figure 1.** Some sources may include textbooks, encyclopedias, government documents, professional journals, science magazines, and the Internet. Always list the sources of your information.

Figure 1 The Internet can be a valuable research tool.

Evaluate Sources of Information Not all sources of information are reliable. You should evaluate all of your sources of information, and use only those you know to be dependable. For example, if you are researching ways to make homes more energy efficient, a site written by the U.S. Department of Energy would be more reliable than a site written by a company that is trying to sell a new type of weatherproofing material. Also, remember that research always is changing. Consult the most current resources available to you. For example, a 1985 resource about saving energy would not reflect the most recent findings.

Sometimes scientists use data that they did not collect themselves, or conclusions drawn by other researchers. This data must be evaluated carefully. Ask questions about how the data were obtained, if the investigation was carried out properly, and if it has been duplicated exactly with the same results. Would you reach the same conclusion from the data? Only when you have confidence in the data can you believe it is true and feel comfortable using it.

Interpret Scientific Illustrations As you research a topic in science, you will see drawings, diagrams, and photographs to help you understand what you read. Some illustrations are included to help you understand an idea that you can't see easily by yourself, like the tiny particles in an atom in **Figure 2.** A drawing helps many people to remember details more easily and provides examples that clarify difficult concepts or give additional information about the topic you are studying. Most illustrations have labels or a caption to identify or to provide more information.

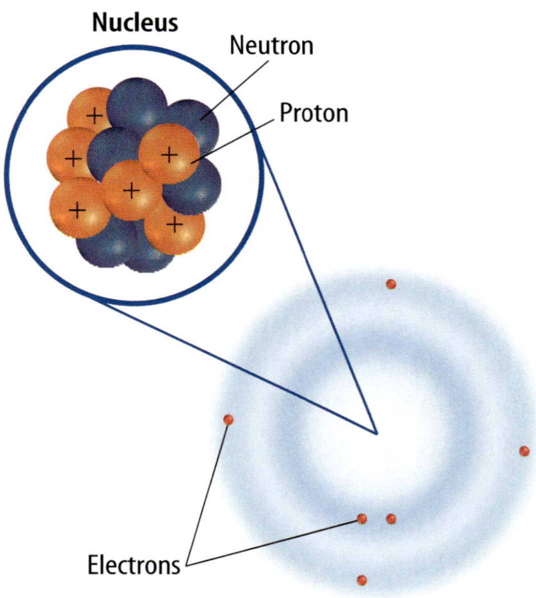

Figure 2 This drawing shows an atom of carbon with its six protons, six neutrons, and six electrons.

Concept Maps One way to organize data is to draw a diagram that shows relationships among ideas (or concepts). A concept map can help make the meanings of ideas and terms more clear, and help you understand and remember what you are studying. Concept maps are useful for breaking large concepts down into smaller parts, making learning easier.

Network Tree A type of concept map that not only shows a relationship, but how the concepts are related is a network tree, shown in **Figure 3.** In a network tree, the words are written in the ovals, while the description of the type of relationship is written across the connecting lines.

When constructing a network tree, write down the topic and all major topics on separate pieces of paper or notecards. Then arrange them in order from general to specific. Branch the related concepts from the major concept and describe the relationship on the connecting line. Continue to more specific concepts until finished.

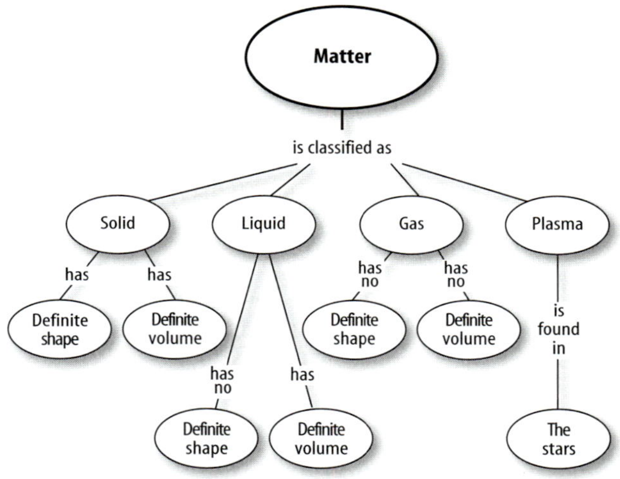

Figure 3 A network tree shows how concepts or objects are related.

Events Chain Another type of concept map is an events chain. Sometimes called a flow chart, it models the order or sequence of items. An events chain can be used to describe a sequence of events, the steps in a procedure, or the stages of a process.

When making an events chain, first find the one event that starts the chain. This event is called the initiating event. Then, find the next event and continue until the outcome is reached, as shown in **Figure 4.**

Science Skill Handbook

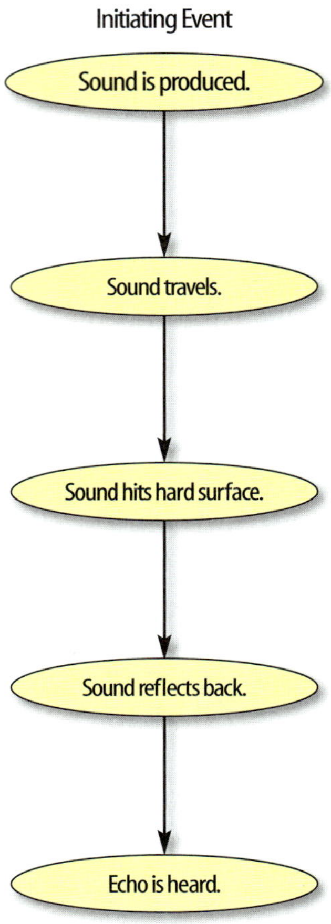

Figure 4 Events-chain concept maps show the order of steps in a process or event. This concept map shows how a sound makes an echo.

Cycle Map

Cycle Map A specific type of events chain is a cycle map. It is used when the series of events do not produce a final outcome, but instead relate back to the beginning event, such as in **Figure 5.** Therefore, the cycle repeats itself.

To make a cycle map, first decide what event is the beginning event. This is also called the initiating event. Then list the next events in the order that they occur, with the last event relating back to the initiating event. Words can be written between the events that describe what happens from one event to the next. The number of events in a cycle map can vary, but usually contain three or more events.

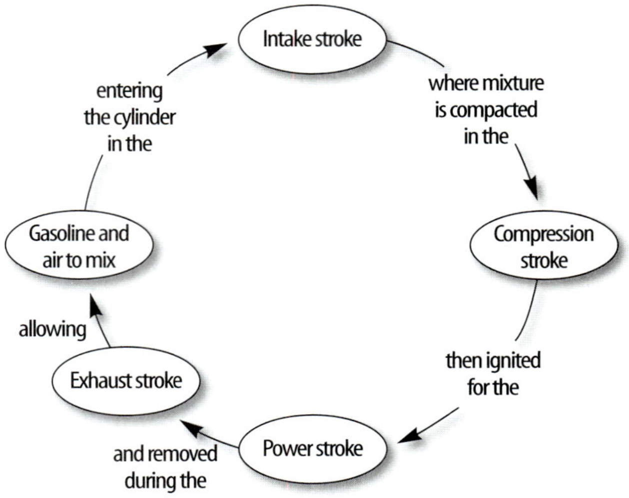

Figure 5 A cycle map shows events that occur in a cycle.

Spider Map A type of concept map that you can use for brainstorming is the spider map. When you have a central idea, you might find that you have a jumble of ideas that relate to it but are not necessarily clearly related to each other. The spider map on sound in **Figure 6** shows that if you write these ideas outside the main concept, then you can begin to separate and group unrelated terms so they become more useful.

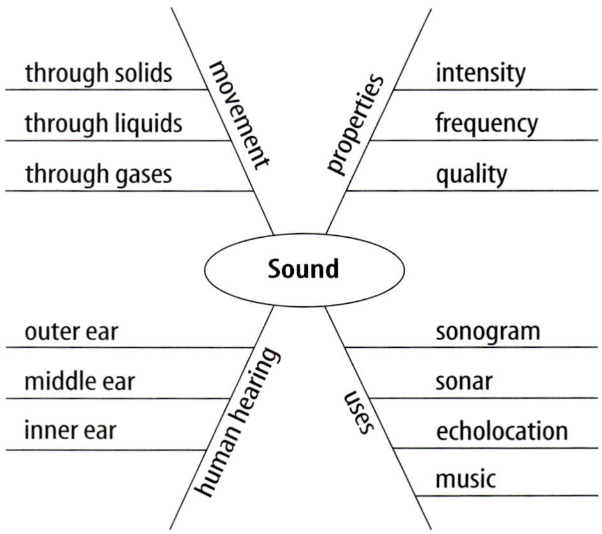

Figure 6 A spider map allows you to list ideas that relate to a central topic but not necessarily to one another.

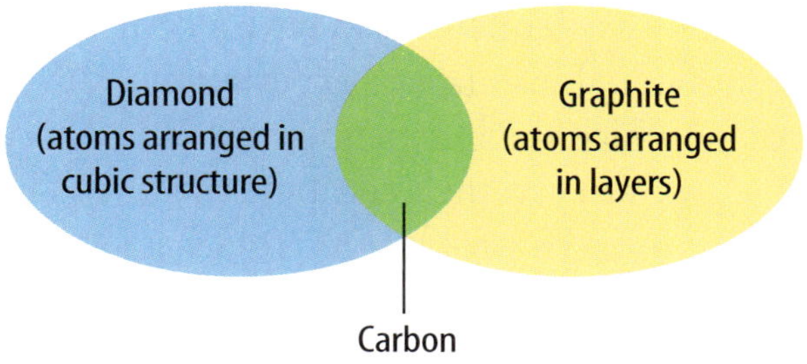

Figure 7 This Venn diagram compares and contrasts two substances made from carbon.

Venn Diagram To illustrate how two subjects compare and contrast you can use a Venn diagram. You can see the characteristics that the subjects have in common and those that they do not, shown in **Figure 7.**

To create a Venn diagram, draw two overlapping ovals that that are big enough to write in. List the characteristics unique to one subject in one oval, and the characteristics of the other subject in the other oval. The characteristics in common are listed in the overlapping section.

Make and Use Tables One way to organize information so it is easier to understand is to use a table. Tables can contain numbers, words, or both.

To make a table, list the items to be compared in the first column and the characteristics to be compared in the first row. The title should clearly indicate the content of the table, and the column or row heads should be clear. Notice that in **Table 1** the units are included.

Table 1 Recyclables Collected During Week

Day of Week	Paper (kg)	Aluminum (kg)	Glass (kg)
Monday	5.0	4.0	12.0
Wednesday	4.0	1.0	10.0
Friday	2.5	2.0	10.0

Make a Model One way to help you better understand the parts of a structure, the way a process works, or to show things too large or small for viewing is to make a model. For example, an atomic model made of a plastic-ball nucleus and pipe-cleaner electron shells can help you visualize how the parts of an atom relate to each other. Other types of models can by devised on a computer or represented by equations.

Form a Hypothesis

A possible explanation based on previous knowledge and observations is called a hypothesis. After researching gasoline types and recalling previous experiences in your family's car you form a hypothesis—our car runs more efficiently because we use premium gasoline. To be valid, a hypothesis has to be something you can test by using an investigation.

Predict When you apply a hypothesis to a specific situation, you predict something about that situation. A prediction makes a statement in advance, based on prior observation, experience, or scientific reasoning. People use predictions to make everyday decisions. Scientists test predictions by performing investigations. Based on previous observations and experiences, you might form a prediction that cars are more efficient with premium gasoline. The prediction can be tested in an investigation.

Design an Experiment A scientist needs to make many decisions before beginning an investigation. Some of these include: how to carry out the investigation, what steps to follow, how to record the data, and how the investigation will answer the question. It also is important to address any safety concerns.

Test the Hypothesis

Now that you have formed your hypothesis, you need to test it. Using an investigation, you will make observations and collect data, or information. This data might either support or not support your hypothesis. Scientists collect and organize data as numbers and descriptions.

Follow a Procedure In order to know what materials to use, as well as how and in what order to use them, you must follow a procedure. **Figure 8** shows a procedure you might follow to test your hypothesis.

Procedure

1. Use regular gasoline for two weeks.
2. Record the number of kilometers between fill-ups and the amount of gasoline used.
3. Switch to premium gasoline for two weeks.
4. Record the number of kilometers between fill-ups and the amount of gasoline used.

Figure 8 A procedure tells you what to do step by step.

Identify and Manipulate Variables and Controls In any experiment, it is important to keep everything the same except for the item you are testing. The one factor you change is called the independent variable. The change that results is the dependent variable. Make sure you have only one independent variable, to assure yourself of the cause of the changes you observe in the dependent variable. For example, in your gasoline experiment the type of fuel is the independent variable. The dependent variable is the efficiency.

Many experiments also have a control—an individual instance or experimental subject for which the independent variable is not changed. You can then compare the test results to the control results. To design a control you can have two cars of the same type. The control car uses regular gasoline for four weeks. After you are done with the test, you can compare the experimental results to the control results.

Collect Data

Whether you are carrying out an investigation or a short observational experiment, you will collect data, as shown in **Figure 9.** Scientists collect data as numbers and descriptions and organize it in specific ways.

Observe Scientists observe items and events, then record what they see. When they use only words to describe an observation, it is called qualitative data. Scientists' observations also can describe how much there is of something. These observations use numbers, as well as words, in the description and are called quantitative data. For example, if a sample of the element gold is described as being "shiny and very dense" the data are qualitative. Quantitative data on this sample of gold might include "a mass of 30 g and a density of 19.3 g/cm^3."

Figure 9 Collecting data is one way to gather information directly.

Figure 10 Record data neatly and clearly so it is easy to understand.

When you make observations you should examine the entire object or situation first, and then look carefully for details. It is important to record observations accurately and completely. Always record your notes immediately as you make them, so you do not miss details or make a mistake when recording results from memory. Never put unidentified observations on scraps of paper. Instead they should be recorded in a notebook, like the one in **Figure 10.** Write your data neatly so you can easily read it later. At each point in the experiment, record your observations and label them. That way, you will not have to determine what the figures mean when you look at your notes later. Set up any tables that you will need to use ahead of time, so you can record any observations right away. Remember to avoid bias when collecting data by not including personal thoughts when you record observations. Record only what you observe.

Estimate Scientific work also involves estimating. To estimate is to make a judgment about the size or the number of something without measuring or counting. This is important when the number or size of an object or population is too large or too difficult to accurately count or measure.

Sample Scientists may use a sample or a portion of the total number as a type of estimation. To sample is to take a small, representative portion of the objects or organisms of a population for research. By making careful observations or manipulating variables within that portion of the group, information is discovered and conclusions are drawn that might apply to the whole population. A poorly chosen sample can be unrepresentative of the whole. If you were trying to determine the rainfall in an area, it would not be best to take a rainfall sample from under a tree.

Measure You use measurements everyday. Scientists also take measurements when collecting data. When taking measurements, it is important to know how to use measuring tools properly. Accuracy also is important.

Length To measure length, the distance between two points, scientists use meters. Smaller measurements might be measured in centimeters or millimeters.

Length is measured using a metric ruler or meter stick. When using a metric ruler, line up the 0-cm mark with the end of the object being measured and read the number of the unit where the object ends. Look at the metric ruler shown in **Figure 11.** The centimeter lines are the long, numbered lines, and the shorter lines are millimeter lines. In this instance, the length would be 4.50 cm.

Figure 11 This metric ruler has centimeter and millimeter divisions.

Mass The SI unit for mass is the kilogram (kg). Scientists can measure mass using units formed by adding metric prefixes to the unit gram (g), such as milligram (mg). To measure mass, you might use a triple-beam balance similar to the one shown in **Figure 12.** The balance has a pan on one side and a set of beams on the other side. Each beam has a rider that slides on the beam.

When using a triple-beam balance, place an object on the pan. Slide the largest rider along its beam until the pointer drops below zero. Then move it back one notch. Repeat the process for each rider proceeding from the larger to smaller until the pointer swings an equal distance above and below the zero point. Sum the masses on each beam to find the mass of the object. Move all riders back to zero when finished.

Instead of putting materials directly on the balance, scientists often take a tare of a container. A tare is the mass of a container into which objects or substances are placed for measuring their masses. To mass objects or substances, find the mass of a clean container. Remove the container from the pan, and place the object or substances in the container. Find the mass of the container with the materials in it. Subtract the mass of the empty container from the mass of the filled container to find the mass of the materials you are using.

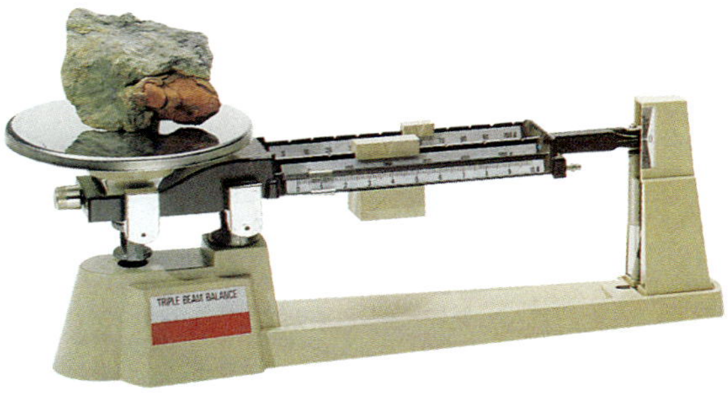

Figure 12 A triple-beam balance is used to determine the mass of an object.

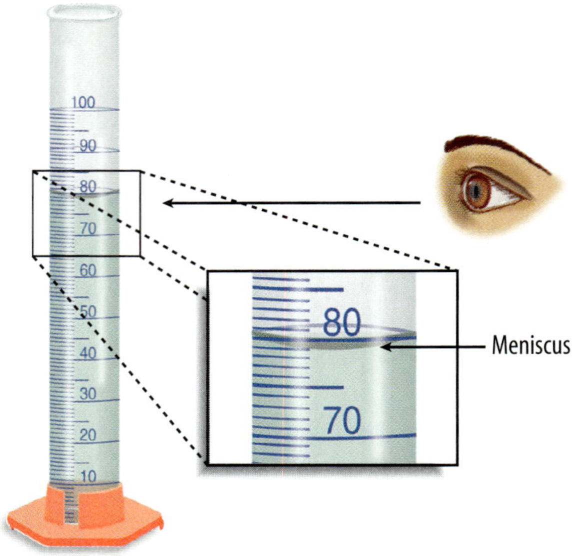

Figure 13 Graduated cylinders measure liquid volume.

Liquid Volume To measure liquids, the unit used is the liter. When a smaller unit is needed, scientists might use a milliliter. Because a milliliter takes up the volume of a cube measuring 1 cm on each side it also can be called a cubic centimeter ($cm^3 = cm \times cm \times cm$).

You can use beakers and graduated cylinders to measure liquid volume. A graduated cylinder, shown in **Figure 13,** is marked from bottom to top in milliliters. In lab, you might use a 10-mL graduated cylinder or a 100-mL graduated cylinder. When measuring liquids, notice that the liquid has a curved surface. Look at the surface at eye level, and measure the bottom of the curve. This is called the meniscus. The graduated cylinder in **Figure 13** contains 79.0 mL, or 79.0 cm^3, of a liquid.

Temperature Scientists often measure temperature using the Celsius scale. Pure water has a freezing point of 0°C and boiling point of 100°C. The unit of measurement is degrees Celsius. Two other scales often used are the Fahrenheit and Kelvin scales.

Figure 14 A thermometer measures the temperature of an object.

Scientists use a thermometer to measure temperature. Most thermometers in a laboratory are glass tubes with a bulb at the bottom end containing a liquid such as colored alcohol. The liquid rises or falls with a change in temperature. To read a glass thermometer like the thermometer in **Figure 14,** rotate it slowly until a red line appears. Read the temperature where the red line ends.

Form Operational Definitions An operational definition defines an object by how it functions, works, or behaves. For example, when you are playing hide and seek and a tree is home base, you have created an operational definition for a tree.

Objects can have more than one operational definition. For example, a ruler can be defined as a tool that measures the length of an object (how it is used). It can also be a tool with a series of marks used as a standard when measuring (how it works).

Analyze the Data

To determine the meaning of your observations and investigation results, you will need to look for patterns in the data. Then you must think critically to determine what the data mean. Scientists use several approaches when they analyze the data they have collected and recorded. Each approach is useful for identifying specific patterns.

Interpret Data The word *interpret* means "to explain the meaning of something." When analyzing data from an experiement, try to find out what the data show. Identify the control group and the test group to see whether or not changes in the independent variable have had an effect. Look for differences in the dependent variable between the control and test groups.

Classify Sorting objects or events into groups based on common features is called classifying. When classifying, first observe the objects or events to be classified. Then select one feature that is shared by some members in the group, but not by all. Place those members that share that feature in a subgroup. You can classify members into smaller and smaller subgroups based on characteristics. Remember that when you classify, you are grouping objects or events for a purpose. Keep your purpose in mind as you select the features to form groups and subgroups.

Compare and Contrast Observations can be analyzed by noting the similarities and differences between two more objects or events that you observe. When you look at objects or events to see how they are similar, you are comparing them. Contrasting is looking for differences in objects or events.

Science Skill Handbook

Recognize Cause and Effect A cause is a reason for an action or condition. The effect is that action or condition. When two events happen together, it is not necessarily true that one event caused the other. Scientists must design a controlled investigation to recognize the exact cause and effect.

Draw Conclusions

When scientists have analyzed the data they collected, they proceed to draw conclusions about the data. These conclusions are sometimes stated in words similar to the hypothesis that you formed earlier. They may confirm a hypothesis, or lead you to a new hypothesis.

Infer Scientists often make inferences based on their observations. An inference is an attempt to explain observations or to indicate a cause. An inference is not a fact, but a logical conclusion that needs further investigation. For example, you may infer that a fire has caused smoke. Until you investigate, however, you do not know for sure.

Apply When you draw a conclusion, you must apply those conclusions to determine whether the data supports the hypothesis. If your data do not support your hypothesis, it does not mean that the hypothesis is wrong. It means only that the result of the investigation did not support the hypothesis. Maybe the experiment needs to be redesigned, or some of the initial observations on which the hypothesis was based were incomplete or biased. Perhaps more observation or research is needed to refine your hypothesis. A successful investigation does not always come out the way you originally predicted.

Avoid Bias Sometimes a scientific investigation involves making judgments. When you make a judgment, you form an opinion. It is important to be honest and not to allow any expectations of results to bias your judgments. This is important throughout the entire investigation, from researching to collecting data to drawing conclusions.

Communicate

The communication of ideas is an important part of the work of scientists. A discovery that is not reported will not advance the scientific community's understanding or knowledge. Communication among scientists also is important as a way of improving their investigations.

Scientists communicate in many ways, from writing articles in journals and magazines that explain their investigations and experiments, to announcing important discoveries on television and radio. Scientists also share ideas with colleagues on the Internet or present them as lectures, like the student is doing in **Figure 15.**

Figure 15 A student communicates to his peers about his investigation.

Science Skill Handbook

SAFETY SYMBOLS

SAFETY SYMBOLS	HAZARD	EXAMPLES	PRECAUTION	REMEDY
DISPOSAL	Special disposal procedures need to be followed.	certain chemicals, living organisms	Do not dispose of these materials in the sink or trash can.	Dispose of wastes as directed by your teacher.
BIOLOGICAL	Organisms or other biological materials that might be harmful to humans	bacteria, fungi, blood, unpreserved tissues, plant materials	Avoid skin contact with these materials. Wear mask or gloves.	Notify your teacher if you suspect contact with material. Wash hands thoroughly.
EXTREME TEMPERATURE	Objects that can burn skin by being too cold or too hot	boiling liquids, hot plates, dry ice, liquid nitrogen	Use proper protection when handling.	Go to your teacher for first aid.
SHARP OBJECT	Use of tools or glassware that can easily puncture or slice skin	razor blades, pins, scalpels, pointed tools, dissecting probes, broken glass	Practice common-sense behavior and follow guidelines for use of the tool.	Go to your teacher for first aid.
FUME	Possible danger to respiratory tract from fumes	ammonia, acetone, nail polish remover, heated sulfur, moth balls	Make sure there is good ventilation. Never smell fumes directly. Wear a mask.	Leave foul area and notify your teacher immediately.
ELECTRICAL	Possible danger from electrical shock or burn	improper grounding, liquid spills, short circuits, exposed wires	Double-check setup with teacher. Check condition of wires and apparatus.	Do not attempt to fix electrical problems. Notify your teacher immediately.
IRRITANT	Substances that can irritate the skin or mucous membranes of the respiratory tract	pollen, moth balls, steel wool, fiberglass, potassium permanganate	Wear dust mask and gloves. Practice extra care when handling these materials.	Go to your teacher for first aid.
CHEMICAL	Chemicals can react with and destroy tissue and other materials	bleaches such as hydrogen peroxide; acids such as sulfuric acid, hydrochloric acid; bases such as ammonia, sodium hydroxide	Wear goggles, gloves, and an apron.	Immediately flush the affected area with water and notify your teacher.
TOXIC	Substance may be poisonous if touched, inhaled, or swallowed.	mercury, many metal compounds, iodine, poinsettia plant parts	Follow your teacher's instructions.	Always wash hands thoroughly after use. Go to your teacher for first aid.
FLAMMABLE	Flammable chemicals may be ignited by open flame, spark, or exposed heat.	alcohol, kerosene, potassium permanganate	Avoid open flames and heat when using flammable chemicals.	Notify your teacher immediately. Use fire safety equipment if applicable.
OPEN FLAME	Open flame in use, may cause fire.	hair, clothing, paper, synthetic materials	Tie back hair and loose clothing. Follow teacher's instruction on lighting and extinguishing flames.	Notify your teacher immediately. Use fire safety equipment if applicable.

Eye Safety
Proper eye protection should be worn at all times by anyone performing or observing science activities.

Clothing Protection
This symbol appears when substances could stain or burn clothing.

Animal Safety
This symbol appears when safety of animals and students must be ensured.

Handwashing
After the lab, wash hands with soap and water before removing goggles.

Safety in the Science Laboratory

The science laboratory is a safe place to work if you follow standard safety procedures. Being responsible for your own safety helps to make the entire laboratory a safer place for everyone. When performing any lab, read and apply the caution statements and safety symbol listed at the beginning of the lab.

General Safety Rules

1. Obtain your teacher's permission to begin all investigations and use laboratory equipment.
2. Study the procedure. Ask your teacher any questions. Be sure you understand safety symbols shown on the page.
3. Notify your teacher about allergies or other health conditions which can affect your participation in a lab.
4. Learn and follow use and safety procedures for your equipment. If unsure, ask your teacher.
5. Never eat, drink, chew gum, apply cosmetics, or do any personal grooming in the lab. Never use lab glassware as food or drink containers. Keep your hands away from your face and mouth.
6. Know the location and proper use of the safety shower, eye wash, fire blanket, and fire alarm.

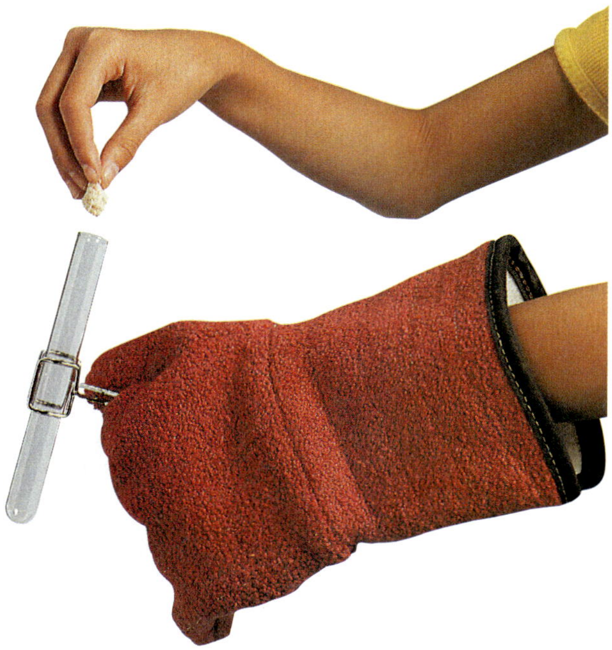

Prevent Accidents

1. Use the safety equipment provided to you. Goggles and a safety apron should be worn during investigations.
2. Do NOT use hair spray, mousse, or other flammable hair products. Tie back long hair and tie down loose clothing.
3. Do NOT wear sandals or other open-toed shoes in the lab.
4. Remove jewelry on hands and wrists. Loose jewelry, such as chains and long necklaces, should be removed to prevent them from getting caught in equipment.
5. Do not taste any substances or draw any material into a tube with your mouth.
6. Proper behavior is expected in the lab. Practical jokes and fooling around can lead to accidents and injury.
7. Keep your work area uncluttered.

Laboratory Work

1. Collect and carry all equipment and materials to your work area before beginning a lab.
2. Remain in your own work area unless given permission by your teacher to leave it.

3. Always slant test tubes away from yourself and others when heating them, adding substances to them, or rinsing them.
4. If instructed to smell a substance in a container, hold the container a short distance away and fan vapors towards your nose.
5. Do NOT substitute other chemicals/substances for those in the materials list unless instructed to do so by your teacher.
6. Do NOT take any materials or chemicals outside of the laboratory.
7. Stay out of storage areas unless instructed to be there and supervised by your teacher.

Laboratory Cleanup

1. Turn off all burners, water, and gas, and disconnect all electrical devices.
2. Clean all pieces of equipment and return all materials to their proper places.
3. Dispose of chemicals and other materials as directed by your teacher. Place broken glass and solid substances in the proper containers. Never discard materials in the sink.
4. Clean your work area.
5. Wash your hands with soap and water thoroughly BEFORE removing your goggles.

Emergencies

1. Report any fire, electrical shock, glassware breakage, spill, or injury, no matter how small, to your teacher immediately. Follow his or her instructions.
2. If your clothing should catch fire, STOP, DROP, and ROLL. If possible, smother it with the fire blanket or get under a safety shower. NEVER RUN.
3. If a fire should occur, turn off all gas and leave the room according to established procedures.
4. In most instances, your teacher will clean up spills. Do NOT attempt to clean up spills unless you are given permission and instructions to do so.
5. If chemicals come into contact with your eyes or skin, notify your teacher immediately. Use the eyewash or flush your skin or eyes with large quantities of water.
6. The fire extinguisher and first-aid kit should only be used by your teacher unless it is an extreme emergency and you have been given permission.
7. If someone is injured or becomes ill, only a professional medical provider or someone certified in first aid should perform first-aid procedures.

1 Space Probe Flights

Time Required 20 minutes

Materials Students can also use clay balls to represent the probes.

Safety Precautions
- Caution students to be careful when using the toothpicks.
- Caution students to avoid areas with traffic when completing their activities.

Teaching Strategies
- The two *Voyager* crafts continue traveling into deep space. Have students research how far away these probes are from Earth at this moment.
- Have students research other probes such as *Magellan* and *Galileo* and have students determine the comparative distances they traveled.

Conclude and Apply
1. The time line should relate the distance traveled to the following information:: *Mariner 2* (1962/Venus), *Pioneer 10* (1972/Jupiter), *Mariner 10* (1974/Mercury), *Viking 1* (1975/Mars), and *Voyager 2* (1977/Neptune).
2. 1 cm: 1 million km

Extra Try at Home Labs

Extra Try at Home Labs

EXTRA Try at Home Labs

From Your Kitchen, Junk Drawer, or Yard

1 Space Probe Flights

Real-World Question
How can we compare the distances traveled by space probes to their destinations?

Possible Materials
- polystyrene balls (5)
- toothpicks (5)
- small stick-on labels (5)
- tennis ball
- meterstick

Procedure
1. Write the names *Mariner 2, Pioneer 10, Mariner 10, Viking 1,* and *Voyager 2* on the five labels and stick each label on a toothpick. Stick a labeled toothpick into each of the polystyrene balls to represent these five United States space probes.
2. Place the tennis ball in an open space such as a basketball court or field.
3. Measure a distance of 0.42 m from the tennis ball and place the *Mariner 2* probe in that spot. Place the *Pioneer 10* probe 6.28 m away, the *Mariner 10* probe 0.92 m from the ball, the *Viking 1* probe 0.78 m away, and the *Voyager 2* probe 43.47 m from the tennis ball.

Conclude and Apply
1. Create a timeline showing the year each probe was launched and its destination and relate this information to the distance traveled.
2. Mercury is 58 million km from the sun and Earth is 150 million km. Use this information to calculate the scale used for this activity.

2 Creating Craters

Real-World Question
Why does the Moon have craters?

Possible Materials
- drink mix or powdered baby formula
- black pepper or paprika
- large, deep cooking tray or large bowl
- marbles
- small, round candies
- aquarium gravel
- tweezers
- bag of cotton balls

Procedure
1. Pour a 3-cm layer of powder over the bottom of a large, deep cooking tray.
2. Sprinkle a fine layer of black pepper over the powder.
3. Lay a 2–3 cm layer of cotton over half of the powder.
4. Drop marbles and other small objects into the powder not covered by the cotton. Carefully remove the objects with tweezers and observe the craters and impact patterns they make.
5. Drop objects on to the half of the tray covered by cotton.
6. Remove the objects and cotton and observe the marks made by objects in the powder.

Conclude and Apply
1. Compare the impacts made by the objects in the powder not covered by cotton with the impacts in the powder covered by cotton.
2. Infer why the Moon has many craters on its surface but Earth does not.

 Adult supervision required for all labs.

These labs are available at bookj.msscience.com.

2 Creating Craters

Time Required 25 minutes

Materials Baby formula powder covered with a fine layer of black pepper works best for creating craters and impact patterns.

Safety Precaution Avoid using eye-irritating powders or spices.

Teaching Strategy In the area not protected by the cotton, students can experiment with throwing objects into the powder at different angles. They should not throw the objects into the powder with a lot of force.

Conclude and Apply
1. Students should observe a variety of different shaped craters surrounded by impact patterns made by the pepper, but there will be few if any marks on the powder covered by cotton.
2. Like the cotton protecting the powder, Earth's atmosphere prevents most incoming meteors from reaching the surface.

Extra Try at Home Labs

3 Many Moons

Real-World Question
How do the number of moons of the nine planets compare?

Possible Materials
- golf balls (5)
- softballs (4)
- colored construction paper
- hole puncher
- pennies (10)
- quarters (8)
- meterstick

Procedure
1. Lay the golf balls and softballs on the floor in a row to represent the nine planets. The golf balls should represent the terrestrial planets and the softballs the gas planets.
2. Next to the golf ball representing Earth place one quarter. A quarter represents a moon with a diameter greater than 1,000 km. Research which planets have moons this size and place quarters next to them.
3. Use pennies to represent moons with a diameter between 200–1,000 km. Place pennies next to the planets with moons this size.
4. Use a hole punch to punch out holes from colored construction paper. These holes represent moons smaller than 200 km in diameter. Research which planets have moons this size and place the holes next to them.

Conclude and Apply
1. Infer why terrestrial planets have fewer moons than gas planets.
2. Infer why astronomers do not believe all the moons in the solar system have been discovered.

4 Big Stars

Real-World Question
How does the size of Earth compare to the size of stars?

Possible Materials
- metric ruler
- meterstick
- tape measure
- masking tape
- white paper
- black marker

Procedure
1. Tape a sheet of white paper to the floor.
2. Draw a dot in the center to the paper. Measure a 1-mm distance from the dot and draw a second dot. This distance represents the diameter of Earth.
3. Measure a distance of 10.9 cm from the first dot and draw a third dot. This distance represents the diameter of the Sun.
4. Measure a distance of 5 m from the first dot and mark the location on the floor with a piece of masking tape. This distance represents the average diameter of a red giant star.
5. Measure a distance of 30 m from the first dot and mark the location on the floor with a piece of masking tape. This distance represents the diameter of the supergiant star Antares.

Conclude and Apply
1. The diameter of Earth is 12,756 km. What is the diameter of the Sun?
2. What is the diameter of an average red giant?

Adult supervision required for all labs.

Extra Try at Home Labs

3 Many Moons

Time Required 40 minutes

Materials
- Provide students with materials and time to research the number of moons.
- To get a better idea of the size comparison between the terrestrial and gas planets, replace the softballs with basketballs.
- Large washers can replace the quarters.

Safety Precaution Caution students to use the metersticks and hole punchers with care.

Teaching Strategy Clear away floor space in your classroom before the lab or reserve the gym to complete this lab.

Conclude and Apply
1. The larger gas planets have more gravity to attract satellites and hold them in orbit.
2. Small moons around the gas planets are constantly being discovered as new telescopes and probes explore the outer planets.

4 Big Stars

Time Required 25 minutes

Materials Instead of using masking tape, tennis balls or small orange cones can be used to mark the comparative diameters of the giant stars.

Safety Precaution Students should use the metersticks with care.

Teaching Strategy Plan to complete this lab outside or in a gymnasium if you want students to complete the final measurement.

Conclude and Apply
1. 1,390,404 km
2. 63,780,000 km

Computer Skills

People who study science rely on computers, like the one in **Figure 16,** to record and store data and to analyze results from investigations. Whether you work in a laboratory or just need to write a lab report with tables, good computer skills are a necessity.

Using the computer comes with responsibility. Issues of ownership, security, and privacy can arise. Remember, if you did not author the information you are using, you must provide a source for your information. Also, anything on a computer can be accessed by others. Do not put anything on the computer that you would not want everyone to know. To add more security to your work, use a password.

Use a Word Processing Program

A computer program that allows you to type your information, change it as many times as you need to, and then print it out is called a word processing program. Word processing programs also can be used to make tables.

Figure 16 A computer will make reports neater and more professional looking.

Learn the Skill To start your word processing program, a blank document, sometimes called "Document 1," appears on the screen. To begin, start typing. To create a new document, click the *New* button on the standard tool bar. These tips will help you format the document.

- The program will automatically move to the next line; press *Enter* if you wish to start a new paragraph.
- Symbols, called non-printing characters, can be hidden by clicking the *Show/Hide* button on your toolbar.
- To insert text, move the cursor to the point where you want the insertion to go, click on the mouse once, and type the text.
- To move several lines of text, select the text and click the *Cut* button on your toolbar. Then position your cursor in the location that you want to move the cut text and click *Paste.* If you move to the wrong place, click *Undo.*
- The spell check feature does not catch words that are misspelled to look like other words, like "cold" instead of "gold." Always reread your document to catch all spelling mistakes.
- To learn about other word processing methods, read the user's manual or click on the *Help* button.
- You can integrate databases, graphics, and spreadsheets into documents by copying from another program and pasting it into your document, or by using desktop publishing (DTP). DTP software allows you to put text and graphics together to finish your document with a professional look. This software varies in how it is used and its capabilities.

Technology Skill Handbook

Use a Database

A collection of facts stored in a computer and sorted into different fields is called a database. A database can be reorganized in any way that suits your needs.

Learn the Skill A computer program that allows you to create your own database is a database management system (DBMS). It allows you to add, delete, or change information. Take time to get to know the features of your database software.

- Determine what facts you would like to include and research to collect your information.
- Determine how you want to organize the information.
- Follow the instructions for your particular DBMS to set up fields. Then enter each item of data in the appropriate field.
- Follow the instructions to sort the information in order of importance.
- Evaluate the information in your database, and add, delete, or change as necessary.

Use the Internet

The Internet is a global network of computers where information is stored and shared. To use the Internet, like the students in **Figure 17,** you need a modem to connect your computer to a phone line and an Internet Service Provider account.

Learn the Skill To access internet sites and information, use a "Web browser," which lets you view and explore pages on the World Wide Web. Each page is its own site, and each site has its own address, called a URL. Once you have found a Web browser, follow these steps for a search (this also is how you search a database).

Figure 17 The Internet allows you to search a global network for a variety of information.

- Be as specific as possible. If you know you want to research "gold," don't type in "elements." Keep narrowing your search until you find what you want.
- Web sites that end in *.com* are commercial Web sites; *.org*, *.edu*, and *.gov* are non-profit, educational, or government Web sites.
- Electronic encyclopedias, almanacs, indexes, and catalogs will help locate and select relevant information.
- Develop a "home page" with relative ease. When developing a Web site, NEVER post pictures or disclose personal information such as location, names, or phone numbers. Your school or community usually can host your Web site. A basic understanding of HTML (hypertext mark-up language), the language of Web sites, is necessary. Software that creates HTML code is called authoring software, and can be downloaded free from many Web sites. This software allows text and pictures to be arranged as the software is writing the HTML code.

Use a Spreadsheet

A spreadsheet, shown in **Figure 18,** can perform mathematical functions with any data arranged in columns and rows. By entering a simple equation into a cell, the program can perform operations in specific cells, rows, or columns.

Learn the Skill Each column (vertical) is assigned a letter, and each row (horizontal) is assigned a number. Each point where a row and column intersect is called a cell, and is labeled according to where it is located—Column A, Row 1 (A1).

- Decide how to organize the data, and enter it in the correct row or column.
- Spreadsheets can use standard formulas or formulas can be customized to calculate cells.
- To make a change, click on a cell to make it activate, and enter the edited data or formula.
- Spreadsheets also can display your results in graphs. Choose the style of graph that best represents the data.

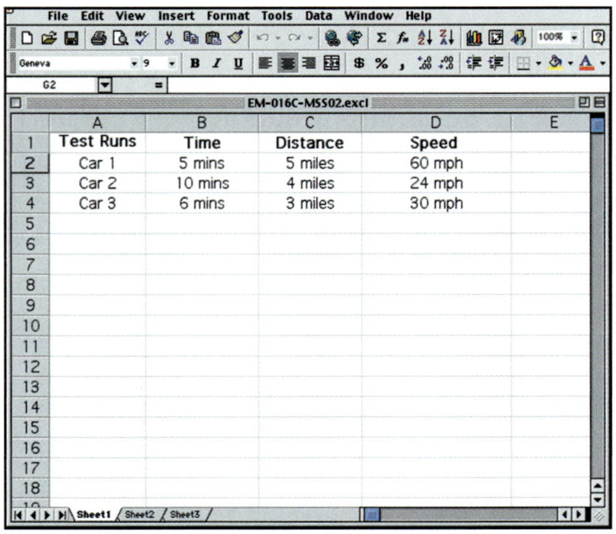

	A	B	C	D	E
1	Test Runs	Time	Distance	Speed	
2	Car 1	5 mins	5 miles	60 mph	
3	Car 2	10 mins	4 miles	24 mph	
4	Car 3	6 mins	3 miles	30 mph	

Figure 18 A spreadsheet allows you to perform mathematical operations on your data.

Use Graphics Software

Adding pictures, called graphics, to your documents is one way to make your documents more meaningful and exciting. This software adds, edits, and even constructs graphics. There is a variety of graphics software programs. The tools used for drawing can be a mouse, keyboard, or other specialized devices. Some graphics programs are simple. Others are complicated, called computer-aided design (CAD) software.

Learn the Skill It is important to have an understanding of the graphics software being used before starting. The better the software is understood, the better the results. The graphics can be placed in word-processing document.

- Clip art can be found on a variety of internet sites, and on CDs. These images can be copied and pasted into your document.
- When beginning, try editing existing drawings, then work up to creating drawings.
- The images are made of tiny rectangles of color called pixels. Each pixel can be altered.
- Digital photography is another way to add images. The photographs in the memory of a digital camera can be downloaded into a computer, then edited and added to the document.
- Graphics software also can allow animation. The software allows drawings to have the appearance of movement by connecting basic drawings automatically. This is called in-betweening, or tweening.
- Remember to save often.

Presentation Skills

Develop Multimedia Presentations

Most presentations are more dynamic if they include diagrams, photographs, videos, or sound recordings, like the one shown in **Figure 19.** A multimedia presentation involves using stereos, overhead projectors, televisions, computers, and more.

Learn the Skill Decide the main points of your presentation, and what types of media would best illustrate those points.

- Make sure you know how to use the equipment you are working with.
- Practice the presentation using the equipment several times.
- Enlist the help of a classmate to push play or turn lights out for you. Be sure to practice your presentation with him or her.
- If possible, set up all of the equipment ahead of time, and make sure everything is working properly.

Figure 19 These students are engaging the audience using a variety of tools.

Computer Presentations

There are many different interactive computer programs that you can use to enhance your presentation. Most computers have a compact disc (CD) drive that can play both CDs and digital video discs (DVDs). Also, there is hardware to connect a regular CD, DVD, or VCR. These tools will enhance your presentation.

Another method of using the computer to aid in your presentation is to develop a slide show using a computer program. This can allow movement of visuals at the presenter's pace, and can allow for visuals to build on one another.

Learn the Skill In order to create multimedia presentations on a computer, you need to have certain tools. These may include traditional graphic tools and drawing programs, animation programs, and authoring systems that tie everything together. Your computer will tell you which tools it supports. The most important step is to learn about the tools that you will be using.

- Often, color and strong images will convey a point better than words alone. Use the best methods available to convey your point.
- As with other presentations, practice many times.
- Practice your presentation with the tools you and any assistants will be using.
- Maintain eye contact with the audience. The purpose of using the computer is not to prompt the presenter, but to help the audience understand the points of the presentation.

Reduce Fractions

$\frac{66 \div 6}{90 \div 6} = \frac{11}{15}$

Add and Subtract Fractions

$\frac{4}{9} + \frac{2}{9} = \frac{6}{9}$

$\frac{6 \div 3}{9 \div 3} = \frac{2}{3}$

Math Review

Use Fractions

A fraction compares a part to a whole. In the fraction $\frac{2}{3}$, the 2 represents the part and is the numerator. The 3 represents the whole and is the denominator.

Reduce Fractions To reduce a fraction, you must find the largest factor that is common to both the numerator and the denominator, the greatest common factor (GCF). Divide both numbers by the GCF. The fraction has then been reduced, or it is in its simplest form.

Example Twelve of the 20 chemicals in the science lab are in powder form. What fraction of the chemicals used in the lab are in powder form?

Step 1 Write the fraction.

$$\frac{\text{part}}{\text{whole}} = \frac{12}{20}$$

Step 2 To find the GCF of the numerator and denominator, list all of the factors of each number.

Factors of 12: 1, 2, 3, 4, 6, 12 (the numbers that divide evenly into 12)

Factors of 20: 1, 2, 4, 5, 10, 20 (the numbers that divide evenly into 20)

Step 3 List the common factors.

1, 2, 4.

Step 4 Choose the greatest factor in the list.

The GCF of 12 and 20 is 4.

Step 5 Divide the numerator and denominator by the GCF.

$$\frac{12 \div 4}{20 \div 4} = \frac{3}{5}$$

In the lab, $\frac{3}{5}$ of the chemicals are in powder form.

Practice Problem At an amusement park, 66 of 90 rides have a height restriction. What fraction of the rides, in its simplest form, has a height restriction?

Add and Subtract Fractions To add or subtract fractions with the same denominator, add or subtract the numerators and write the sum or difference over the denominator. After finding the sum or difference, find the simplest form for your fraction.

Example 1 In the forest outside your house, $\frac{1}{8}$ of the animals are rabbits, $\frac{3}{8}$ are squirrels, and the remainder are birds and insects. How many are mammals?

Step 1 Add the numerators.

$$\frac{1}{8} + \frac{3}{8} = \frac{(1 + 3)}{8} = \frac{4}{8}$$

Step 2 Find the GCF.

$$\frac{4}{8} \text{ (GCF, 4)}$$

Step 3 Divide the numerator and denominator by the GCF.

$$\frac{4}{4} = 1, \ \frac{8}{4} = 2$$

$\frac{1}{2}$ of the animals are mammals.

Example 2 If $\frac{7}{16}$ of the Earth is covered by freshwater, and $\frac{1}{16}$ of that is in glaciers, how much freshwater is not frozen?

Step 1 Subtract the numerators.

$$\frac{7}{16} - \frac{1}{16} = \frac{(7 - 1)}{16} = \frac{6}{16}$$

Step 2 Find the GCF.

$$\frac{6}{16} \text{ (GCF, 2)}$$

Step 3 Divide the numerator and denominator by the GCF.

$$\frac{6}{2} = 3, \ \frac{16}{2} = 8$$

$\frac{3}{8}$ of the freshwater is not frozen.

Practice Problem A bicycle rider is going 15 km/h for $\frac{4}{9}$ of his ride, 10 km/h for $\frac{2}{9}$ of his ride, and 8 km/h for the remainder of the ride. How much of his ride is he going over 8 km/h?

Math Skill Handbook

154 ◆ J STUDENT RESOURCES

Unlike Denominators To add or subtract fractions with unlike denominators, first find the least common denominator (LCD). This is the smallest number that is a common multiple of both denominators. Rename each fraction with the LCD, and then add or subtract. Find the simplest form if necessary.

Example 1 A chemist makes a paste that is $\frac{1}{2}$ table salt (NaCl), $\frac{1}{3}$ sugar ($C_6H_{12}O_6$), and the rest water (H_2O). How much of the paste is a solid?

Step 1 Find the LCD of the fractions.

$\frac{1}{2} + \frac{1}{3}$ (LCD, 6)

Step 2 Rename each numerator and each denominator with the LCD.

$1 \times 3 = 3,\ 2 \times 3 = 6$

$1 \times 2 = 2,\ 3 \times 2 = 6$

Step 3 Add the numerators.

$\frac{3}{6} + \frac{2}{6} = \frac{(3+2)}{6} = \frac{5}{6}$

$\frac{5}{6}$ of the paste is a solid.

Example 2 The average precipitation in Grand Junction, CO, is $\frac{7}{10}$ inch in November, and $\frac{3}{5}$ inch in December. What is the total average precipitation?

Step 1 Find the LCD of the fractions.

$\frac{7}{10} + \frac{3}{5}$ (LCD, 10)

Step 2 Rename each numerator and each denominator with the LCD.

$7 \times 1 = 7,\ 10 \times 1 = 10$

$3 \times 2 = 6,\ 5 \times 2 = 10$

Step 3 Add the numerators.

$\frac{7}{10} + \frac{6}{10} = \frac{(7+6)}{10} = \frac{13}{10}$

$\frac{13}{10}$ inches total precipitation, or $1\frac{3}{10}$ inches.

Practice Problem On an electric bill, about $\frac{1}{8}$ of the energy is from solar energy and about $\frac{1}{10}$ is from wind power. How much of the total bill is from solar energy and wind power combined?

Example 3 In your body, $\frac{7}{10}$ of your muscle contractions are involuntary (cardiac and smooth muscle tissue). Smooth muscle makes $\frac{3}{15}$ of your muscle contractions. How many of your muscle contractions are made by cardiac muscle?

Step 1 Find the LCD of the fractions.

$\frac{7}{10} - \frac{3}{15}$ (LCD, 30)

Step 2 Rename each numerator and each denominator with the LCD.

$7 \times 3 = 21,\ 10 \times 3 = 30$

$3 \times 2 = 6,\ 15 \times 2 = 30$

Step 3 Subtract the numerators.

$\frac{21}{30} - \frac{6}{30} = \frac{(21-6)}{30} = \frac{15}{30}$

Step 4 Find the GCF.

$\frac{15}{30}$ (GCF, 15)

$\frac{1}{2}$

$\frac{1}{2}$ of all muscle contractions are cardiac muscle.

Example 4 Tony wants to make cookies that call for $\frac{3}{4}$ of a cup of flour, but he only has $\frac{1}{3}$ of a cup. How much more flour does he need?

Step 1 Find the LCD of the fractions.

$\frac{3}{4} - \frac{1}{3}$ (LCD, 12)

Step 2 Rename each numerator and each denominator with the LCD.

$3 \times 3 = 9,\ 4 \times 3 = 12$

$1 \times 4 = 4,\ 3 \times 4 = 12$

Step 3 Subtract the numerators.

$\frac{9}{12} - \frac{4}{12} = \frac{(9-4)}{12} = \frac{5}{12}$

$\frac{5}{12}$ of a cup of flour.

Practice Problem Using the information provided to you in Example 3 above, determine how many muscle contractions are voluntary (skeletal muscle).

Math Skill Handbook

Unlike Denominators

Problem 1

$1 \times 5 = 5, 8 \times 5 = 40$

$1 \times 4 = 4, 10 \times 4 = 40$

$\frac{5}{40} + \frac{4}{40} = \frac{9}{40}$

Problem 2

If $\frac{7}{10}$ are involuntary, the remainder are voluntary.

$\frac{10}{10} - \frac{7}{10} = \frac{3}{10}$

Math Skill Handbook

Multiply Fractions

Multiply Fractions To multiply with fractions, multiply the numerators and multiply the denominators. Find the simplest form if necessary.

Example Multiply $\frac{3}{5}$ by $\frac{1}{3}$.

Step 1 Multiply the numerators and denominators.

$$\frac{3}{5} \times \frac{1}{3} = \frac{(3 \times 1)}{(5 \times 3)} = \frac{3}{15}$$

Step 2 Find the GCF.

$$\frac{3}{15} \text{ (GCF, 3)}$$

Step 3 Divide the numerator and denominator by the GCF.

$$\frac{3}{3} = 1, \ \frac{15}{3} = 5$$

$$\frac{1}{5}$$

$\frac{3}{5}$ multiplied by $\frac{1}{3}$ is $\frac{1}{5}$.

Practice Problem Multiply $\frac{3}{14}$ by $\frac{5}{16}$.

Find a Reciprocal

Find a Reciprocal Two numbers whose product is 1 are called multiplicative inverses, or reciprocals.

Example Find the reciprocal of $\frac{3}{8}$.

Step 1 Inverse the fraction by putting the denominator on top and the numerator on the bottom.

$$\frac{8}{3}$$

The reciprocal of $\frac{3}{8}$ is $\frac{8}{3}$.

Practice Problem Find the reciprocal of $\frac{4}{9}$.

Divide Fractions

Divide Fractions To divide one fraction by another fraction, multiply the dividend by the reciprocal of the divisor. Find the simplest form if necessary.

Example 1 Divide $\frac{1}{9}$ by $\frac{1}{3}$.

Step 1 Find the reciprocal of the divisor.

The reciprocal of $\frac{1}{3}$ is $\frac{3}{1}$.

Step 2 Multiply the dividend by the reciprocal of the divisor.

$$\frac{\frac{1}{9}}{\frac{1}{3}} = \frac{1}{9} \times \frac{3}{1} = \frac{(1 \times 3)}{(9 \times 1)} = \frac{3}{9}$$

Step 3 Find the GCF.

$$\frac{3}{9} \text{ (GCF, 3)}$$

Step 4 Divide the numerator and denominator by the GCF.

$$\frac{3}{3} = 1, \ \frac{9}{3} = 3$$

$$\frac{1}{3}$$

$\frac{1}{9}$ divided by $\frac{1}{3}$ is $\frac{1}{3}$.

Example 2 Divide $\frac{3}{5}$ by $\frac{1}{4}$.

Step 1 Find the reciprocal of the divisor.

The reciprocal of $\frac{1}{4}$ is $\frac{4}{1}$.

Step 2 Multiply the dividend by the reciprocal of the divisor.

$$\frac{\frac{3}{5}}{\frac{1}{4}} = \frac{3}{5} \times \frac{4}{1} = \frac{(3 \times 4)}{(5 \times 1)} = \frac{12}{5}$$

$\frac{3}{5}$ divided by $\frac{1}{4}$ is $\frac{12}{5}$ or $2\frac{2}{5}$.

Practice Problem Divide $\frac{3}{11}$ by $\frac{7}{10}$.

Multiply Fractions

$$\frac{3}{14} \times \frac{5}{16} = \frac{(3 \times 5)}{(14 \times 16)} = \frac{15}{224}$$

Find a Reciprocal

$$\frac{9}{4}$$

Divide Fractions

The reciprocal of $\frac{7}{10}$ is $\frac{10}{7}$.

$$\frac{3}{11} \times \frac{10}{7} = \frac{(3 \times 10)}{(11 \times 7)} = \frac{30}{77}$$

Math Skill Handbook

156 ◆ J STUDENT RESOURCES

Use Ratios

When you compare two numbers by division, you are using a ratio. Ratios can be written 3 to 5, 3:5, or $\frac{3}{5}$. Ratios, like fractions, also can be written in simplest form.

Ratios can represent probabilities, also called odds. This is a ratio that compares the number of ways a certain outcome occurs to the number of outcomes. For example, if you flip a coin 100 times, what are the odds that it will come up heads? There are two possible outcomes, heads or tails, so the odds of coming up heads are 50:100. Another way to say this is that 50 out of 100 times the coin will come up heads. In its simplest form, the ratio is 1:2.

Example 1 A chemical solution contains 40 g of salt and 64 g of baking soda. What is the ratio of salt to baking soda as a fraction in simplest form?

Step 1 Write the ratio as a fraction.

$$\frac{\text{salt}}{\text{baking soda}} = \frac{40}{64}$$

Step 2 Express the fraction in simplest form.

The GCF of 40 and 64 is 8.

$$\frac{40}{64} = \frac{40 \div 8}{64 \div 8} = \frac{5}{8}$$

The ratio of salt to baking soda in the sample is 5:8.

Example 2 Sean rolls a 6-sided die 6 times. What are the odds that the side with a 3 will show?

Step 1 Write the ratio as a fraction.

$$\frac{\text{number of sides with a 3}}{\text{number of sides}} = \frac{1}{6}$$

Step 2 Multiply by the number of attempts.

$$\frac{1}{6} \times 6 \text{ attempts} = \frac{6}{6} \text{ attempts} = 1 \text{ attempt}$$

1 attempt out of 6 will show a 3.

Practice Problem Two metal rods measure 100 cm and 144 cm in length. What is the ratio of their lengths in simplest form?

Use Decimals

A fraction with a denominator that is a power of ten can be written as a decimal. For example, 0.27 means $\frac{27}{100}$. The decimal point separates the ones place from the tenths place.

Any fraction can be written as a decimal using division. For example, the fraction $\frac{5}{8}$ can be written as a decimal by dividing 5 by 8. Written as a decimal, it is 0.625.

Add or Subtract Decimals When adding and subtracting decimals, line up the decimal points before carrying out the operation.

Example 1 Find the sum of 47.68 and 7.80.

Step 1 Line up the decimal places when you write the numbers.

$$\begin{array}{r} 47.68 \\ +\ 7.80 \\ \hline \end{array}$$

Step 2 Add the decimals.

$$\begin{array}{r} 47.68 \\ +\ 7.80 \\ \hline 55.48 \end{array}$$

The sum of 47.68 and 7.80 is 55.48.

Example 2 Find the difference of 42.17 and 15.85.

Step 1 Line up the decimal places when you write the number.

$$\begin{array}{r} 42.17 \\ -15.85 \\ \hline \end{array}$$

Step 2 Subtract the decimals.

$$\begin{array}{r} 42.17 \\ -15.85 \\ \hline 26.32 \end{array}$$

The difference of 42.17 and 15.85 is 26.32.

Practice Problem Find the sum of 1.245 and 3.842.

Math Skill Handbook

Use Ratios

$$\frac{100 \text{ cm}}{144 \text{ cm}} = \frac{100 \div 4}{144 \div 4} = \frac{25}{36}$$

25:36

Add or Subtract Decimals

$$\begin{array}{r} {\scriptstyle 1} \\ 1.245 \\ +\ 3.842 \\ \hline 5.087 \end{array}$$

Math Skill Handbook

MATH SKILL HANDBOOK J ◆ 157

Multiply Decimals

Multiply 4.6 and 2.2 by 10.

$46 \times 22 = 1012$

Each factor had one decimal place.

10.12

Divide Decimals

Multiply both factors by 10.

Divide 756 by 36.

$$\begin{array}{r} 21 \\ 36\overline{)756} \\ \underline{72} \\ 36 \\ \underline{36} \\ 0 \end{array}$$

Use Proportions

$$\frac{3}{18} = \frac{5}{w}$$

$w \times 3 = 5 \times 18$

$$\frac{3w}{3} = \frac{90}{3}$$

$w = 30$

Multiply Decimals To multiply decimals, multiply the numbers like any other number, ignoring the decimal point. Count the decimal places in each factor. The product will have the same number of decimal places as the sum of the decimal places in the factors.

Example Multiply 2.4 by 5.9.

Step 1 Multiply the factors like two whole numbers.
$24 \times 59 = 1416$

Step 2 Find the sum of the number of decimal places in the factors. Each factor has one decimal place, for a sum of two decimal places.

Step 3 The product will have two decimal places.
14.16

The product of 2.4 and 5.9 is 14.16.

Practice Problem Multiply 4.6 by 2.2.

Divide Decimals When dividing decimals, change the divisor to a whole number. To do this, multiply both the divisor and the dividend by the same power of ten. Then place the decimal point in the quotient directly above the decimal point in the dividend. Then divide as you do with whole numbers.

Example Divide 8.84 by 3.4.

Step 1 Multiply both factors by 10.
$3.4 \times 10 = 34,\ 8.84 \times 10 = 88.4$

Step 2 Divide 88.4 by 34.

$$\begin{array}{r} 2.6 \\ 34\overline{)88.4} \\ \underline{-68} \\ 204 \\ \underline{-204} \\ 0 \end{array}$$

8.84 divided by 3.4 is 2.6.

Practice Problem Divide 75.6 by 3.6.

Use Proportions

An equation that shows that two ratios are equivalent is a proportion. The ratios $\frac{2}{4}$ and $\frac{5}{10}$ are equivalent, so they can be written as $\frac{2}{4} = \frac{5}{10}$. This equation is a proportion.

When two ratios form a proportion, the cross products are equal. To find the cross products in the proportion $\frac{2}{4} = \frac{5}{10}$, multiply the 2 and the 10, and the 4 and the 5. Therefore $2 \times 10 = 4 \times 5$, or $20 = 20$.

Because you know that both proportions are equal, you can use cross products to find a missing term in a proportion. This is known as solving the proportion.

Example The heights of a tree and a pole are proportional to the lengths of their shadows. The tree casts a shadow of 24 m when a 6-m pole casts a shadow of 4 m. What is the height of the tree?

Step 1 Write a proportion.

$$\frac{\text{height of tree}}{\text{height of pole}} = \frac{\text{length of tree's shadow}}{\text{length of pole's shadow}}$$

Step 2 Substitute the known values into the proportion. Let h represent the unknown value, the height of the tree.

$$\frac{h}{6} = \frac{24}{4}$$

Step 3 Find the cross products.
$h \times 4 = 6 \times 24$

Step 4 Simplify the equation.
$4h = 144$

Step 5 Divide each side by 4.

$$\frac{4h}{4} = \frac{144}{4}$$

$h = 36$

The height of the tree is 36 m.

Practice Problem The ratios of the weights of two objects on the Moon and on Earth are in proportion. A rock weighing 3 N on the Moon weighs 18 N on Earth. How much would a rock that weighs 5 N on the Moon weigh on Earth?

158 ◆ J STUDENT RESOURCES

Use Percentages

The word *percent* means "out of one hundred." It is a ratio that compares a number to 100. Suppose you read that 77 percent of the Earth's surface is covered by water. That is the same as reading that the fraction of the Earth's surface covered by water is $\frac{77}{100}$. To express a fraction as a percent, first find the equivalent decimal for the fraction. Then, multiply the decimal by 100 and add the percent symbol.

Example Express $\frac{13}{20}$ as a percent.

Step 1 Find the equivalent decimal for the fraction.

$$\begin{array}{r} 0.65 \\ 20\overline{)13.00} \\ \underline{12\,0} \\ 1\,00 \\ \underline{1\,00} \\ 0 \end{array}$$

Step 2 Rewrite the fraction $\frac{13}{20}$ as 0.65.

Step 3 Multiply 0.65 by 100 and add the % sign.

$0.65 \times 100 = 65 = 65\%$

So, $\frac{13}{20} = 65\%$.

This also can be solved as a proportion.

Example Express $\frac{13}{20}$ as a percent.

Step 1 Write a proportion.

$\frac{13}{20} = \frac{x}{100}$

Step 2 Find the cross products.

$1300 = 20x$

Step 3 Divide each side by 20.

$\frac{1300}{20} = \frac{20x}{20}$

$65\% = x$

Practice Problem In one year, 73 of 365 days were rainy in one city. What percent of the days in that city were rainy?

Solve One-Step Equations

A statement that two things are equal is an equation. For example, $A = B$ is an equation that states that A is equal to B.

An equation is solved when a variable is replaced with a value that makes both sides of the equation equal. To make both sides equal the inverse operation is used. Addition and subtraction are inverses, and multiplication and division are inverses.

Example 1 Solve the equation $x - 10 = 35$.

Step 1 Find the solution by adding 10 to each side of the equation.

$x - 10 = 35$

$x - 10 + 10 = 35 + 10$

$x = 45$

Step 2 Check the solution.

$x - 10 = 35$

$45 - 10 = 35$

$35 = 35$

Both sides of the equation are equal, so $x = 45$.

Example 2 In the formula $a = bc$, find the value of c if $a = 20$ and $b = 2$.

Step 1 Rearrange the formula so the unknown value is by itself on one side of the equation by dividing both sides by b.

$a = bc$

$\frac{a}{b} = \frac{bc}{b}$

$\frac{a}{b} = c$

Step 2 Replace the variables a and b with the values that are given.

$\frac{a}{b} = c$

$\frac{20}{2} = c$

$10 = c$

Step 3 Check the solution.

$a = bc$

$20 = 2 \times 10$

$20 = 20$

Both sides of the equation are equal, so $c = 10$ is the solution when $a = 20$ and $b = 2$.

Practice Problem In the formula $h = gd$, find the value of d if $g = 12.3$ and $h = 17.4$.

Use Percentages

$\frac{73}{365} = \frac{x}{100}$

$\frac{7300}{365} = \frac{365x}{365}$

$20\% = x$

Solve One-Step Equations

$h = gd$

$\frac{17.4}{12.3} = \frac{12.3d}{12.3}$

$1.41 = d$

Math Skill Handbook

MATH SKILL HANDBOOK J ◆ 159

Use Statistics

mean

$8 + 4 + 12 + 8 + 11 + 14 + 16 = 73$

$73 \div 7 = 10.4$

median

4, 8, 8, 11, 12, 14, 16

mode

4, 8, 8, 11, 12, 14, 16

range

4, 8, 8, 11, 12, 14, 16

$16 - 4 = 12$

Math Skill Handbook

Use Statistics

The branch of mathematics that deals with collecting, analyzing, and presenting data is statistics. In statistics, there are three common ways to summarize data with a single number—the mean, the median, and the mode.

The **mean** of a set of data is the arithmetic average. It is found by adding the numbers in the data set and dividing by the number of items in the set.

The **median** is the middle number in a set of data when the data are arranged in numerical order. If there were an even number of data points, the median would be the mean of the two middle numbers.

The **mode** of a set of data is the number or item that appears most often.

Another number that often is used to describe a set of data is the range. The **range** is the difference between the largest number and the smallest number in a set of data.

A **frequency table** shows how many times each piece of data occurs, usually in a survey. **Table 2** below shows the results of a student survey on favorite color.

Table 2 Student Color Choice

Color	Tally	Frequency
red	IIII	4
blue	卌	5
black	II	2
green	III	3
purple	卌 II	7
yellow	卌 I	6

Based on the frequency table data, which color is the favorite?

Example The speeds (in m/s) for a race car during five different time trials are 39, 37, 44, 36, and 44.

To find the mean:

Step 1 Find the sum of the numbers.

$39 + 37 + 44 + 36 + 44 = 200$

Step 2 Divide the sum by the number of items, which is 5.

$200 \div 5 = 40$

The mean is 40 m/s.

To find the median:

Step 1 Arrange the measures from least to greatest.

36, 37, 39, 44, 44

Step 2 Determine the middle measure.

36, 37, 39, 44, 44

The median is 39 m/s.

To find the mode:

Step 1 Group the numbers that are the same together.

44, 44, 36, 37, 39

Step 2 Determine the number that occurs most in the set.

44, 44, 36, 37, 39

The mode is 44 m/s.

To find the range:

Step 1 Arrange the measures from largest to smallest.

44, 44, 39, 37, 36

Step 2 Determine the largest and smallest measures in the set.

44, 44, 39, 37, 36

Step 3 Find the difference between the largest and smallest measures.

$44 - 36 = 8$

The range is 8 m/s.

Practice Problem Find the mean, median, mode, and range for the data set 8, 4, 12, 8, 11, 14, 16.

Math Skill Handbook

Use Geometry

The branch of mathematics that deals with the measurement, properties, and relationships of points, lines, angles, surfaces, and solids is called geometry.

Perimeter The **perimeter** (P) is the distance around a geometric figure. To find the perimeter of a rectangle, add the length and width and multiply that sum by two, or $2(l + w)$. To find perimeters of irregular figures, add the length of the sides.

Example 1 Find the perimeter of a rectangle that is 3 m long and 5 m wide.

Step 1 You know that the perimeter is 2 times the sum of the width and length.

$P = 2(3\text{ m} + 5\text{ m})$

Step 2 Find the sum of the width and length.

$P = 2(8\text{ m})$

Step 3 Multiply by 2.

$P = 16\text{ m}$

The perimeter is 16 m.

Example 2 Find the perimeter of a shape with sides measuring 2 cm, 5 cm, 6 cm, 3 cm.

Step 1 You know that the perimeter is the sum of all the sides.

$P = 2 + 5 + 6 + 3$

Step 2 Find the sum of the sides.

$P = 2 + 5 + 6 + 3$

$P = 16$

The perimeter is 16 cm.

Practice Problem Find the perimeter of a rectangle with a length of 18 m and a width of 7 m.

Practice Problem Find the perimeter of a triangle measuring 1.6 cm by 2.4 cm by 2.4 cm.

Area of a Rectangle The **area** (A) is the number of square units needed to cover a surface. To find the area of a rectangle, multiply the length times the width, or $l \times w$. When finding area, the units also are multiplied. Area is given in square units.

Example Find the area of a rectangle with a length of 1 cm and a width of 10 cm.

Step 1 You know that the area is the length multiplied by the width.

$A = (1\text{ cm} \times 10\text{ cm})$

Step 2 Multiply the length by the width. Also multiply the units.

$A = 10\text{ cm}^2$

The area is 10 cm^2.

Practice Problem Find the area of a square whose sides measure 4 m.

Area of a Triangle To find the area of a triangle, use the formula:

$$A = \frac{1}{2}(\text{base} \times \text{height})$$

The base of a triangle can be any of its sides. The height is the perpendicular distance from a base to the opposite endpoint, or vertex.

Example Find the area of a triangle with a base of 18 m and a height of 7 m.

Step 1 You know that the area is $\frac{1}{2}$ the base times the height.

$A = \frac{1}{2}(18\text{ m} \times 7\text{ m})$

Step 2 Multiply $\frac{1}{2}$ by the product of 18×7. Multiply the units.

$A = \frac{1}{2}(126\text{ m}^2)$

$A = 63\text{ m}^2$

The area is 63 m^2.

Practice Problem Find the area of a triangle with a base of 27 cm and a height of 17 cm.

Perimeter

Problem 1

$P = 2(18\text{ m} + 7\text{ m})$

$P = 2(25\text{ m})$

$P = 50\text{ m}$

Problem 2

$P = 1.6\text{ cm} + 2.4\text{ cm} + 2.4\text{ cm}$

$P = 6.4\text{ cm}$

Area of a Rectangle

$A = (4\text{ m} \times 4\text{ m})$

$A = 16\text{ m}^2$

Area of a Triangle

$A = \frac{1}{2}(27\text{ cm} \times 17\text{ cm})$

$A = \frac{1}{2}(459\text{ cm}^2)$

$A = 229.5\text{ cm}^2$

Math Skill Handbook

MATH SKILL HANDBOOK J ◆ 161

Circumference of a Circle

$C = 2\pi r$

$C = 2\pi(19)$

$C = 38\pi$

$C = 119.3$

Area of a Circle

$A = \pi r^2$

$A = \pi(16\text{ m})^2$

$A = \pi\ 256\text{ m}^2$

$A = 803.8\text{ m}^2$

Circumference of a Circle The **diameter** (d) of a circle is the distance across the circle through its center, and the **radius** (r) is the distance from the center to any point on the circle. The radius is half of the diameter. The distance around the circle is called the **circumference** (C). The formula for finding the circumference is:

$$C = 2\pi r \text{ or } C = \pi d$$

The circumference divided by the diameter is always equal to 3.1415926... This nonterminating and nonrepeating number is represented by the Greek letter π (pi). An approximation often used for π is 3.14.

Example 1 Find the circumference of a circle with a radius of 3 m.

Step 1 You know the formula for the circumference is 2 times the radius times π.
$C = 2\pi(3)$

Step 2 Multiply 2 times the radius.
$C = 6\pi$

Step 3 Multiply by π.
$C = 19\text{ m}$

The circumference is 19 m.

Example 2 Find the circumference of a circle with a diameter of 24.0 cm.

Step 1 You know the formula for the circumference is the diameter times π.
$C = \pi(24.0)$

Step 2 Multiply the diameter by π.
$C = 75.4\text{ cm}$

The circumference is 75.4 cm.

Practice Problem Find the circumference of a circle with a radius of 19 cm.

Area of a Circle The formula for the area of a circle is:
$A = \pi r^2$

Example 1 Find the area of a circle with a radius of 4.0 cm.

Step 1 $A = \pi(4.0)^2$

Step 2 Find the square of the radius.
$A = 16\pi$

Step 3 Multiply the square of the radius by π.
$A = 50\text{ cm}^2$

The area of the circle is 50 cm^2.

Example 2 Find the area of a circle with a radius of 225 m.

Step 1 $A = \pi(225)^2$

Step 2 Find the square of the radius.
$A = 50625\pi$

Step 3 Multiply the square of the radius by π.
$A = 158962.5$

The area of the circle is 158,962 m^2.

Example 3 Find the area of a circle whose diameter is 20.0 mm.

Step 1 You know the formula for the area of a circle is the square of the radius times π, and that the radius is half of the diameter.
$A = \pi\left(\frac{20.0}{2}\right)^2$

Step 2 Find the radius.
$A = \pi(10.0)^2$

Step 3 Find the square of the radius.
$A = 100\pi$

Step 4 Multiply the square of the radius by π.
$A = 314\text{ mm}^2$

The area is 314 mm^2.

Practice Problem Find the area of a circle with a radius of 16 m.

Math Skill Handbook

162 ◆ J STUDENT RESOURCES

Volume The measure of space occupied by a solid is the **volume** (V). To find the volume of a rectangular solid multiply the length times width times height, or $V = l \times w \times h$. It is measured in cubic units, such as cubic centimeters (cm^3).

Example Find the volume of a rectangular solid with a length of 2.0 m, a width of 4.0 m, and a height of 3.0 m.

Step 1 You know the formula for volume is the length times the width times the height.
$V = 2.0 \text{ m} \times 4.0 \text{ m} \times 3.0 \text{ m}$

Step 2 Multiply the length times the width times the height.
$V = 24 \text{ m}^3$

The volume is 24 m^3.

Practice Problem Find the volume of a rectangular solid that 8 m long, 4 m wide, and 4 m high.

To find the volume of other solids, multiply the area of the base times the height.

Example 1 Find the volume of a solid that has a triangular base with a length of 8.0 m and a height of 7.0 m. The height of the entire solid is 15.0 m.

Step 1 You know that the base is a triangle, and the area of a triangle is $\frac{1}{2}$ the base times the height, and the volume is the area of the base times the height.
$V = \left[\frac{1}{2}(b \times h)\right] \times 15$

Step 2 Find the area of the base.
$V = \left[\frac{1}{2}(8 \times 7)\right] \times 15$
$V = \left(\frac{1}{2} \times 56\right) \times 15$

Step 3 Multiply the area of the base by the height of the solid.
$V = 28 \times 15$
$V = 420 \text{ m}^3$

The volume is 420 m^3.

Example 2 Find the volume of a cylinder that has a base with a radius of 12.0 cm, and a height of 21.0 cm.

Step 1 You know that the base is a circle, and the area of a circle is the square of the radius times π, and the volume is the area of the base times the height.
$V = (\pi r^2) \times 21$
$V = (\pi 12^2) \times 21$

Step 2 Find the area of the base.
$V = 144\pi \times 21$
$V = 452 \times 21$

Step 3 Multiply the area of the base by the height of the solid.
$V = 9490 \text{ cm}^3$

The volume is 9490 cm^3.

Example 3 Find the volume of a cylinder that has a diameter of 15 mm and a height of 4.8 mm.

Step 1 You know that the base is a circle with an area equal to the square of the radius times π. The radius is one-half the diameter. The volume is the area of the base times the height.
$V = (\pi r^2) \times 4.8$
$V = \left[\pi\left(\frac{1}{2} \times 15\right)^2\right] \times 4.8$
$V = (\pi 7.5^2) \times 4.8$

Step 2 Find the area of the base.
$V = 56.25\pi \times 4.8$
$V = 176.63 \times 4.8$

Step 3 Multiply the area of the base by the height of the solid.
$V = 847.8$

The volume is 847.8 mm^3.

Practice Problem Find the volume of a cylinder with a diameter of 7 cm in the base and a height of 16 cm.

Math Skill Handbook

Volume

Problem 1
$V = 8 \text{ m} \times 4 \text{ m} \times 4 \text{ m}$
$V = 128 \text{ m}^3$

Problem 2
$V = \pi r^2 \times \text{height}$
$V = \left[\pi\left(\frac{1}{2} \times 7\right)^2\right] \times 16$
$V = [\pi(3.5)^2] \times 16$
$V = [\pi(12.25)] \times 16$
$V = 38.46 \times 16$
$V = 615.36$

Math Skill Handbook

Measure in SI

smaller; 1000; one thousandth

Dimensional Analysis

$x\text{ mg} = 1\cancel{\text{kg}} \times \frac{1000\cancel{\text{g}}}{1\cancel{\text{kg}}} \times \frac{1000\text{ mg}}{1\cancel{\text{g}}} =$

1,000,000 mg

1,000,000 mg = 1 kg

Science Applications

Measure in SI

The metric system of measurement was developed in 1795. A modern form of the metric system, called the International System (SI), was adopted in 1960 and provides the standard measurements that all scientists around the world can understand.

The SI system is convenient because unit sizes vary by powers of 10. Prefixes are used to name units. Look at **Table 3** for some common SI prefixes and their meanings.

Table 3 Some SI Prefixes

Prefix	Symbol	Meaning	
kilo-	k	1,000	thousand
hecto-	h	100	hundred
deka-	da	10	ten
deci-	d	0.1	tenth
centi-	c	0.01	hundredth
milli-	m	0.001	thousandth

Example How many grams equal one kilogram?

Step 1 Find the prefix *kilo* in **Table 3.**

Step 2 Using **Table 3,** determine the meaning of *kilo.* According to the table, it means 1,000. When the prefix *kilo* is added to a unit, it means that there are 1,000 of the units in a "*kilo*unit."

Step 3 Apply the prefix to the units in the question. The units in the question are grams. There are 1,000 grams in a kilogram.

Practice Problem Is a milligram larger or smaller than a gram? How many of the smaller units equal one larger unit? What fraction of the larger unit does one smaller unit represent?

Dimensional Analysis

Convert SI Units In science, quantities such as length, mass, and time sometimes are measured using different units. A process called dimensional analysis can be used to change one unit of measure to another. This process involves multiplying your starting quantity and units by one or more conversion factors. A conversion factor is a ratio equal to one and can be made from any two equal quantities with different units. If 1,000 mL equal 1 L then two ratios can be made.

$$\frac{1{,}000\text{ mL}}{1\text{ L}} = \frac{1\text{ L}}{1{,}000\text{ mL}} = 1$$

One can covert between units in the SI system by using the equivalents in **Table 3** to make conversion factors.

Example 1 How many cm are in 4 m?

Step 1 Write conversion factors for the units given. From **Table 3,** you know that 100 cm = 1 m. The conversion factors are

$$\frac{100\text{ cm}}{1\text{ m}} \text{ and } \frac{1\text{ m}}{100\text{ cm}}$$

Step 2 Decide which conversion factor to use. Select the factor that has the units you are converting from (m) in the denominator and the units you are converting to (cm) in the numerator.

$$\frac{100\text{ cm}}{1\text{ m}}$$

Step 3 Multiply the starting quantity and units by the conversion factor. Cancel the starting units with the units in the denominator. There are 400 cm in 4 m.

$$4\cancel{\text{ m}} \times \frac{100\text{ cm}}{1\cancel{\text{ m}}} = 400\text{ cm}$$

Practice Problem How many milligrams are in one kilogram? (Hint: You will need to use two conversion factors from **Table 3.**)

Math Skill Handbook

164 ◆ J STUDENT RESOURCES

Table 4 Unit System Equivalents

Type of Measurement	Equivalent
Length	1 in = 2.54 cm 1 yd = 0.91 m 1 mi = 1.61 km
Mass and Weight*	1 oz = 28.35 g 1 lb = 0.45 kg 1 ton (short) = 0.91 tonnes (metric tons) 1 lb = 4.45 N
Volume	1 in^3 = 16.39 cm^3 1 qt = 0.95 L 1 gal = 3.78 L
Area	1 in^2 = 6.45 cm^2 1 yd^2 = 0.83 m^2 1 mi^2 = 2.59 km^2 1 acre = 0.40 hectares
Temperature	$°C = \frac{(°F - 32)}{1.8}$ K = °C + 273

*Weight is measured in standard Earth gravity.

Convert Between Unit Systems **Table 4** gives a list of equivalents that can be used to convert between English and SI units.

Example If a meterstick has a length of 100 cm, how long is the meterstick in inches?

Step 1 Write the conversion factors for the units given. From **Table 4,** 1 in = 2.54 cm.

$$\frac{1 \text{ in}}{2.54 \text{ cm}} \text{ and } \frac{2.54 \text{ cm}}{1 \text{ in}}$$

Step 2 Determine which conversion factor to use. You are converting from cm to in. Use the conversion factor with cm on the bottom.

$$\frac{1 \text{ in}}{2.54 \text{ cm}}$$

Step 3 Multiply the starting quantity and units by the conversion factor. Cancel the starting units with the units in the denominator. Round your answer based on the number of significant figures in the conversion factor.

$$100 \cancel{\text{cm}} \times \frac{1 \text{ in}}{2.54 \cancel{\text{cm}}} = 39.37 \text{ in}$$

The meterstick is 39.4 in long.

Practice Problem A book has a mass of 5 lbs. What is the mass of the book in kg?

Practice Problem Use the equivalent for in and cm (1 in = 2.54 cm) to show how 1 in^3 = 16.39 cm^3.

Convert Between Unit Systems

$$\frac{(1 \text{ in})^3}{(2.54 \text{ cm})^3} = \frac{1 \text{ in} \times 1 \text{ in} \times 1 \text{ in}}{2.54 \text{ cm} \times 2.54 \text{ cm} \times 2.54 \text{ cm}} = \frac{1 \text{ in}^3}{16.39 \text{ cm}^3}$$

Precision and Significant Digits

Problem 1
7; 4

Problem 2
$5.28 \times 5.2 = 27.456$

5.28 has 3 significant digits.

5.2 has 2 significant digits.

When multiplying and dividing, the answer is rounded to the smallest number of significant digits of the numbers being multiplied or divided—in this case, 2.

27.456 is rounded to 27.

Scientific Notation

Problem 1
4; 4

Problem 2
0.000007

Problem 3
5.39×10^2

Math Skill Handbook

Precision and Significant Digits

When you make a measurement, the value you record depends on the precision of the measuring instrument. This precision is represented by the number of significant digits recorded in the measurement. When counting the number of significant digits, all digits are counted except zeros at the end of a number with no decimal point such as 2,050, and zeros at the beginning of a decimal such as 0.03020. When adding or subtracting numbers with different precision, round the answer to the smallest number of decimal places of any number in the sum or difference. When multiplying or dividing, the answer is rounded to the smallest number of significant digits of any number being multiplied or divided.

Example The lengths 5.28 and 5.2 are measured in meters. Find the sum of these lengths and record your answer using the correct number of significant digits.

Step 1 Find the sum.

5.28 m	2 digits after the decimal
+ 5.2 m	1 digit after the decimal
10.48 m	

Step 2 Round to one digit after the decimal because the least number of digits after the decimal of the numbers being added is 1.

The sum is 10.5 m.

Practice Problem How many significant digits are in the measurement 7,071,301 m? How many significant digits are in the measurement 0.003010 g?

Practice Problem Multiply 5.28 and 5.2 using the rule for multiplying and dividing. Record the answer using the correct number of significant digits.

Scientific Notation

Many times numbers used in science are very small or very large. Because these numbers are difficult to work with scientists use scientific notation. To write numbers in scientific notation, move the decimal point until only one non-zero digit remains on the left. Then count the number of places you moved the decimal point and use that number as a power of ten. For example, the average distance from the Sun to Mars is 227,800,000,000 m. In scientific notation, this distance is 2.278×10^{11} m. Because you moved the decimal point to the left, the number is a positive power of ten.

The mass of an electron is about 0.000 000 000 000 000 000 000 000 000 000 911 kg. Expressed in scientific notation, this mass is 9.11×10^{-31} kg. Because the decimal point was moved to the right, the number is a negative power of ten.

Example Earth is 149,600,000 km from the Sun. Express this in scientific notation.

Step 1 Move the decimal point until one non-zero digit remains on the left.
1.496 000 00

Step 2 Count the number of decimal places you have moved. In this case, eight.

Step 3 Show that number as a power of ten, 10^8.

The Earth is 1.496×10^8 km from the Sun.

Practice Problem How many significant digits are in 149,600,000 km? How many significant digits are in 1.496×10^8 km?

Practice Problem Parts used in a high performance car must be measured to 7×10^{-6} m. Express this number as a decimal.

Practice Problem A CD is spinning at 539 revolutions per minute. Express this number in scientific notation.

Math Skill Handbook

Make and Use Graphs

Data in tables can be displayed in a graph—a visual representation of data. Common graph types include line graphs, bar graphs, and circle graphs.

Line Graph A line graph shows a relationship between two variables that change continuously. The independent variable is changed and is plotted on the x-axis. The dependent variable is observed, and is plotted on the y-axis.

Example Draw a line graph of the data below from a cyclist in a long-distance race.

Table 5 Bicycle Race Data

Time (h)	Distance (km)
0	0
1	8
2	16
3	24
4	32
5	40

Step 1 Determine the x-axis and y-axis variables. Time varies independently of distance and is plotted on the x-axis. Distance is dependent on time and is plotted on the y-axis.

Step 2 Determine the scale of each axis. The x-axis data ranges from 0 to 5. The y-axis data ranges from 0 to 40.

Step 3 Using graph paper, draw and label the axes. Include units in the labels.

Step 4 Draw a point at the intersection of the time value on the x-axis and corresponding distance value on the y-axis. Connect the points and label the graph with a title, as shown in **Figure 20**.

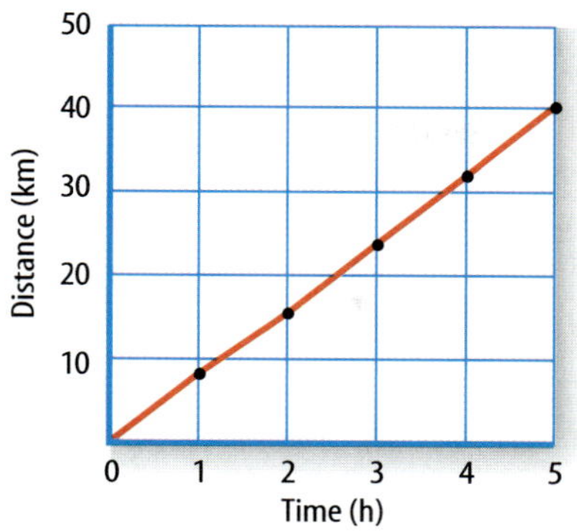

Figure 20 This line graph shows the relationship between distance and time during a bicycle ride.

Practice Problem A puppy's shoulder height is measured during the first year of her life. The following measurements were collected: (3 mo, 52 cm), (6 mo, 72 cm), (9 mo, 83 cm), (12 mo, 86). Graph this data.

Find a Slope The slope of a straight line is the ratio of the vertical change, rise, to the horizontal change, run.

$$\text{Slope} = \frac{\text{vertical change (rise)}}{\text{horizontal change (run)}} = \frac{\text{change in } y}{\text{change in } x}$$

Example Find the slope of the graph in **Figure 20.**

Step 1 You know that the slope is the change in y divided by the change in x.

$$\text{Slope} = \frac{\text{change in } y}{\text{change in } x}$$

Step 2 Determine the data points you will be using. For a straight line, choose the two sets of points that are the farthest apart.

$$\text{Slope} = \frac{(40-0)\text{ km}}{(5-0)\text{ hr}}$$

Step 3 Find the change in y and x.

$$\text{Slope} = \frac{40\text{ km}}{5\text{h}}$$

Step 4 Divide the change in y by the change in x.

$$\text{Slope} = \frac{8\text{ km}}{\text{h}}$$

The slope of the graph is 8 km/h.

Line Graph

x	y
3	52
6	72
9	83
12	86

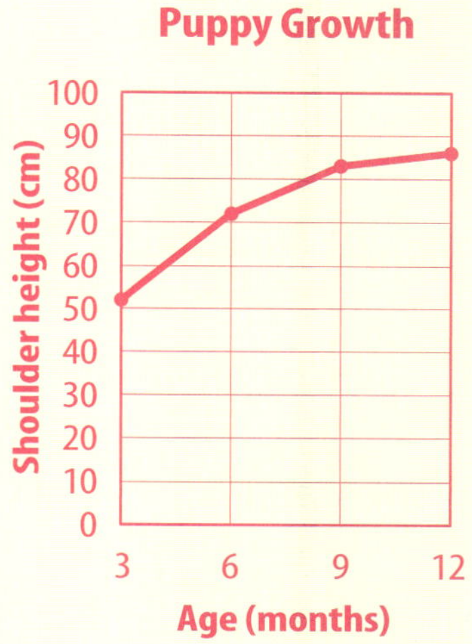

Bar Graph

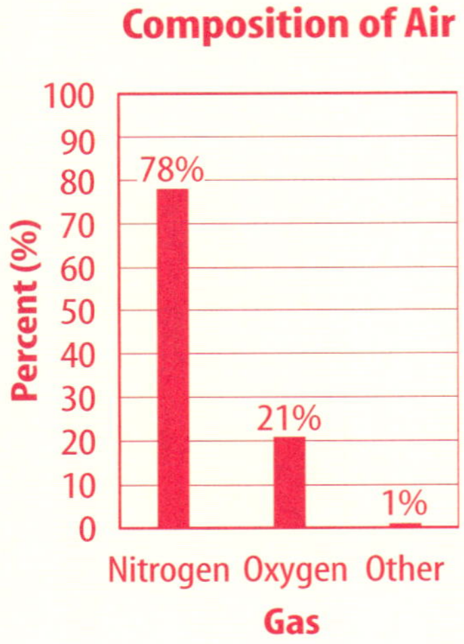

Circle Graph

The total amount of aluminum collected is:

$4.0 \text{ kg} + 1.0 \text{ kg} + 2.0 \text{ kg} = 7.0 \text{ kg}$

$\frac{4.0 \text{ kg}}{7.0 \text{ kg}} = \frac{x}{360°}$; $x = 206°$

$\frac{1.0 \text{ kg}}{7.0 \text{ kg}} = \frac{x}{360°}$; $x = 51°$

$\frac{2.0 \text{ kg}}{7.0 \text{ kg}} = \frac{x}{360°}$; $x = 103°$

Bar Graph To compare data that does not change continuously you might choose a bar graph. A bar graph uses bars to show the relationships between variables. The *x*-axis variable is divided into parts. The parts can be numbers such as years, or a category such as a type of animal. The *y*-axis is a number and increases continuously along the axis.

Example A recycling center collects 4.0 kg of aluminum on Monday, 1.0 kg on Wednesday, and 2.0 kg on Friday. Create a bar graph of this data.

Step 1 Select the *x*-axis and *y*-axis variables. The measured numbers (the masses of aluminum) should be placed on the *y*-axis. The variable divided into parts (collection days) is placed on the *x*-axis.

Step 2 Create a graph grid like you would for a line graph. Include labels and units.

Step 3 For each measured number, draw a vertical bar above the *x*-axis value up to the *y*-axis value. For the first data point, draw a vertical bar above Monday up to 4.0 kg.

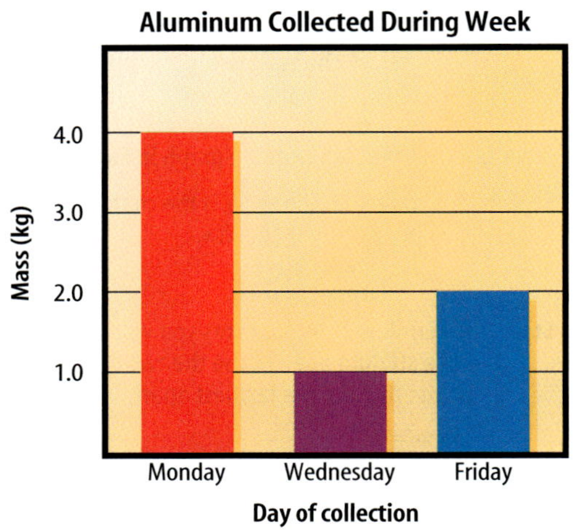

Practice Problem Draw a bar graph of the gases in air: 78% nitrogen, 21% oxygen, 1% other gases.

Circle Graph To display data as parts of a whole, you might use a circle graph. A circle graph is a circle divided into sections that represent the relative size of each piece of data. The entire circle represents 100%, half represents 50%, and so on.

Example Air is made up of 78% nitrogen, 21% oxygen, and 1% other gases. Display the composition of air in a circle graph.

Step 1 Multiply each percent by 360° and divide by 100 to find the angle of each section in the circle.

$$78\% \times \frac{360°}{100} = 280.8°$$

$$21\% \times \frac{360°}{100} = 75.6°$$

$$1\% \times \frac{360°}{100} = 3.6°$$

Step 2 Use a compass to draw a circle and to mark the center of the circle. Draw a straight line from the center to the edge of the circle.

Step 3 Use a protractor and the angles you calculated to divide the circle into parts. Place the center of the protractor over the center of the circle and line the base of the protractor over the straight line.

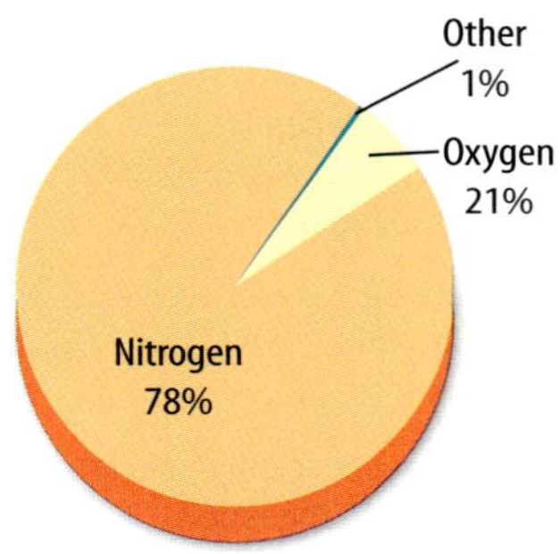

Practice Problem Draw a circle graph to represent the amount of aluminum collected during the week shown in the bar graph to the left.

Reference Handbooks

Weather Map Symbols

Sample Station Model

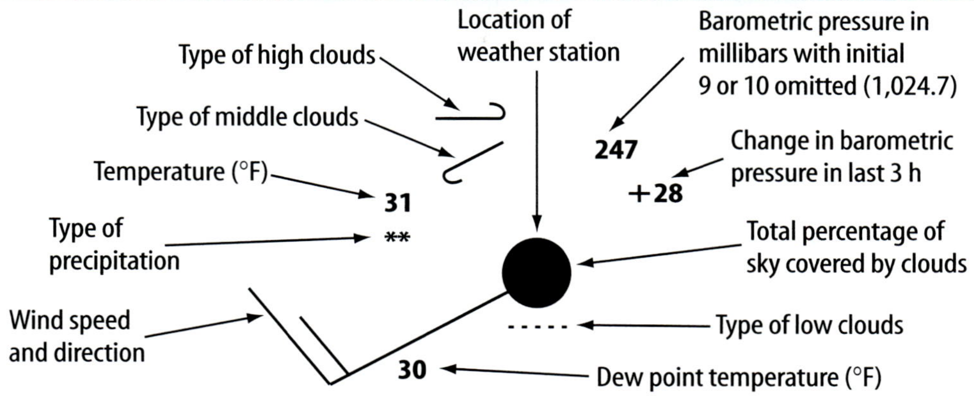

Sample Plotted Report at Each Station

Precipitation	Wind Speed and Direction	Sky Coverage	Some Types of High Clouds
Fog	0 calm	No cover	Scattered cirrus
Snow	1–2 knots	1/10 or less	Dense cirrus in patches
Rain	3–7 knots	2/10 to 3/10	Veil of cirrus covering entire sky
Thunderstorm	8–12 knots	4/10	Cirrus not covering entire sky
Drizzle	13–17 knots	–	
Showers	18–22 knots	6/10	
	23–27 knots	7/10	
	48–52 knots	Overcast with openings	
	1 knot = 1.852 km/h	Completely overcast	

Some Types of Middle Clouds	Some Types of Low Clouds	Fronts and Pressure Systems	
Thin altostratus layer	Cumulus of fair weather	(H) or High (L) or Low	Center of high- or low-pressure system
Thick altostratus layer	Stratocumulus		Cold front
Thin altostratus in patches	Fractocumulus of bad weather		Warm front
Thin altostratus in bands	Stratus of fair weather		Occluded front
			Stationary front

Minerals

Minerals					
Mineral (formula)	**Color**	**Streak**	**Hardness**	**Breakage Pattern**	**Uses and Other Properties**
Graphite (C)	black to gray	black to gray	1–1.5	basal cleavage (scales)	pencil lead, lubricants for locks, rods to control some small nuclear reactions, battery poles
Galena (PbS)	gray	gray to black	2.5	cubic cleavage perfect	source of lead, used for pipes, shields for X rays, fishing equipment sinkers
Hematite (Fe_2O_3)	black or reddish-brown	reddish-brown	5.5–6.5	irregular fracture	source of iron; converted to pig iron, made into steel
Magnetite (Fe_3O_4)	black	black	6	conchoidal fracture	source of iron, attracts a magnet
Pyrite (FeS_2)	light, brassy, yellow	greenish-black	6–6.5	uneven fracture	fool's gold
Talc ($Mg_3 Si_4O_{10} (OH)_2$)	white, greenish	white	1	cleavage in one direction	used for talcum powder, sculptures, paper, and tabletops
Gypsum ($CaSO_4 \cdot 2H_2O$)	colorless, gray, white, brown	white	2	basal cleavage	used in plaster of paris and dry wall for building construction
Sphalerite (ZnS)	brown, reddish-brown, greenish	light to dark brown	3.5–4	cleavage in six directions	main ore of zinc; used in paints, dyes, and medicine
Muscovite ($KAl_3Si_3 O_{10}(OH)_2$)	white, light gray, yellow, rose, green	colorless	2–2.5	basal cleavage	occurs in large, flexible plates; used as an insulator in electrical equipment, lubricant
Biotite ($K(Mg,Fe)_3 (AlSi_3O_{10}) (OH)_2$)	black to dark brown	colorless	2.5–3	basal cleavage	occurs in large, flexible plates
Halite (NaCl)	colorless, red, white, blue	colorless	2.5	cubic cleavage	salt; soluble in water; a preservative

Minerals

Minerals					
Mineral (formula)	**Color**	**Streak**	**Hardness**	**Breakage Pattern**	**Uses and Other Properties**
Calcite ($CaCO_3$)	colorless, white, pale blue	colorless, white	3	cleavage in three directions	fizzes when HCl is added; used in cements and other building materials
Dolomite ($CaMg(CO_3)_2$)	colorless, white, pink, green, gray, black	white	3.5–4	cleavage in three directions	concrete and cement; used as an ornamental building stone
Fluorite (CaF_2)	colorless, white, blue, green, red, yellow, purple	colorless	4	cleavage in four directions	used in the manufacture of optical equipment; glows under ultraviolet light
Hornblende ($(CaNa)_{2-3}(Mg,Al,Fe)_5$-$(Al,Si)_2Si_6O_{22}(OH)_2$)	green to black	gray to white	5–6	cleavage in two directions	will transmit light on thin edges; 6-sided cross section
Feldspar ($(KAlSi_3O_8)$ $(NaAlSi_3O_8)$, $(CaAl_2Si_2O_8)$)	colorless, white to gray, green	colorless	6	two cleavage planes meet at 90° angle	used in the manufacture of ceramics
Augite ($((Ca,Na)(Mg,Fe,Al)(Al,Si)_2O_6)$)	black	colorless	6	cleavage in two directions	square or 8-sided cross section
Olivine ($((Mg,Fe)_2SiO_4)$)	olive, green	none	6.5–7	conchoidal fracture	gemstones, refractory sand
Quartz (SiO_2)	colorless, various colors	none	7	conchoidal fracture	used in glass manufacture, electronic equipment, radios, computers, watches, gemstones

Rocks

Rocks		
Rock Type	**Rock Name**	**Characteristics**
Igneous (intrusive)	Granite	Large mineral grains of quartz, feldspar, hornblende, and mica. Usually light in color.
	Diorite	Large mineral grains of feldspar, hornblende, and mica. Less quartz than granite. Intermediate in color.
	Gabbro	Large mineral grains of feldspar, augite, and olivine. No quartz. Dark in color.
Igneous (extrusive)	Rhyolite	Small mineral grains of quartz, feldspar, hornblende, and mica, or no visible grains. Light in color.
	Andesite	Small mineral grains of feldspar, hornblende, and mica or no visible grains. Intermediate in color.
	Basalt	Small mineral grains of feldspar, augite, and possibly olivine or no visible grains. No quartz. Dark in color.
	Obsidian	Glassy texture. No visible grains. Volcanic glass. Fracture looks like broken glass.
	Pumice	Frothy texture. Floats in water. Usually light in color.
Sedimentary (detrital)	Conglomerate	Coarse grained. Gravel or pebble-size grains.
	Sandstone	Sand-sized grains 1/16 to 2 mm.
	Siltstone	Grains are smaller than sand but larger than clay.
	Shale	Smallest grains. Often dark in color. Usually platy.
Sedimentary (chemical or organic)	Limestone	Major mineral is calcite. Usually forms in oceans and lakes. Often contains fossils.
	Coal	Forms in swampy areas. Compacted layers of organic material, mainly plant remains.
Sedimentary (chemical)	Rock Salt	Commonly forms by the evaporation of seawater.
Metamorphic (foliated)	Gneiss	Banding due to alternate layers of different minerals, of different colors. Parent rock often is granite.
	Schist	Parallel arrangement of sheetlike minerals, mainly micas. Forms from different parent rocks.
	Phyllite	Shiny or silky appearance. May look wrinkled. Common parent rocks are shale and slate.
	Slate	Harder, denser, and shinier than shale. Common parent rock is shale.
Metamorphic (nonfoliated)	Marble	Calcite or dolomite. Common parent rock is limestone.
	Soapstone	Mainly of talc. Soft with greasy feel.
	Quartzite	Hard with interlocking quartz crystals. Common parent rock is sandstone.

Topographic Map Symbols

Topographic Map Symbols	
Primary highway, hard surface	Index contour
Secondary highway, hard surface	Supplementary contour
Light-duty road, hard or improved surface	Intermediate contour
Unimproved road	Depression contours
Railroad: single track	
Railroad: multiple track	Boundaries: national
Railroads in juxtaposition	State
	County, parish, municipal
Buildings	Civil township, precinct, town, barrio
cem Schools, church, and cemetery	Incorporated city, village, town, hamlet
Buildings (barn, warehouse, etc.)	Reservation, national or state
Wells other than water (labeled as to type)	Small park, cemetery, airport, etc.
Tanks: oil, water, etc. (labeled only if water)	Land grant
Located or landmark object; windmill	Township or range line, U.S. land survey
Open pit, mine, or quarry; prospect	
	Township or range line, approximate location
Marsh (swamp)	
Wooded marsh	Perennial streams
Woods or brushwood	Elevated aqueduct
Vineyard	Water well and spring
Land subject to controlled inundation	Small rapids
Submerged marsh	Large rapids
Mangrove	Intermittent lake
Orchard	Intermittent stream
Scrub	Aqueduct tunnel
Urban area	Glacier
	Small falls
x7369 Spot elevation	Large falls
670 Water elevation	Dry lake bed

Reference Handbooks

PERIODIC TABLE OF THE ELEMENTS

Columns of elements are called groups. Elements in the same group have similar chemical properties.

Element — Hydrogen
Atomic number — 1
Symbol — H
Atomic mass — 1.008
State of matter

Gas
Liquid
Solid
Synthetic

The first three symbols tell you the state of matter of the element at room temperature. The fourth symbol identifies elements that are not present in significant amounts on Earth. Useful amounts are made synthetically.

Period	1	2	3	4	5	6	7	8	9
1	Hydrogen 1 H 1.008								
2	Lithium 3 Li 6.941	Beryllium 4 Be 9.012							
3	Sodium 11 Na 22.990	Magnesium 12 Mg 24.305							
4	Potassium 19 K 39.098	Calcium 20 Ca 40.078	Scandium 21 Sc 44.956	Titanium 22 Ti 47.867	Vanadium 23 V 50.942	Chromium 24 Cr 51.996	Manganese 25 Mn 54.938	Iron 26 Fe 55.845	Cobalt 27 Co 58.933
5	Rubidium 37 Rb 85.468	Strontium 38 Sr 87.62	Yttrium 39 Y 88.906	Zirconium 40 Zr 91.224	Niobium 41 Nb 92.906	Molybdenum 42 Mo 95.94	Technetium 43 Tc (98)	Ruthenium 44 Ru 101.07	Rhodium 45 Rh 102.906
6	Cesium 55 Cs 132.905	Barium 56 Ba 137.327	Lanthanum 57 La 138.906	Hafnium 72 Hf 178.49	Tantalum 73 Ta 180.948	Tungsten 74 W 183.84	Rhenium 75 Re 186.207	Osmium 76 Os 190.23	Iridium 77 Ir 192.217
7	Francium 87 Fr (223)	Radium 88 Ra (226)	Actinium 89 Ac (227)	Rutherfordium 104 Rf (261)	Dubnium 105 Db (262)	Seaborgium 106 Sg (266)	Bohrium 107 Bh (264)	Hassium 108 Hs (277)	Meitnerium 109 Mt (268)

The number in parentheses is the mass number of the longest-lived isotope for that element.

Rows of elements are called periods. Atomic number increases across a period.

The arrow shows where these elements would fit into the periodic table. They are moved to the bottom of the table to save space.

Lanthanide series	Cerium 58 Ce 140.116	Praseodymium 59 Pr 140.908	Neodymium 60 Nd 144.24	Promethium 61 Pm (145)	Samarium 62 Sm 150.36
Actinide series	Thorium 90 Th 232.038	Protactinium 91 Pa 231.036	Uranium 92 U 238.029	Neptunium 93 Np (237)	Plutonium 94 Pu (244)

Metal

Metalloid

Nonmetal

The color of an element's block tells you if the element is a metal, nonmetal, or metalloid.

Science Online

Visit bookj.msscience.com for updates to the periodic table.

10	11	12	13	14	15	16	17	18
								Helium 2 **He** 4.003
			Boron 5 **B** 10.811	Carbon 6 **C** 12.011	Nitrogen 7 **N** 14.007	Oxygen 8 **O** 15.999	Fluorine 9 **F** 18.998	Neon 10 **Ne** 20.180
			Aluminum 13 **Al** 26.982	Silicon 14 **Si** 28.086	Phosphorus 15 **P** 30.974	Sulfur 16 **S** 32.065	Chlorine 17 **Cl** 35.453	Argon 18 **Ar** 39.948
Nickel 28 **Ni** 58.693	Copper 29 **Cu** 63.546	Zinc 30 **Zn** 65.409	Gallium 31 **Ga** 69.723	Germanium 32 **Ge** 72.64	Arsenic 33 **As** 74.922	Selenium 34 **Se** 78.96	Bromine 35 **Br** 79.904	Krypton 36 **Kr** 83.798
Palladium 46 **Pd** 106.42	Silver 47 **Ag** 107.868	Cadmium 48 **Cd** 112.411	Indium 49 **In** 114.818	Tin 50 **Sn** 118.710	Antimony 51 **Sb** 121.760	Tellurium 52 **Te** 127.60	Iodine 53 **I** 126.904	Xenon 54 **Xe** 131.293
Platinum 78 **Pt** 195.078	Gold 79 **Au** 196.967	Mercury 80 **Hg** 200.59	Thallium 81 **Tl** 204.383	Lead 82 **Pb** 207.2	Bismuth 83 **Bi** 208.980	Polonium 84 **Po** (209)	Astatine 85 **At** (210)	Radon 86 **Rn** (222)
Darmstadtium 110 **Ds** (281)	Unununium * 111 **Uuu** (272)	Ununbium * 112 **Uub** (285)		Ununquadium * 114 **Uuq** (289)		** 116		** 118

* The names and symbols for elements 111–114 are temporary. Final names will be selected when the elements' discoveries are verified.

** Elements 116 and 118 were thought to have been created. The claim was retracted because the experimental results could not be repeated.

Europium 63 **Eu** 151.964	Gadolinium 64 **Gd** 157.25	Terbium 65 **Tb** 158.925	Dysprosium 66 **Dy** 162.500	Holmium 67 **Ho** 164.930	Erbium 68 **Er** 167.259	Thulium 69 **Tm** 168.934	Ytterbium 70 **Yb** 173.04	Lutetium 71 **Lu** 174.967
Americium 95 **Am** (243)	Curium 96 **Cm** (247)	Berkelium 97 **Bk** (247)	Californium 98 **Cf** (251)	Einsteinium 99 **Es** (252)	Fermium 100 **Fm** (257)	Mendelevium 101 **Md** (258)	Nobelium 102 **No** (259)	Lawrencium 103 **Lr** (262)

Glossary/Glosario

Cómo usar el glosario en español:
1. Busca el término en inglés que desees encontrar.
2. El término en español, junto con la definición, se encuentran en la columna de la derecha.

Pronunciation Key

Use the following key to help you sound out words in the glossary.

a	**b**ack (BAK)	**ew**	**food** (FEWD)
ay	**day** (DAY)	**yoo**	**pure** (PYOOR)
ah	**fa**ther (FAH thur)	**yew**	**few** (FYEW)
ow	**flow**er (FLOW ur)	**uh**	**comma** (CAH muh)
ar	**car** (CAR)	**u** (+ con)	**rub** (RUB)
e	**le**ss (LES)	**sh**	**sh**elf (SHELF)
ee	**lea**f (LEEF)	**ch**	**na**ture (NAY chur)
ih	**tri**p (TRIHP)	**g**	**g**ift (GIHFT)
i (i + con + e)	**i**dea (i DEE uh)	**j**	**g**em (JEM)
oh	**go** (GOH)	**ing**	**sing** (SING)
aw	**so**ft (SAWFT)	**zh**	**vi**sion (VIH zhun)
or	**orb**it (OR buht)	**k**	**c**ake (KAYK)
oy	**coi**n (COYN)	**s**	**s**eed, **c**ent (SEED, SENT)
oo	**foo**t (FOOT)	**z**	**z**one, rai**s**e (ZOHN, RAYZ)

English — A — Español

absolute magnitude: measure of the amount of light a star actually gives off. (p. 106)

apparent magnitude: measure of the amount of light from a star that is received on Earth. (p. 106)

asteroid: a piece of rock or metal made up of material similar to that which formed the planets; mostly found in the asteroid belt between the orbits of Mars and Jupiter. (p. 92)

axis: imaginary vertical line that cuts through the center of Earth and around which Earth spins. (p. 41)

magnitud absoluta: medida de la cantidad real de luz que genera una estrella. (p. 106)

magnitud aparente: medida de la cantidad de luz recibida en la Tierra desde una estrella. (p. 106)

asteroide: pedazo de roca o metal formado de material similar al que forma los planetas; se encuentran principalmente en el cinturón de asteroides entre las órbitas de Marte y Júpiter. (p. 92)

eje: línea vertical imaginaria que atraviesa el centro de la Tierra y alrededor de la cual gira ésta. (p. 41)

B

Big Bang theory: states that about 13.7 billion years ago, the universe began with a huge, fiery explosion. (p. 125)

black hole: final stage in the evolution of a very massive star, where the core's mass collapses to a point that it's gravity is so strong that not even light can escape. (p. 118)

teoría de la Gran Explosión: establece que hace aproximadamente 13.7 billones de años el universo se originó con una enorme explosión. (p. 125)

agujero negro: etapa final en la evolución de una estrella masiva, en donde la masa del núcleo se colapsa hasta el punto de que su gravedad es tan fuerte que ni siquiera la luz puede escapar. (p. 118)

C

chromosphere: layer of the Sun's atmosphere above the photosphere. (p. 109)

cromosfera: capa de la atmósfera del sol que se encuentra sobre la fotosfera. (p. 109)

comet: space object made of dust and rock particles mixed with frozen water, methane, and ammonia that forms a bright coma as it approaches the Sun. (p. 90)

cometa: objeto espacial formado por partículas de polvo y roca mezcladas con agua congelada, metano y amoníaco que forman una cola brillante cuando se aproxima al sol. (p. 90)

constellation: group of stars that forms a pattern in the sky that looks like a familiar object (Libra), animal (Pegasus), or character (Orion). (p. 104)

constelación: grupo de estrellas que forma un patrón en el cielo y que semeja un objeto (Libra), un animal (Pegaso) o un personaje familiar (Orión). (p. 104)

corona: outermost, largest layer of the Sun's atmosphere; extends millions of kilometers into space and has temperatures up to 2 million K. (p. 109)

corona: capa más externa y más grande de la atmósfera solar; se extiende millones de kilómetros dentro del espacio y tiene una temperatura hasta de 2 millones de grados Kelvin. (p. 109)

E

Earth: third planet from the Sun; has an atmosphere that protects life and surface temperatures that allow water to exist as a solid, liquid, and gas. (p. 78)

Tierra: tercer planeta más cercano al sol; tiene una atmósfera que protege la vida y temperaturas en su superficie que permiten la presencia de agua en estado sólido, líquido y gaseoso. (p. 78)

electromagnetic spectrum: arrangement of electromagnetic waves according to their wavelengths. (p. 9)

espectro electromagnético: ordenamiento de las ondas electromagnéticas de acuerdo con su longitud de onda. (p. 9)

ellipse (ee LIHPS): elongated, closed curve that describes Earth's yearlong orbit around the Sun. (p. 43)

elipse: curva cerrada y elongada que describe la órbita anual de la Tierra alrededor del sol. (p. 43)

equinox (EE kwuh nahks): twice-yearly time—each spring and fall—when the Sun is directly over the equator and the number of daylight and nighttime hours are equal worldwide. (p. 45)

equinoccio: dos veces al año—en primavera y otoño—cuando el sol está posicionado directamente sobre el ecuador y el número de horas del día y de la noche son iguales en todo el mundo. (p. 45)

F

full moon: phase that occurs when all of the Moon's surface facing Earth reflects light. (p. 47)

luna llena: fase que ocurre cuando toda la superficie de la luna frente a la Tierra refleja la luz del sol. (p. 47)

G

galaxy: large group of stars, dust, and gas held together by gravity; can be elliptical, spiral, or irregular. (p. 120)

galaxia: grupo grande de estrellas, polvo y gas en donde todo está unido por gravedad; puede ser elíptica, espiral o irregular. (p. 120)

giant: late stage in the life of comparatively low-mass main sequence star in which hydrogen in the core is deleted, the core contracts and temperatures inside the star increase, causing its outer layers to expand and cool. (p. 117)

gigante: etapa tardía en la vida de una estrella de secuencia principal, de relativamente poca masa, en la que el hidrógeno en el núcleo está agotado, el núcleo se contrae y la temperatura en el interior de la estrella aumenta, causando que las capas externas se expandan y enfríen. (p. 117)

Great Red Spot: giant, high-pressure storm in Jupiter's atmosphere. (p. 82)

La Gran Mancha Roja: tormenta gigante de alta presión en la atmósfera de Júpiter. (p. 82)

I

impact basin: a hollow left on the surface of the Moon caused by an object striking its surface. (p. 57)

cráter de impacto: un hueco dejado en la superficie de la luna causada por un objeto que chocó contra su superficie. (p. 57)

J

Jupiter: largest and fifth planet from the Sun; contains more mass than all the other planets combined, has continuous storms of high-pressure gas, and an atmosphere mostly of hydrogen and helium. (p. 82)

Júpiter: el quinto planeta más cercano al sol, y también el más grande; contiene más masa que todos los otros planetas en conjunto, tiene tormentas continuas de gas a alta presión y una atmósfera compuesta principalmente por hidrógeno y helio. (p. 82)

light-year: unit representing the distance light travels in one year—about 9.5 trillion km—used to record distances between stars and galaxies. (p. 107)

año luz: unidad que representa la distancia que la luz viaja en un año—cerca de 9.5 trillones de kilómetros—usada para registrar las distancias entre las estrellas y las galaxias. (p. 107)

lunar eclipse: occurs when Earth's shadow falls on the Moon. (p. 50)

eclipse lunar: ocurre cuando la sombra de la Tierra cubre la luna. (p. 50)

M

maria (MAHR ee uh): dark-colored, relatively flat regions of the Moon formed when ancient lava reached the surface and filled craters on the Moon's surface. (p. 51)

mares: regiones de la Luna relativamente planas y de color oscuro que se formaron cuando la lava alcanzó la superficie y llenó los cráteres en la seperficie lunar. (p. 51)

Mars: fourth planet from the Sun; has polar ice caps, a thin atmosphere, and a reddish appearance caused by iron oxide in weathered rocks and soil. (p. 78)

Marte: cuarto planeta más cercano al sol; tiene casquetes de hielo polar, una atmósfera delgada y una apariencia rojiza causada por el óxido de hierro presente en las rocas y suelo de su superficie. (p. 78)

Mercury: smallest planet, closest to the Sun; does not have a true atmosphere; has a surface with many craters and high cliffs. (p. 76)

Mercurio: el planeta más pequeño y más cercano al sol; no tiene una atmósfera verdadera; tiene una superficie con muchos cráteres y grandes acantilados. (p. 76)

meteor: a meteoroid that burns up in Earth's atmosphere. (p. 91)

meteoro: un meteoroide que se incinera en la atmósfera de la Tierra. (p. 91)

meteorite: a meteoroid that strikes the surface of a moon or planet. (p. 92)

meteorito: un meteoroide que choca contra la superficie de la luna o de algún planeta. (p. 92)

moon phase: change in appearance of the Moon as viewed from the Earth, due to the relative positions of the Moon, Earth, and Sun. (p. 47)

fase lunar: cambio en la apariencia de la luna según es vista desde la Tierra; se debe a las posiciones relativas de la luna, la Tierra y el sol. (p. 47)

N

nebula: large cloud of gas and dust that contracts under gravitational force and breaks apart into smaller pieces, each of which might collapse to form a star. (p. 116)

nebulosa: nube grande de polvo y gas que se contrae bajo la fuerza gravitacional y se descompone en pedazos más pequeños, cada uno de los cuales se puede colapsar para formar una estrella. (p. 116)

Neptune: usually the eighth planet from the Sun; is large and gaseous, has rings that vary in thickness, and is bluish-green in color. (p. 86)

Neptuno: el octavo planeta desde el sol; es grande y gaseoso, tiene anillos que varían en espesor y tiene un color verde-azulado. (p. 86)

neutron star: collapsed core of a supernova that can shrink to about 20 km in diameter and contains only neutrons in the dense core. (p. 118)

estrella de neutrones: núcleo colapsado de una supernova que puede contraerse hasta tener un diámetro de 20 kilómetros y contiene sólo neutrones en su denso núcleo. (p. 118)

new moon: moon phase that occurs when the Moon is between Earth and the Sun, at which point the Moon cannot be seen because its lighted half is facing the Sun and its dark side faces Earth. (p. 47)

luna nueva: fase lunar que ocurre cuando la luna se encuentra entre la Tierra y el sol, punto en el cual la luna no puede verse porque su mitad iluminada está frente al sol y su lado oscuro frente a la Tierra. (p. 47)

O

observatory: building that can house an optical telescope; often has a dome-shaped roof that can be opened for viewing. (p. 10)

orbit: curved path followed by a satellite as it revolves around an object. (p. 17)

observatorio: edificación que puede albergar un telescopio óptico; a menudo tiene un techo en forma de domo que puede abrirse para la observación. (p. 10)

órbita: trayectoria curva seguida por un satélite conforme gira alrededor de un objeto. (p. 17)

P

photosphere: lowest layer of the Sun's atmosphere; gives off light and has temperatures of about 6,000K. (p. 109)

Pluto: considered to be the ninth planet from the Sun; has a solid icy-rock surface and a single moon, Charon. (p. 87)

Project Apollo: final stage in the U.S. program to reach the Moon, in which Neil Armstrong was the first human to step onto the Moon's surface. (p. 22)

Project Gemini: second stage in the U.S. program to reach the Moon, in which an astronaut team connected with another spacecraft in orbit. (p. 21)

Project Mercury: first step in the U.S. program to reach the Moon; orbited a piloted spacecraft around Earth and brought it back safely. (p. 21)

fotosfera: capa más interna de la atmósfera del sol; emite luz y tiene temperaturas de cerca de 6,000 grados Kelvin. (p. 109)

Plutón: considerado como el noveno planeta desde el sol; tiene una superficie sólida de roca congelada y una luna, Caronte. (p. 87)

Proyecto Apolo: etapa final en el proyecto norteamericano para llegar a la luna en el que Neil Armstrong fue el primer ser humano en caminar sobre la superficie lunar. (p. 22)

Proyecto Géminis: segunda etapa del proyecto norteamericano para llegar a la luna en el que un grupo de astronautas se conectó con otra nave espacial en órbita. (p. 21)

Proyecto Mercurio: primera etapa del proyecto norteamericano para llegar a la luna en el que una nave espacial tripulada recorrió la órbita de la Tierra y regresó de manera segura. (p. 21)

R

radio telescope: collects and records radio waves traveling through space; can be used day or night under most weather conditions. (p. 13)

radiotelescopio: recolecta y registra ondas de radio que viajan a través del espacio; puede usarse de día o de noche en la mayoría de condiciones climáticas. (p. 13)

reflecting telescope: optical telescope that uses a concave mirror to focus light and form an image at the focal point. (p. 10)

refracting telescope: optical telescope that uses a double convex lens to bend light and form an image at the focal point. (p. 10)

revolution: Earth's yearlong elliptical orbit around the Sun. (p. 43)

rocket: special engine that can work in space and burns liquid or solid fuel. (p. 15)

rotation: spinning of Earth on its imaginary axis, which takes about 24 hours to complete and causes day and night to occur. (p. 41)

telescopio reflectante: telescopio óptico que utiliza un espejo cóncavo para enfocar la luz y formar una imagen en el punto focal. (p. 10)

telescopio de refracción: telescopio óptico que utiliza un lente doble convexo para formar una imagen en el punto focal. (p. 10)

revolución: órbita elíptica de un año de duración que la Tierra recorre alrededor del sol. (p. 43)

cohete: máquina especial que puede funcionar en el espacio y quema combustible sólido o líquido. (p. 15)

rotación: rotación de la Tierra sobre su eje imaginario, lo cual toma cerca de 24 horas para completarse y causa la alternancia entre el día y la noche. (p. 41)

S

satellite: any natural or artificial object that revolves around another object. (p. 17)

Saturn: second-largest and sixth planet from the Sun; has a complex ring system, at least 31 moons, and a thick atmosphere made mostly of hydrogen and helium. (p. 84)

solar eclipse: occurs when the Moon passes directly between the Sun and Earth and casts a shadow over part of Earth. (p. 49)

solar system: system of nine planets, including Earth, and other objects that revolve around the Sun. (p. 71)

solstice: twice-yearly point at which the Sun reaches its greatest distance north or south of the equator. (p. 44)

space probe: instrument that travels far into the solar system and gathers data to send back to Earth. (p. 18)

space shuttle: reusable spacecraft that can carry cargo, astronauts, and satellites to and from space. (p. 23)

space station: large facility with living quarters, work and exercise areas, and equipment and support systems for humans to live and work in space and conduct research. (p. 24)

sphere (SFIHR): a round, three-dimensional object whose surface is the same distance from its center at all points; Earth is a sphere that bulges somewhat at the equator and is slightly flattened at the poles. (p. 40)

satélite: cualquier objeto natural o artificial que gire alrededor de otro objeto. (p. 17)

Saturno: además de ser el sexto planeta más cercano al sol, también es el segundo en tamaño; tiene un sistema de anillos complejo, por lo menos 31 lunas y una atmósfera gruesa compuesta principalmente de hidrógeno y helio. (p. 84)

eclipse solar: ocurre cuando la luna pasa directamente entre el sol y la Tierra y se genera una sombra sobre una parte de la Tierra. (p. 49)

sistema solar: sistema de nueve planetas, incluyendo a la Tierra y otros objetos que giran alrededor del sol. (p. 71)

solsticio: punto en el cual dos veces al año el sol alcanza su mayor distancia al norte o al sur del ecuador. (p. 44)

sonda espacial: instrumento que viaja grandes distancias en el sistema solar, recopila datos y los envía a la Tierra. (p. 18)

trasbordador espacial: nave espacial reutilizable que puede llevar carga, astronautas y satélites hacia y desde el espacio. (p. 23)

estación espacial: instalación grande con áreas para hospedarse, trabajar y hacer ejercicio; tiene equipos y sistemas de apoyo para que los seres humanos vivan, trabajen y lleven a cabo investigaciones en el espacio. (p. 24)

esfera: un objeto tridimensional y redondo donde cualquier punto de su superficie está a la misma distancia del centro; la Tierra es una esfera algo abultada en el ecuador y ligeramente achatada en los polos. (p. 40)

sunspots: areas on the Sun's surface that are cooler and less bright than surrounding areas, are caused by the Sun's magnetic field, and occur in cycles. (p. 110)

supergiant: late stage in the life cycle of a massive star in which the core heats up, heavy elements form by fusion, and the star expands; can eventually explode to form a supernova. (p. 119)

manchas solares: áreas en la superficie solar que son más frías y menos brillantes que las áreas circundantes, son causadas por el campo magnético solar y ocurren en ciclos. (p. 110)

supergigante: etapa tardía en el ciclo de vida de una estrella masiva en la que el núcleo se calienta, se forman elementos pesados por fusión y la estrella se expande; eventualmente puede explotar para formar una supernova. (p. 119)

U

Uranus (YOOR uh nus): seventh planet from the Sun; is large and gaseous, has a distinct bluish-green color, and rotates on an axis nearly parallel to the plane of its orbit. (p. 85)

Urano: séptimo planeta desde el sol; es grande y gaseoso, tiene un color verde-azulado distintivo y gira sobre un eje casi paralelo al plano de su órbita. (p. 85)

V

Venus: second planet from the Sun; similar to Earth in mass and size; has a thick atmosphere and a surface with craters, faultlike cracks, and volcanoes. (p. 77)

Venus: segundo planeta más cercano al sol; similar a la Tierra en masa y tamaño; tiene una atmósfera gruesa y una superficie con cráteres, grietas similares a fallas y volcanes. (p. 77)

W

waning: describes phases that occur after a full moon, as the visible lighted side of the Moon grows smaller. (p. 48)

waxing: describes phases following a new moon, as more of the Moon's lighted side becomes visible. (p. 48)

white dwarf: late stage in the life cycle of a comparatively low-mass main sequence star; formed when its core depletes its helium and its outer layers escape into space, leaving behind a hot, dense core. (p. 118)

menguante: describe las fases posteriores a la luna llena, de manera que el lado iluminado de la luna es cada vez menos visible. (p. 48)

creciente: describe las fases posteriores a la luna nueva, de manera que el lado iluminado de la luna es cada vez más visible. (p. 48)

enana blanca: etapa tardía en el ciclo de vida de una estrella de secuencia principal, de relativamente poca masa, formada cuando el núcleo agota su helio y sus capas externas escapan al espacio, dejando atrás un núcleo denso y caliente. (p. 118)

Index

Italic numbers = *illustration/photo* **Bold numbers = vocabulary term**
lab = a page on which the entry is used in a lab
act = a page on which the entry is used in an activity

Index

S

T

U

V

W

Magnification Key: Magnifications listed are the magnifications at which images were originally photographed.
LM–Light Microscope
SEM–Scanning Electron Microscope
TEM–Transmission Electron Microscope

Acknowledgments: Glencoe would like to acknowledge the artists and agencies who participated in illustrating this program: Absolute Science Illustration; Andrew Evansen; Argosy; Articulate Graphics; Craig Attebery represented by Frank & Jeff Lavaty; CHK America; John Edwards and Associates; Gagliano Graphics; Pedro Julio Gonzalez represented by Melissa Turk & The Artist Network; Robert Hynes represented by Mendola Ltd.; Morgan Cain & Associates; JTH Illustration; Laurie O'Keefe; Matthew Pippin represented by Beranbaum Artist's Representative; Precision Graphics; Publisher's Art; Rolin Graphics, Inc.; Wendy Smith represented by Melissa Turk & The Artist Network; Kevin Torline represented by Berendsen and Associates, Inc.; WILDlife ART; Phil Wilson represented by Cliff Knecht Artist Representative; Zoo Botanica.

Photo Credits

Cover (tl)NASA/Science Photo Library/Photo Researchers, (tr)Billy & Sally Fletcher/Tom Stack & Assoc., (b)Photodisc; **i ii** (tl)NASA/Science Photo Library/Photo Researchers, (tr)Billy & Sally Fletcher/Tom Stack & Assoc., (b)Photodisc; **iv** (bkgd)John Evans, cover: (tl)NASA/Science Photo Library/Photo Researchers, (tr)Billy & Sally Fletcher/Tom Stack & Assoc., (b)Photodisc; **v** (t)PhotoDisc, (b)John Evans **vi** (l)John Evans, (r)Geoff Butler; **vii** (l)John Evans, (r)PhotoDisc; **viii** PhotoDisc; **ix** Aaron Haupt Photography; **x** (t)Julian Baum/Science Photo Library/Photo Researchers, (b)NASA/Science Photo Library/Photo Researchers; **xi** AFP/CORBIS; **xii** NASA; **1** Pekka Parviainen/Science Photo Library/Photo Researchers; **2** (t)David J. Phillip/AP/Wide World Photos; **2–3** Malin Space Science Systems/NASA; **3** (t)Malin Space Science Systems/NASA/JPL, (br)courtesy DC Golden; **5** Geco UK/Science Photo Library/Photo Researchers; **6–7** TSADO/NASA/Tom Stack & Assoc.; **8** (l)Weinberg-Clark/The Image Bank/Getty Images, (r)Stephen Marks/The Image Bank/Getty Images; **9** (l)PhotoEdit, Inc., (r)Wernher Krutein/Liaison Agency/Getty Images; **10** Chuck Place/Stock Boston; **11** NASA; **12** (t)Roger Ressmeyer/CORBIS, (b)Simon Fraser/Science Photo Library/Photo Researchers; **13** Raphael Gaillarde/Liaison Agency/Getty Images; **14** (t)Icon Images, (b)Diane Graham-Henry & Kathleen Culbert-Aguilar; **15** NASA; **16** NASA/Science Photo Library/Photo Researchers; **17** NASA; **18** (Mariner 2, Pioneer 10)NASA/Science Source/Photo Researchers, (Viking 1)M. Salaber/Liaison Agency/Getty Images, (Magellan)Julian Baum/Science Photo Library/Photo Researchers; **19** (Venera 8)Dorling Kindersley Images, (Surface of Venus)TASS from Sovfoto, (Mercury, Venus) NASA/JPL, (Voyager 2, Neptune)NASA/JPL/Caltech, (others) NASA; **20** AFP/CORBIS; **21** NASA; **22** NASA/Science Source/Photo Researchers; **23** NASA/Liaison Agency/Getty Images; **24** (t)NASA, (b)NASA/Liaison Agency/Getty Images; **25** NASA/Science Source/Photo Researchers; **26** NASA/JPL/Malin Space Science Systems; **27** NASA/JPL/Liaison Agency/Getty Images; **28** (t)David Ducros/Science Photo Library/Photo Researchers, (b)NASA; **29** The Cover Story/CORBIS; **30** Roger Ressmeyer/CORBIS; **31** Doug Martin; **32** Robert McCall; **33** (l)Novosti/Science Photo Library/Photo Researchers, (c)Roger K. Burnard, (r)NASA; **36** Tom Steyer/Getty Images; **37** NASA/Science Photo Library/Photo Researchers; **38–39** Chad Ehlers/Stone/Getty Images; **48** (bl)Richard J. Wainscoat/Peter Arnold, Inc., (others)Lick Observatory; **50** Dr. Fred Espenak/Science Photo Library/Photo Researchers; **51** Bettmann/CORBIS; **52** NASA; **54** Roger Ressmeyer/CORBIS; **57** BMDO/NRL/LLNL/Science Photo Library/Photo Researchers; **58** (t)Zuber et al/Johns Hopkins University/NASA/Photo Researchers, (b)NASA; **59** NASA; **61** Matt Meadows; **62** Cosmo Condina/Stone; **64** Lick Observatory; **65** NASA; **68–69** Roger Ressmeyer/CORBIS; **69** Matt Meadows; **72** European Southern Observatory/Photo Researchers; **74** Bettmann/CORBIS; **76** USGS/Science Photo Library/Photo Researchers; **77** (t)NASA/Photo Researchers, (b)JPL/TSADO/Tom Stack & Assoc.; **78** (t)Science Photo Library/Photo Researchers, (bl)USGS/TSADO/Tom Stack & Assoc., (bc)USGS/Tom Stack & Assoc., (br)USGS/Tom Stack & Assoc.; **79** NASA/JPL/Malin Space Science Systems; **81** Science Photo Library/Photo Researchers; **82** (l)NASA/Science Photo Library/Photo Researchers, (r)CORBIS; **83** (Io)USGS/TSADO/Tom Stack & Assoc., (Europa)NASA/JPL/Photo Researchers, (Ganymede) NASA/TSADO/Tom Stack & Assoc., (Callisto)JPL, (b)NASA; **84** JPL; **85** Heidi Hammel/NASA; **86** (l)NASA/Science Source/Photo Researchers, (r)NASA/JPL/TSADO/Tom Stack & Assoc.; **87** CORBIS; **88** (Mercury)NASA/JPL/TSADO/Tom Stack & Assoc., (Venus)NASA/Science Source/Photo Researchers, (Earth)CORBIS, (Mars)NASA/USGS/TSADO/Tom Stack & Assoc.; **89** (Jupiter)NASA/Science Photo Library/Photo Researchers, (Saturn)NASA/Science Source/Photo Researchers, (Uranus)ASP/Science Source/Photo Researchers, (Neptune)W. Kaufmann/JPL/Science Source/Photo Researchers, (Pluto)CORBIS; **90** Pekka Parviainen/Science Photo Library/Photo Researchers; **91** Pekka Parviainen/Science Photo Library/Photo Researchers; **92** Georg Gerster/Photo Researchers; **93** JPL/TSADO/Tom Stack & Assoc.; **95** Bettmann/CORBIS; **96** (t b)Museum of Natural History/Smithsonian Institution; **97** (t)NASA, (bl)JPL/NASA, (br)file photo; **98** NASA/Science Source/Photo Researchers; **100** John R. Foster/Photo Researchers; **102–103** TSADO/ESO/Tom Stack & Assoc.; **107** Bob Daemmrich; **110** (t)Carnegie Institution of Washington, (b)NSO/SEL/Roger Ressmeyer/CORBIS;**111** (l)NASA, (r)Picture Press/CORBIS, (b)Bryan & Cherry Alexander/Photo Researchers; **112** Celestial Image Co./Science Photo Library/Photo Researchers; **113** Tim Courlas; **115** Luke Dodd/Science Photo Library/Photo Researchers; **118** AFP/CORBIS; **119** NASA; **121** (t)Kitt Peak National Observatory, (b)CORBIS; **125** R. Williams (ST Scl)/NASA; **126** Matt Meadows; **128** Dennis Di Cicco/Peter Arnold, Inc.; **129** (l)file photo, (r)AFP/CORBIS; **134** PhotoDisc; **136** Tom Pantages; **140** Michell D. Bridwell/PhotoEdit, Inc.; **141** (t)Mark Burnett, (b)Dominic Oldershaw; **142** StudiOhio; **143** Timothy Fuller; **144** Aaron Haupt; **146** KS Studios; **147** Matt Meadows; **150** Amanita Pictures; **151** Bob Daemmrich; **153** Davis Barber/PhotoEdit, Inc.; **169** Matt Meadows; **170** (l)Dr. Richard Kessel, (c)NIBSC/Science Photo Library/Photo Researchers, (r)David John/Visuals Unlimited; **171** (t)Runk/Schoenberger from Grant Heilman, (bl)Andrew Syred/Science Photo Library/Photo Researchers, (br)Rich Brommer; **172** (tr)G.R. Roberts, (l)Ralph Reinhold/Earth Scenes, (br)Scott Johnson/Animals Animals; **173** Martin Harvey/DRK Photo.

PERIODIC TABLE OF THE ELEMENTS

Columns of elements are called groups. Elements in the same group have similar chemical properties.

Gas
Liquid
Solid
Synthetic

The first three symbols tell you the state of matter of the element at room temperature. The fourth symbol identifies elements that are not present in significant amounts on Earth. Useful amounts are made synthetically.

Element — Hydrogen
Atomic number — 1
Symbol — H
Atomic mass — 1.008
State of matter

	1	2	3	4	5	6	7	8	9
1	Hydrogen 1 **H** 1.008								
2	Lithium 3 **Li** 6.941	Beryllium 4 **Be** 9.012							
3	Sodium 11 **Na** 22.990	Magnesium 12 **Mg** 24.305							
4	Potassium 19 **K** 39.098	Calcium 20 **Ca** 40.078	Scandium 21 **Sc** 44.956	Titanium 22 **Ti** 47.867	Vanadium 23 **V** 50.942	Chromium 24 **Cr** 51.996	Manganese 25 **Mn** 54.938	Iron 26 **Fe** 55.845	Cobalt 27 **Co** 58.933
5	Rubidium 37 **Rb** 85.468	Strontium 38 **Sr** 87.62	Yttrium 39 **Y** 88.906	Zirconium 40 **Zr** 91.224	Niobium 41 **Nb** 92.906	Molybdenum 42 **Mo** 95.94	Technetium 43 **Tc** (98)	Ruthenium 44 **Ru** 101.07	Rhodium 45 **Rh** 102.906
6	Cesium 55 **Cs** 132.905	Barium 56 **Ba** 137.327	Lanthanum 57 **La** 138.906	Hafnium 72 **Hf** 178.49	Tantalum 73 **Ta** 180.948	Tungsten 74 **W** 183.84	Rhenium 75 **Re** 186.207	Osmium 76 **Os** 190.23	Iridium 77 **Ir** 192.217
7	Francium 87 **Fr** (223)	Radium 88 **Ra** (226)	Actinium 89 **Ac** (227)	Rutherfordium 104 **Rf** (261)	Dubnium 105 **Db** (262)	Seaborgium 106 **Sg** (266)	Bohrium 107 **Bh** (264)	Hassium 108 **Hs** (277)	Meitnerium 109 **Mt** (268)

The number in parentheses is the mass number of the longest-lived isotope for that element.

Rows of elements are called periods. Atomic number increases across a period.

Lanthanide series	Cerium 58 **Ce** 140.116	Praseodymium 59 **Pr** 140.908	Neodymium 60 **Nd** 144.24	Promethium 61 **Pm** (145)	Samarium 62 **Sm** 150.36
Actinide series	Thorium 90 **Th** 232.038	Protactinium 91 **Pa** 231.036	Uranium 92 **U** 238.029	Neptunium 93 **Np** (237)	Plutonium 94 **Pu** (244)

The arrow shows where these elements would fit into the periodic table. They are moved to the bottom of the table to save space.